Pathogen Detection and Identification in Wastewater

Pathogen Detection and Identification in Wastewater

Editors

Guangming Jiang
Ryo Honda
Sudipti Arora

MDPI

Basel • Beijing • Wuhan • Barcelona • Belgrade • Novi Sad • Cluj • Manchester

Editors

Guangming Jiang
School of Civil, Mining, Environmental and Architectural Engineering
University of Wollongong
Wollongong
Australia

Ryo Honda
School of Geosciences and Civil Engineering
Kanazawa University
Kanazawa
Japan

Sudipti Arora
Dr. B. Lal Institute of Biotechnology
Jaipur
India

Editorial Office
MDPI
St. Alban-Anlage 66
4052 Basel, Switzerland

This is a reprint of articles from the Special Issue published online in the open access journal *Water* (ISSN 2073-4441) (available at: www.mdpi.com/journal/water/special_issues/pathogen_wastewater).

For citation purposes, cite each article independently as indicated on the article page online and as indicated below:

Lastname, A.A.; Lastname, B.B. Article Title. *Journal Name* **Year**, *Volume Number*, Page Range.

ISBN 978-3-7258-0638-6 (Hbk)
ISBN 978-3-7258-0637-9 (PDF)
doi.org/10.3390/books978-3-7258-0637-9

Contents

Contents

Editors
Guangming Jiang
School of Civil, Mining, Environmental and Architectural Engineering
University of Wollongong
Wollongong
Australia

Ryo Honda
School of Geosciences and Civil Engineering
Kanazawa University
Kanazawa
Japan

Sudipti Arora
Dr. B. Lal Institute of Biotechnology
Jaipur
India

Editorial Office
MDPI
St. Alban-Anlage 66
4052 Basel, Switzerland

This is a reprint of articles from the Special Issue published online in the open access journal *Water* (ISSN 2073-4441) (available at: www.mdpi.com/journal/water/special_issues/pathogen_wastewater).

For citation purposes, cite each article independently as indicated on the article page online and as indicated below:

Lastname, A.A.; Lastname, B.B. Article Title. *Journal Name* **Year**, *Volume Number*, Page Range.

ISBN 978-3-7258-0638-6 (Hbk)
ISBN 978-3-7258-0637-9 (PDF)
doi.org/10.3390/books978-3-7258-0637-9

Pathogen Detection and Identification in Wastewater

Editors

Guangming Jiang
Ryo Honda
Sudipti Arora

MDPI

Basel • Beijing • Wuhan • Barcelona • Belgrade • Novi Sad • Cluj • Manchester

About the Editors

Guangming Jiang

Guangming Jiang is an Associate Professor of Environmental Engineering at the University of Wollongong, Australia. He was awarded his Ph.D. at The University of Queensland (Australia). He also completed a Master of Applied Science in Environmental Microbiology at Lincoln University (New Zealand), and a Bachelor of Engineering in Environmental Engineering at Harbin Institute of Technology (China).

He has broad research interests in wastewater treatment processes, environmental health, wastewater pathogens, wastewater-based epidemiology, microbiologically influenced corrosion, bioconcrete and modelling. He has published 1 book, 13 book chapters and more than 160 SCI journal papers, primarily in top-ranked journals such as *Water Research, Environmental Science* and *Technology*. His previous research outcome has led to fully commercialized technology through USP Technologies Ltd. in the USA. Moreover, he provided consulting and testing services to some major wastewater infrastructure projects such as the Singapore Deep Tunnel Sewerage System and Auckland Interception Sewer.

He is an editor or associate editor of multiple top-tier journals, including *Water Research, Journal of Water Process Engineering, iScience*, and *Journal of Environmental Engineering (ASCE)*, in addition to being an Editorial Board Member of the Journal of *Hazardous Materials, Environment International*, and *International Journal of Hygiene and Environmental Health.*

Ryo Honda

Ryo Honda is a Professor at the Faculty of Geosciences and Civil Engineering at Kanazawa University. His expertise lies in membrane-applied biological processes for water treatment and resource recovery, antimicrobial resistance in water environments, and wastewater-based epidemiology. He obtained his Ph.D. from The University of Tokyo in 2005 and has progressed through various academic roles, culminating in his current professorship since 2021. Dr. Honda holds board memberships in the Japan Wastewater Surveillance Association and Japan National Council for Promoting Wastewater Surveillance. His accolades include the Excellent Research Award from the Kurita Water and Environment Foundation in 2022, and WET Excellent Paper Award from the Japan Society on Water Environment in 2018.

Sudipti Arora

Dr. Sudipti Arora is an Environmental Biotechnology Research Scientist and Assistant Director at the Dr. B. Lal Institute of Biotechnology in Jaipur. She is also the President of Energy Swaraj Foundation Jaipur. With expertise in waste management, she holds a patent for a nature-based wastewater treatment solution using vermifiltration technology. As the founder of "Prakrit: a Centre of Excellence in Environmental Biotechnology" and Director of the NGO "Prakrit Sustainable Integrated Foundation", Dr. Arora is dedicated to transformative education and innovative solutions for sustainability, aligning with the UN's Sustainable Development Goals. She earned her Ph.D. from IIT Roorkee and has actively contributed to numerous prestigious organizations. Recognized for her outstanding research, Dr. Arora has received awards such as the "Prof. G. P. Talwar young scientist award" and "Young Water Professional Award" from the Ministry of Jal Shakti, Government of India. Her commitment to societal impact is evident through her involvement in various projects, publications and awards, showcasing her leadership and dedication to making a positive difference in environmental and community well-being.

water

MDPI

Editorial

Pathogen Detection and Identification in Wastewater

Guangming Jiang [1,*], Ryo Honda [2] and Sudipti Arora [3]

1 School of Civil, Mining, Environmental and Architectural Engineering, University of Wollongong, Wollongong, NSW 2522, Australia
2 School of Geosciences and Civil Engineering, Kanazawa University, Kakumamachi, Kanazawa 920-1192, Ishikawa, Japan; rhonda@se.kanazawa-u.ac.jp
3 Dr. B. Lal Institute of Biotechnology, Jaipur 302017, India; sudiptiarora@gmail.com
* Correspondence: gjiang@uow.edu.au; Tel.: +61-02-4221-3792

1. Introduction

The COVID-19 pandemic has renewed research needs for the detection and monitoring of various pathogens in urban wastewater systems including sewerage systems and wastewater treatment or recycling plants [1]. Not only have the detection methods evolved from culture based to molecular based technology over the years, but also the purposes and applications have expanded from water quality monitoring to wastewater-based epidemiology (WBE) [2–5]. To ensure the effectiveness of wastewater treatment and the protection of public health, much research has been completed to enhance wastewater sampling, pathogen recovery and concentration, detection sensitivity and specificity, data analysis, results interpretation, and mathematical modelling [6]. These recent developments allow the more accurate and timely identification of the occurrence of pathogenic bacteria, fungi, and viruses in wastewater. Meanwhile, it has been a great challenge to detect new pathogens, such as SARS-CoV-2, at very low levels in wastewater. Significant research is still needed to improve the detection and identification capability for any emerging pathogens. Genome sequencing and other molecular-based methods are promising in providing comprehensive information including the identification of pathogen variants, resistance genes, and the design of PCR probes.

The challenges of public health and safety impose an ever-increasing importance on the fate and spread of pathogens in wastewater systems. WBE has been developed and widely applied as an important supplementary tool for pathogen surveillance through wastewater analysis. It is widely recognized for its unique advantages, in comparison to conventional epidemiological surveillance, of low-cost, nearly real-time detection, and early warning capacity [7]. It thus finds fit-for-purpose use in tracking COVID-19 outbreaks for low-middle income regions and countries, although it is not limited to such regions or countries. The inherent limitations and associated uncertainties of the WBE approach are partly due to the pathogen detection methodologies [8,9]. Thus, it is essential to improve current pathogen detection in wastewater for the higher accuracy of WBE back-calculation of infections. Eventually, WBE can be designed as a useful tool in managing wastewater treatment and the interventions of pandemic through process optimization and public health intervention and policymaking [10]. Looking into the future, wastewater can be used to monitor and track a range of pathogens of significance to population health.

This Special Issue "*Pathogen Detection and Identification in Wastewater*" has captured the above-mentioned new opportunities and challenges for advancing the detection of pathogens and its application, focusing on methodological advancements and innovations, and their applications for WBE. Many manuscripts were submitted for consideration for this Special Issue, and all of them were subject to the rigorous review process instituted by the journal, *Water*. In total, 11 papers were finally accepted for publication and inclusion in this Special Issue (eight articles and three reviews). These papers were the contributions of researchers from Australia, Canda, China, Finland, India, Italy, Japan, Mexico, Sweden, and

Citation: Jiang, G.; Honda, R.; Arora, S. Pathogen Detection and Identification in Wastewater. *Water* **2024**, *16*, 611. https://doi.org/10.3390/w16040611

Received: 21 January 2024
Revised: 15 February 2024
Accepted: 17 February 2024
Published: 19 February 2024

USA. This Editorial will briefly summarize the articles published within this Special Issue, with the main purpose of guiding and encouraging readers in exploring the collection of studies.

2. An Overview of Published Articles

The first review article (contribution 1) by Zhang et al., as the feature paper of this Special Issue, provided a comprehensive overview of pathogens in wastewater and the range of available molecular detection methods, including nucleic acid targeting methods, immunology-based methods, biosensors, and paper-based detection. It also discussed the recent developments in wastewater sampling, DNA/RNA extraction, detection and quantification of pathogens, and the profiling of multiple pathogens. It is thus a starting point for novice research students or researchers to efficiently obtain an overview of the information and grasp some key background knowledge of pathogen detection and identification in wastewater.

Contribution 2 by Tiwari et al. provided a systematic review on the tracing of COVID-19 viral variants in wastewater. Following the PRISMA guideline, it identified and summarized 80 studies for the variants detected, detection methods, and the geographical distribution of studies. It is valuable in understanding the use and effectiveness of WBE for different SARS-CoV-2 variants.

Contribution 3 by Guo et al. is another review on how temperature and water types (i.e., wastewater, freshwater, and seawater) affect the decay of coronaviruses in terms of both infectivity and RNA. The article stated that the sensitivity of the WBE back-estimation of enveloped SARS-CoV-2 was higher than non-enveloped enteric viruses, which were less degradable in wastewater. Also, wastewater dilution by stormwater inflow or seawater infiltration has implications for the decay of coronaviruses and thus the transmission risk and WBE data interpretation.

Four papers of this Special Issue have focused on the development and evaluation of pathogen detection methods in wastewater.

Sajid et al. (contribution 4) demonstrated the use and effectiveness of membrane-inlet mass spectrometry (MIMS) as a low-impact method to differentiate between the pathogenic and non-pathogenic bacterial strains, through analyzing the volatile compounds produced by bacteria.

Hasing et al. (contribution 5) developed and optimized a method for SARS-CoV-2 detection in wastewater using moderate-speed centrifuged solids. This method showed advantages by having a similar sensitivity when compared to the ultrafiltration reference method but with lower PCR inhibition, lower costs, fewer processing steps, and a shorter turnaround time.

Pellegrinelli et al. (contribution 6) reported on an inter-laboratory proficiency test to evaluate the performance of different pre-analytical (three viral concentration methods in combination with three RNA extraction protocols) and analytical methods (two primer/probe sets in three master mixes) for identifying the presence of SARS-CoV-2 in untreated municipal wastewaters samples. The results indicated that the PEG-8000 precipitation in combination with real-time RT-PCR targeting the N gene was the best performing workflow in detecting SARS-CoV-2 in raw wastewater.

Zhang et al. (contribution 7) studied the impact of sewer biofilms on the dynamics of SARS-CoV-2 RNA concentration in naturally contaminated real wastewater using lab-scale reactors. A significant reduction in viral concentration in the water phase and its accumulation in biofilms were revealed. The findings indicated the complex role of sewer compartments and the implications for the interpretation of WBE data.

Four articles in this Special Issue have reported on the surveillance of SARS-CoV-2, its variants, and human adenovirus (HAdVs) through wastewater analysis in different countries. They highlighted the diverse applications (case studies), effectiveness, and challenges of WBE as a supplementary epidemiological surveillance tool.

Sosa-Hernández et al. (contribution 8) reported on the first long-term WBE study of COVID-19 in Mexico and demonstrated its value in providing information regarding community infections in low–middle-income populations. It indicated the importance of strategically placing WBE control centers in target communities.

Arora et al. (contribution 9) investigated the SARS-CoV-2 detection at each treatment stage of 14 wastewater treatment plants (WWTPs) in Northern India. It reported that aerobic biological wastewater treatment can be effective in removing SARS-CoV-2 viruses due to the evidence of no-detection in secondary- or tertiary-treated samples.

Nag et al. (contribution 10) reported a case study on monitoring SARS-CoV-2 variants from 11 distinct wastewater treatment plants across Jaipur (India). Wastewater surveillance for variant characterization was shown to be a reliable and practical method for tracking the diversity of SARS-CoV-2 strains in the community that is considerably faster than clinical genomic surveillance.

Maniah et al. (contribution 11) detected human adenoviruses in three wastewater treatment plants in Saudi Arabia. Sequencing showed that the F species of HAdVs, especially serotype 41, dominated in wastewater. Seasonal variations were found to be negligible, with relative humidity being a significant factor, rather than wind speed.

3. Conclusions

This compilation of articles devoted to the pathogen detection and identification in wastewater encompasses a diverse range of research, with a topical focus on its extensive use in WBE. It is a topic of enduring importance and research interest, as reflected by the different existing methodologies, their optimization, and the emergence of novel methods. The COVID-19 pandemic provided researchers with ample opportunities in conducting research on viral detection in wastewater. The profound knowledge and experiences generated over this time will have long-lasting and widespread positive effects for this research field in the future.

The challenges of current pathogen detection in wastewater lie in the fact that there are highly varying performances and very low comparability of different wastewater surveillance studies, largely due to the highly diverse sampling and analytical methods being used to detect pathogen concentrations in wastewater. Large variations in reproducibility, sensitivity, and efficiency of different analytical protocols lead to high analytical uncertainties. Nevertheless, future research should aim to improve the reproducibility and comparability by standardizing wastewater sampling, analysis, and data reporting protocols.

Acknowledgments: As Guest Editor of the Special Issue "*Pathogen Detection and Identification in Wastewater*", we would like to express our deep appreciation to all authors whose valuable work was published under this issue and thus contributed to the success of the edition.

Conflicts of Interest: The authors declare no conflicts of interest.

List of Contributions:

1. Zhang, S.; Li, X.; Wu, J.; Coin, L.; O'Brien, J.; Hai, F.; Jiang, G. Molecular Methods for Pathogenic Bacteria Detection and Recent Advances in Wastewater Analysis. *Water* **2021**, *13*, 3551. linebreak https://doi.org/10.3390/w13243551.
2. Tiwari, A.; Adhikari, S.; Zhang, S.; Solomon, T.; Lipponen, A.; Islam, M.; Thakali, O.; Sangkham, S.; Shaheen, M.; Jiang, G.; et al. Tracing COVID-19 Trails in Wastewater: A Systematic Review of SARS-CoV-2 Surveillance with Viral Variants. *Water* **2023**, *15*, 1018. https://doi.org/10.3390/w15061018.
3. Guo, Y.; Liu, Y.; Gao, S.; Zhou, X.; Sivakumar, M.; Jiang, G. Effects of Temperature and Water Types on the Decay of Coronavirus: A Review. *Water* **2023**, *15*, 1051. https://doi.org/10.3390/w15061051.
4. Sajid, S.; Aryal, I.; Chaudhri, S.; Lauritsen, F.; Jørgensen, M.; Jenssen, H.; Prabhala, B. MIMS as a Low-Impact Tool to Identify Pathogens in Water. *Water* **2023**, *15*, 184. https://doi.org/10.3390/w15010184.
5. Hasing, M.; Yu, J.; Qiu, Y.; Maal-Bared, R.; Bhavanam, S.; Lee, B.; Hrudey, S.; Pang, X. Comparison of Detecting and Quantitating SARS-CoV-2 in Wastewater Using Moderate-

Speed Centrifuged Solids versus an Ultrafiltration Method. *Water* **2021**, *13*, 2166. https://doi.org/10.3390/w13162166.
6. Pellegrinelli, L.; Castiglioni, S.; Cocuzza, C.; Bertasi, B.; Primache, V.; Schiarea, S.; Salmoiraghi, G.; Franzetti, A.; Musumeci, R.; Tilola, M.; et al. Evaluation of Pre-Analytical and Analytical Methods for Detecting SARS-CoV-2 in Municipal Wastewater Samples in Northern Italy. *Water* **2022**, *14*, 833. https://doi.org/10.3390/w14050833.
7. Zhang, S.; Sharma, E.; Tiwari, A.; Chen, Y.; Sherchan, S.; Gao, S.; Zhou, X.; Shi, J.; Jiang, G. The Reduction of SARS-CoV-2 RNA Concentration in the Presence of Sewer Biofilms. *Water* **2023**, *15*, 2132. https://doi.org/10.3390/w15112132.
8. Sosa-Hernández, J.; Oyervides-Muñoz, M.; Melchor-Martínez, E.; Driver, E.; Bowes, D.; Kraberger, S.; Lucero-Saucedo, S.; Fontenele, R.; Parra-Arroyo, L.; Holland, L.; et al. Extensive Wastewater-Based Epidemiology as a Resourceful Tool for SARS-CoV-2 Surveillance in a Low-to-Middle-Income Country through a Successful Collaborative Quest: WBE, Mobility, and Clinical Tests. *Water* **2022**, *14*, 1842. https://doi.org/10.3390/w14121842.
9. Arora, S.; Nag, A.; Rajpal, A.; Tyagi, V.; Tiwari, S.; Sethi, J.; Sutaria, D.; Rajvanshi, J.; Saxena, S.; Shrivastava, S.; et al. Imprints of Lockdown and Treatment Processes on the Wastewater Surveillance of SARS-CoV-2: A Curious Case of Fourteen Plants in Northern India. *Water* **2021**, *13*, 2265. https://doi.org/10.3390/w13162265.
10. Nag, A.; Arora, S.; Sinha, V.; Meena, E.; Sutaria, D.; Gupta, A.; Medicherla, K. Monitoring of SARS-CoV-2 Variants by Wastewater-Based Surveillance as a Sustainable and Pragmatic Approach: A Case Study of Jaipur (India). *Water* **2022**, *14*, 297. https://doi.org/10.3390/w14030297.
11. Maniah, K.; Nour, I.; Hanif, A.; Yassin, M.; Alkathiri, A.; Al-Ashkar, I.; Eifan, S. Molecular Identification of Human Adenovirus Isolated from Different Wastewater Treatment Plants in Riyadh, Saudi Arabia: Surveillance and Meteorological Impacts. *Water* **2023**, *15*, 1367. https://doi.org/10.3390/w15071367.

References

1. Bivins, A.; North, D.; Ahmad, A.; Ahmed, W.; Alm, E.; Been, F.; Bhattacharya, P.; Bijlsma, L.; Boehm, A.B.; Brown, J.; et al. Wastewater-Based Epidemiology: Global Collaborative to Maximize Contributions in the Fight Against COVID-19. *Environ. Sci. Technol.* **2020**, *54*, 7754–7757. [CrossRef] [PubMed]
2. Botes, M.; de Kwaadsteniet, M.; Cloete, T.E. Application of quantitative PCR for the detection of microorganisms in water. *Anal. Bioanal. Chem.* **2013**, *405*, 91–108. [CrossRef] [PubMed]
3. Gerardi, M.H.; Zimmerman, M.C. *Wastewater Pathogens*; John Wiley & Sons, Inc.: Hoboken, NJ, USA, 2004.
4. Sinclair, R.G.; Choi, C.Y.; Riley, M.R.; Gerba, C.P. Pathogen surveillance through monitoring of sewer systems. *Adv. Appl. Microbiol.* **2008**, *65*, 249–269. [PubMed]
5. Toze, S. PCR and the detection of microbial pathogens in water and wastewater. *Water Res.* **1999**, *33*, 3545–3556. [CrossRef]
6. Gilbride, K.A.; Lee, D.Y.; Beaudette, L.A. Molecular techniques in wastewater: Understanding microbial communities, detecting pathogens, and real-time process control. *J. Microbiol. Methods* **2006**, *66*, 1–20. [CrossRef] [PubMed]
7. Singer, A.C.; Thompson, J.R.; Filho, C.R.M.; Street, R.; Li, X.; Castiglioni, S.; Thomas, K.V. A world of wastewater-based epidemiology. *Nat. Water* **2023**, *1*, 408–415. [CrossRef]
8. Jiang, G.; Liu, Y.; Tang, S.; Kitajima, M.; Haramoto, E.; Arora, S.; Choi, P.M.; Jackson, G.; D'Aoust, P.M.; Delatolla, R.; et al. Moving forward with COVID-19: Future research prospects of wastewater-based epidemiology methodologies and applications. *Curr. Opin. Environ. Sci. Health* **2023**, *33*, 100458. [CrossRef] [PubMed]
9. Li, X.; Zhang, S.; Shi, J.; Luby, S.P.; Jiang, G. Uncertainties in estimating SARS-CoV-2 prevalence by wastewater-based epidemiology. *Chem. Eng. J.* **2021**, *415*, 129039. [CrossRef] [PubMed]
10. Diamond, M.B.; Keshaviah, A.; Bento, A.I.; Conroy-Ben, O.; Driver, E.M.; Ensor, K.B.; Halden, R.U.; Hopkins, L.P.; Kuhn, K.G.; Moe, C.L.; et al. Wastewater surveillance of pathogens can inform public health responses. *Nat. Med.* **2022**, *28*, 1992–1995. [CrossRef] [PubMed]

water

MDPI

Review

Molecular Methods for Pathogenic Bacteria Detection and Recent Advances in Wastewater Analysis

Shuxin Zhang [1], Xuan Li [1], Jiangping Wu [1], Lachlan Coin [2], Jake O'Brien [3], Faisal Hai [1] and Guangming Jiang [1,4,*]

1 School of Civil, Mining and Environmental Engineering, University of Wollongong, Wollongong 2522, Australia; sz877@uowmail.edu.au (S.Z.); xuanl@uow.edu.au (X.L.); jw130@uowmail.edu.au (J.W.); faisal@uow.edu.au (F.H.)
2 Department of Microbiology and Immunology at the University of Melbourne, Melbourne 3010, Australia; lachlan.coin@unimelb.edu.au
3 Queensland Alliance for Environmental Health Sciences (QAEHS), University of Queensland, Woolloongabba 4102, Australia; j.obrien2@uq.edu.au
4 Illawarra Health and Medical Research Institute (IHMRI), University of Wollongong, Wollongong 2522, Australia
* Correspondence: gjiang@uow.edu.au; Tel.: +61-02-4221-3792

Abstract: With increasing concerns about public health and the development of molecular techniques, new detection tools and the combination of existing approaches have increased the abilities of pathogenic bacteria monitoring by exploring new biomarkers, increasing the sensitivity and accuracy of detection, quantification, and analyzing various genes such as functional genes and antimicrobial resistance genes (ARG). Molecular methods are gradually emerging as the most popular detection approach for pathogens, in addition to the conventional culture-based plate enumeration methods. The analysis of pathogens in wastewater and the back-estimation of infections in the community, also known as wastewater-based epidemiology (WBE), is an emerging methodology and has a great potential to supplement current surveillance systems for the monitoring of infectious diseases and the early warning of outbreaks. However, as a complex matrix, wastewater largely challenges the analytical performance of molecular methods. This review synthesized the literature of typical pathogenic bacteria in wastewater, types of biomarkers, molecular methods for bacterial analysis, and their recent advances in wastewater analysis. The advantages and limitation of these molecular methods were evaluated, and their prospects in WBE were discussed to provide insight for future development.

Keywords: pathogenic bacteria; biomarkers; molecular methods; wastewater-based epidemiology; antimicrobial resistance

Citation: Zhang, S.; Li, X.; Wu, J.; Coin, L.; O'Brien, J.; Hai, F.; Jiang, G. Molecular Methods for Pathogenic Bacteria Detection and Recent Advances in Wastewater Analysis. *Water* **2021**, *13*, 3551. https://doi.org/10.3390/w13243551

Academic Editor: Alejandro Gonzalez-Martinez

Received: 15 November 2021
Accepted: 9 December 2021
Published: 12 December 2021

1. Introduction

Diseases induced by human pathogens are a major threat to public health worldwide. According to an estimation by the World Health Organization (WHO), 600 million people (almost 1 in 10 people) fall ill from eating contaminated food, and among them 420,000 people die every year, leading to the loss of 33 million healthy lives [1]. Unsafe food causes an annual loss of USD 110 billion in productivity and medical expenses in low- and middle-income countries. In addition, the ever-increasing waterborne diseases become a global burden, causing a financial loss of almost USD 12 billion and over 2.2 million deaths each year. Water and foodborne diseases overburden health care systems and affect global economies, thus hindering socio-economic development.

Among various waterborne and foodborne pathogens, pathogenic bacteria are the largest and most common group. Food and waterborne pathogenic bacteria, including *Escherichia coli* (*E. coli*), *Campylobacter* spp., (*C. jejuni* and *C. coli*), *Legionella* spp., *Salmonella* spp., and *Shigella* spp. are responsible for most infection cases, sometimes with severe and fatal outcomes. In recent years, although water and foodborne disease outbreaks have been declining with the increasing efforts in improving public health, the burden of infectious water and foodborne diseases is still a pressing global issue [2,3]. Furthermore, the growing antimicrobial resistance (AMR) of pathogens threatens the effective prevention and treatment of an ever-increasing range of infections, reminding us of the urgency and importance of strengthening our capacity to monitor and prevent the increasing risk of these human pathogens.

The traditional detection method of pathogenic bacteria mainly relies on microbial culturing. Culture-based methods are low-cost, easy to operate, and highly standardized and thus are broadly used for the regulatory purposes of pathogenic bacteria monitoring such as enumerating FIB in bathing water [4]. However, the main limitations of these methods are the lack of differentiation between the target and other non-target endogenous microorganisms of the same samples, false negative/positive results, time and labor-consuming procedures, and the inability to detect viable but nonculturable (VBNC) cells [5]. Moreover, in the application of quantitative research, culture-based methods often underestimate the number of bacteria. This affects the quantification accuracy of targets and underestimates the prevalence of pathogens in the human community. In addition, VBNC can become viable and may cause disease and increase the public health risk [4]. Therefore, as a rapid analyzing tool with high accuracy and specificity, molecular methods have quickly become the mainstream detection technique of pathogenic bacteria.

According to the biological markers being used, molecular methods can be divided in to two major groups, i.e., the nucleic acid targeting method and protein/antigen targeting method [6]. The nucleic acid targeting method includes fluorescence amplification-based methods, such as the polymerase chain reaction (PCR), quantitative or real-time PCR (qPCR), digital PCR (dPCR), deoxyribonucleic acid (DNA) microarray, fluorescence in situ hybridization (FISH), and molecular beacon, and sequencing-based methods such as pyrosequencing, Illumina sequencing, and nanopore sequencing. The protein and antigen targeting method includes a traditional antibody–antigen interaction method similar to immunological methods (lateral flow tests (LFTs)) and enzyme-linked immunosorbent assays (ELISA) [7]. Moreover, by combining the basic molecular detection approaches with a metal and paper platform, biosensor-based and paper-based devices have become a rapid, cheap, and portable on-site method for pathogenic bacteria detection [8–10].

Wastewater-based epidemiology (WBE) is a method to obtain qualitative and quantitative data on the chemical use/exposure and infectious cases of residents within a given wastewater catchment area based on the analysis of chemical compounds, pathogens, and certain biomarkers in raw sewage [11,12]. Sewage collected from a wastewater treatment plant (WWTP) or sewers can be regarded as the pooled urine and stool sample within a community and can be used to evaluate the health status of the whole community [11]. Studies based on wastewater have showed that sewage can reveal not only illicit drug use, diet, and lifestyles, but also disease outbreaks within a community [11–13]. In addition, different from clinical testing, wastewater analysis can include pathogens shed by asymptomatic and presymptomatic individuals. This makes it a powerful tool for the early warning and timely intervention of infectious disease [14–16]. Wastewater can be monitored for pathogenic and benign microbes through a variety of technologies. Traditional techniques usually include filtering, staining, and examining samples under a microscope. More sensitive and specific methods such as PCR-based methods and DNA sequencing have also been employed in the analysis of human pathogens in wastewater. However, the concentration of pathogenic bacteria is usually lower than indicator microorganisms and thus requires a highly sensitive detection method. The complex wastewater matrix often causes false-negative results because of the presence of various inhibitors. In addition, sample processing methods vary for different downstream analytical methodologies. Therefore, it is still challenging to utilize molecular detection methods in the accurate and quantitative detection of pathogenic bacteria in raw wastewater. This review focuses on the molecular analytical techniques for pathogenic bacteria in wastewater and summarizes recent advances of these approaches in wastewater-based epidemiology. The prospects of these approaches in wastewater analysis are discussed to provide insights for further application.

2. Pathogenic Bacteria in Wastewater

Human pathogens, causing serious infections and even death, are one of the leading threats to global public health. Various human pathogens can be grouped as bacteria (e.g., enterohemorrhagic *E. coli* (EHEC), *Campylobacter* spp., and *Salmonella* spp., etc.), viruses (e.g., influenza viruses, hepatitis virus, rotaviruses, and Norwalk viruses, etc.), protozoa (e.g., *Giardia lamblia* and *Cryptosporidium parvum*, etc.), and parasites (e.g., ascaris, Ancylostoma, etc.) [17]. Currently, there are approximately 538 species of pathogenic bacteria infecting human beings. This number is much higher than the overall number of other pathogens such as viruses (around 208 types), parasitic protozoa (around 57 species), and some fungi and helminths [17]. The bacteria listed in Table 1 are bacterial human pathogens detected in raw wastewaters, which are reported as the most common species that can cause global health concern (e.g., gastroenteritis or pneumonia). Most pathogens in wastewater are shed by human patients, although some might originate from other sources such as animals.

Table 1. Pathogenic bacteria detected in wastewater with high significance to public health.

Pathogenic Bacteria		Related Disease	Infectivity [a]	Persistence [b]	Density in WWTP Influent	Reference
Enteric pathogens	*E. coli* O157:H7	Gastroenteritis	High	Moderate	10^1–10^6 CFU/100 mL	[18]
	Campylobacter spp.	Gastroenteritis	Moderate	Moderate	10^2–10^5 MPN/100 mL	[19]
	Shigella spp.	Shigellosis	High	Short	10–10^7 MPN/100 mL	[20,21]
	Salmonella spp.	Salmonellosis; Typhoid fever	Low	May multiply	1–10^7 MPN/100 mL	[20,22]
	Clostridioides difficile	Severe diarrhea and colitis	High	Long	-	[23,24]
Non-enteric pathogens	*Legionella* spp.	Acute respiratory illness, legionellosis	Moderate	May multiply	10^7–10^{10} cells/100 mL	[25,26]
	Mycobacterium spp.	Pulmonary disease, skin infection	Low	May multiply	10^5 gene copies/100 mL	[27]

Note: [a] [6]; [b] [17]; MPN: most probable number.

2.1. Enteric Pathogenic Bacteria

2.1.1. *E. coli* O157:H7

The strain of *E. coli* that causes diarrhea is called diarrheagenic *E. coli*. Among several pathotypes of diarrheagenic *E. coli*, enterohemorrhagic *E. coli* (EHEC) is different because of its ability to produce Shiga toxin. Both of *E. coli* O157:H7 and other non-O157 STEC (Shiga toxin producing *E. coli*) such as O26:H11, O111:H8, and O118:H16 can release Shiga toxins, but only O157:H7 genotypes can induce disease in humans. Others commonly reside in cattle without causing diseases [28]. *E. coli* O157:H7 is the predominant and most virulent serotype in the pathogenic subgroup of *E. coli*. It could cause not only diarrhea and hemorrhagic colitis, but also hemolytic uremic syndrome, which is a serious long-term complication, mainly affecting children, that leads to kidney failure and death. Virulence factors of *E. coli* O157:H7 include the type III secretion system, Shiga-like toxin 1 and 2, acid tolerance response system, hemolysin, and extracellular serine protease [29,30]. *E. coli* is widely used as the fecal contamination indicator, and the O157:H7 genotype is often employed as a model for pathogenic bacteria study in wastewater [18]. The density of *E. coli* O157:H7 in raw wastewater was found as 10–10^6 CFU/100 mL [18]

2.1.2. *Campylobacter* spp.

Campylobacter spp. is one of four major causes of diarrhea, and it is also regarded as the most common cause of human gastroenteritis worldwide. There are 13 pathogenic *Campylobacter* species known to be associated with human infections such as *C. jejuni*, *C. coli*, *C. lari*, *C. concisus*, *C. rectus*, *C. hyointestinalis*, *C. insulaenigrae*, *C. sputorum*, *C. helveticus*, *C. fetus*, *C. mucosalis*, *C. upsaliensis*, and *C. ureolyticus*. Among 17 species and six subspecies of *Campylobacter*, *C. jejuni* and *C. coli* are the most related to infections, accounting for 80–85% and 10–15% of total infections, respectively [31]. *C. jejuni* and *C. coli* are also the main species widely detected and isolated from wastewater [19,32,33]. Pathogenic Campylobacter is responsible for 400–500 million infections annually [34]. In Europe, nearly 230 thousand cases have been reported every year since 2015 [35]. Presumably, the infection dose of campylobacteriosis is very low, with 360 colony-forming units (CFU) being adequate to cause the illness. *Campylobacter* spp. possess different virulence factors (VFs) related to motility, adhesion, invasion, toxin-activity, immune evasion, and iron-uptake [36]. VFs, such as the cadF gene and iam locus, are involved in different invasion steps [37,38]. Other VFs, such as the tripartite toxin encoded in the cdtA, cdtB, and cdtC genes [39], block the CDC2 kinase, inducing progressive cellular distension, which causes cell death [36]. Therefore, those genes are widely used in the diagnosis of *Campylobacter* spp. since thermotolerant *Campylobacter* spp. is difficult to culture [19,40,41].

2.1.3. *Salmonella* spp.

Salmonellosis is one of the most important zoonotic diseases caused by *Salmonella* spp. and transmits to humans through raw food products. A few serotypes, such as *S. typhimurium* and *S. enteritidis*, can cause human infection with poultry as the main host. Two foods that are most commonly associated with *Salmonella* infection are eggs and poultry meat [42]. According to the Centers for Disease Control and Prevention, from 2006 to 2017, *Salmonella* was responsible for about 53.4% of all foodborne disease outbreaks in the USA, and approximately 32.7% of these outbreaks were related to produce consumption [43]. Additionally, *S. Typhi* and *S. Paratyphi* are the main causes of typhoid fever and paratyphoid fever, respectively. Both are human-specific, Gram-negative, and human-restricted bacterial pathogens. Transmission can occur from person to person by eating contaminated food or water or by contact with an acute or chronic infected person [44]. Although a few culturing methods have been developed to isolate and culture those two *Salmonella* species in various samples, the cultivation is still a difficult task since they are fastidious microorganisms. The molecular approach is considered better for the detection and quantification of *Salmonella* spp. than the culture-based approach. Many protocols have been developed to target different genes or gene regions specific to *Salmonella* spp. The most popular gene targets are invA and flagellin genes (fliC-d for *Salmonella Typhi*, fliC-a for *Salmonella Paratyphi* A). Some tests are multiplexed to improve sensitivity and specificity [45–49]. Protein markers such as membrane vesicle protein PagC have also been employed in the detection of pathogenic *Salmonella* as a novel biomarker [50].

2.1.4. *Shigella* spp.

Shigella is a Gram-negative bacterium, a facultative anaerobe of the Enterobacteriaceae family. It is thought to be responsible for shigellosis or bacillary dysentery [51]. *Shigella* spp. includes four serotypes: *S. dysenteriae*, *S. flexneri*, *S. boydii*, and *S. sonnei*. Shigellosis is an invasive disease of the colon that is mainly caused by *S. sonnei* and *S. flexneri*. The main mode of transmission is fecal–oral infection, with an infection dose as low as 10 bacterial cells. The continuous transmission in humans must be passed from one person to another, as the bacterium does not survive long outside the body during the plankton-like phase [52]. About 165 million cases of *Shigella* disease are recorded worldwide each year, resulting in 1 million deaths, particularly in developing countries. It is reported that the phenotypic and genotypic characteristics of *Shigella* species are too close to be distinguished from diarrheagenic *E. coli*. This close genetic relationship between *Shigella* and *E. coli* leads to the widespread presence of virulence genes, making it difficult or even impossible to distinguish the virulence types of *Shigella* from *E. coli*. In recent years, in order to achieve rapid and reliable identification of the four *Shigella* species, a series of PCR-based methods has been developed by targeting various genes in plasmid DNA, including invasion plasmid antigen H (ipaH) [53], ial [54], virA [55], the she pathogenicity island (spi) [56] and tuf [57].

2.1.5. *Clostridioides difficile*

Clostridioides difficile (homotypic synonym *Clostridium difficile*, also known as *C. difficile*, or *C. diff*) is a kind of Gram-positive spore-forming bacterium [58]. Pathogenic *C. difficile* strains can induce diarrhea and life-threatening pseudomembranous colitis, often requiring antibiotic treatment. These strains are mainly identified by their ability to produce the enterotoxin A (TcdA) and/or the cytotoxin B (TcdB) [59]. *C. difficile*-associated diarrhea

(CDAD) is a very common nosocomial infection related with high morbidity and mortality, which imposes a huge financial burden to healthcare facilities [60]. In the last two decades, toxigenic *C. difficile* became one of the most important causes of hospital infections, with many infections leading to diarrhea and potentially fatal pseudomembranous colitis [61]. It has been recognized that community-acquired *C. difficile* infections are increasing among people with no apparent contact with healthcare facilities and without any known risk factors for *C. difficile* infection (CDI) [62]. Since *C. difficile* is a spore-forming bacterium, it is considered as an environmentally resistant pathogenic bacterium with the ability to prolong survival under environmental conditions. Therefore, *C. difficile* in the feces of both symptomatic and asymptomatic CDI patients can enter hospitals and domestic wastewater, which can be the possible approaches for CDI transmission within a community [63]. Moreover, AMR of *C. difficile* raises a major threat to the global health care system, not only because of the treatment of CDI, but also because it can be a reservoir of AMR genes to spread them to other pathogens [64]. These facts raise concerns of *C. difficile* infection and transmission and entail its surveillance based on WBE. Several studies have investigated the prevalence of *C. difficile* in wastewater by cultivation and PCR methods [59,65,66].

2.2. *Non-Enteric Pathogenic Bacteria*

Many of the pathogenic bacteria detected in wastewater are enteric in origin. However, a few of the pathogenic bacteria, which cause non-enteric diseases such as *Legionella* spp. and *Mycobacterium* spp., have also been detected in wastewater [67–69].

2.2.1. *Legionella* spp.

Pneumonia caused by *Legionella* spp. is a life-threatening pulmonary infection that is mostly caused by *Legionella pneumophila* [70]. In addition, another 19 species have also been confirmed as human pathogens based on results isolated from clinical specimens [71]. Infections could be spread not only in communities, but also in hospitals. Moreover, Legionnaires' disease (LD) is clinically and radiologically indistinguishable from community-acquired pneumonia (CAP) caused by other bacterial pathogens [72]. For the treatment, *Legionella* spp. are unaffected by β-lactam antibiotics since they are obligatory intracellular bacteria. The treatment of infections thus requires a high dose of quinolones or macrolides [73]. Early diagnosis of LD is essential for the monitoring of outbreak and treatment in hospitals [74,75]. *Legionella* has been found in wastewater with concentrations up to 10^8 CFU/L. A recent study found that exposure to aerosols dispersed from WWTPs caused LD in residents living near WWTPs during 2013–2018 in the Netherlands [26].

2.2.2. *Mycobacterium* spp.

The *Mycobacterium* genus includes more than 170 species [76], of which at least two, *Mycobacterium tuberculosis* and *Mycobacterium leprosy*, are regarded as obligate human pathogens. Most others are opportunistic organisms that cause disease both in humans and animals when conditions are favorable. Generally, *Mycobacteria* are classified into two main groups, the genetically related *M. tuberculosis* complex (MTC) organisms and nontuberculous mycobacteria (NTM). The NTM are also known as environmental mycobacteria due to their widespread presence in soil and water [77]. Tuberculosis (TB) is a disease caused by infection with *M. tuberculosis*, which caused 1.4 million deaths in 2019. TB became one of the top ten causes of death and is the leading cause from a single infectious agent (ahead of HIV/AIDS). In addition, multidrug-resistant TB (MDR-TB) is also a public health crisis and a health security threat. In 2019, 206,030 patients with multidrug-resistant or rifampicin tuberculosis (MDR/RR-TB) were detected and reported globally, an increase of 10% from 186,883 in 2018 [78]. However, nontuberculous mycobacteria (NTM) have never been quantified in wastewaters before Radomski's study because of the inefficient analytical approaches [27]. More wastewater studies should be conducted for its environmental surveillance.

3. Molecular Methods for Pathogenic Bacteria Detection

3.1. Biomarkers of Pathogenic Bacteria

Biomarkers including nucleic acids, proteins, antigens, adenosine triphosphate (ATP), and metabolic products [10] are employed in the analysis of microorganisms. To differentiate microorganisms within one sample, nucleic acids (DNA/RNA), proteins, and antigens are usually selected as biomarkers because of their special physical and chemical characteristics within different pathogens. The detection of DNA/RNA is based on the specific hybridization and amplification of targets, thus enabling good specificity and accuracy. In case of pathogenic bacteria in wastewater, the most important biomarker is the pathogenic DNA or RNA residues from these bacteria. The biomarkers include the genus/species-specific genes, functional genes, and antimicrobial resistance genes [11]. Moreover, in the analysis of antimicrobial resistance, gene transfer is another significant point. Various mobile genetic elements, including plasmids, transposons, bacteriophages, integrons, and combinations of them, are notable nucleic acid targets for investigating the prevalence and spread of resistance genes in bacteria [79].

ATP assay, enzymatic activity tests, and metabolic products are mostly used to assess the activity of living cells. Due to the linearity between the total number of ATP and the total colony-forming units, the metabolically active cells could be directly quantified using the amount of ATP [80,81]. Ions and some organic acids are the metabolic products of microorganisms. These metabolites could be detected by electrochemical methods, thus were utilized to reflect the metabolic states of microorganisms [82].

Microbial surfaces contain a variety of proteins that are expressed by specific DNA/RNA in different pathogens. By screening these proteins using antibodies and nucleic acids, new biomarkers can be discovered, and pathogens can be specifically detected. Antigens are another kind of molecules on the cell surface of pathogens. They can be specifically bound to antibodies and induce immune responses of the host. Each type of pathogen carries one or more unique antigens on their surface, even within strains. It thus enables the specific identification of pathogens using antibodies [83]. Moreover, by analyzing the specific antigens of each strain, the subtypes of the strain can be determined.

Aptamers are single-stranded DNA or RNA oligonucleotides with high affinities and specificities that can bind a variety of targets, from single molecules to whole cells [84]. They can form diverse, complex secondary structures such as multi-branched loops and G-quadruplexes, which can specifically target the surface proteins of microorganisms or cells. In environmental monitoring, aptamers are superior to antibodies due to their chemical stability, easy chemical modification, relative ease of synthesis, and biocompatibility. With the systematic evolution of ligands by exponential enrichment (SELEX) method, many aptamers have been successfully employed to detect various pathogens in environmental samples [8,85,86].

3.2. Molecular Methods

For a long time, the culture and colony counting-based method has been the dominant method in the detection of pathogens (i.e., the 'gold standard'). It can assess live microbes or viable cells in samples. However, these methods may produce false-positive or false-negative results when evaluating highly aggregated microbial cells. Furthermore, not all microbial cultures can be grown under laboratory conditions. For example, a study of *Campylobacter* indicated that the culture-based method failed to correctly detect *Campylobacter* in 30% of positive patient stool samples compared to non-cultural methods, including PCR and enzyme immunoassay (EIA) [87]. Moreover, the culture methods are time and resource intensive, which are gradually replaced by more rapid and specific molecular methods. Therefore, in order to meet the requirements for reliable analysis of pathogenic bacteria, including high specificity, high sensitivity, good reproducibility, automation, and cost effectivity, molecular methods have gradually emerged to replace the dominant position of culture methods. In recent decades, various rapid, sensitive, and specific molecular methods have been developed. These molecular methods are discussed below and listed in Table 2.

Table 2. Molecular detection methods and example of applications in the analysis of pathogenic bacteria.

Molecular Method				Target Bacteria/Genes	Sample Type	Limit of Detection	References
Nucleic acid targeting methods	**Polymerase chain reaction (PCR)-based method**	**Multiplex-PCR (mPCR)**		*Enteropathogens*	Wastewater	-	[88]
		Single qPCR		*Salmonella*	Salmonella Isolates	-	[89]
		Multiplex qPCR	**Taqman method**	invA of *Salmonella* spp.; the paratose synthase (prt) gene, and the tyvelose epimerase (tyv) gene of group D and group A *Salmonella*, the *Salmonella*-differentiating fragment 1 (Sdf-1) sequence of *S. Enteritidis*	Environmental Samples	1 copies/reaction [b]	[90]
			SYBR green	mcr-1 gene	Wastewater	12 copies/reaction [b]	[91]
		Microfluidic quantitative PCR		Antibiotic resistance and heavy metal resistance genes	Wastewater	-	[92]
		Droplet digital PCR (ddPCR)		VBNC *E. coli* O157:H7/rfbE	Food	5–6 copies/μL [b]	[93]
	DNA Microarray			*Salmonella enterica*, *Shigella flexneri*, *E. coli* O157:H7, and *Listeria monocytogenes*	Food	10^2 CFU/mL [a]	[94]
	LAMP			*V. parahaemolyticus*	Flatfish	1 CFU/mL in buffer [b]; 10 CFU/g in fish sample [a]	[95]
	FISH			*Salmonella*	Minced lamb meat	10 CFU/g [a]	[96]
	Sequencing	**Pyrosequencing**		Bacterial communities	Sputum	-	[97]
		Illumina technology		16S rRNA gene	Well-characterized bacterial reference sample	-	[98]
		Oxford Nanopore Technologies					
		Whole-genome sequencing (WGS)		381 different resistance genes	Wastewater	-	[99]
Immunology-based methods	**Enzyme-linked immunosorbent assay (ELISA)**			*S. enterica typhimurium*.	River water	9.2×10^3 CFU/mL [a]	[100]
Biosensor-based methods	**Cross-linking reaction between antibody and water-soluble cf-GQDs (carboxyl functionalized graphene quantum dots)**			*E. coli* O157: H7	Milk	10^2 CFU/mL [a,b]	[101]
Paper-based device	**Origami paper-based device**			*E. coli*	Bacteria culture	10^3 CFU/mL [b] within 35 min	[102]
	Paper-based ELISA			*E. coli* O157:H7	Food	1×10^4 CFU/mL [a,b]	[9]

Note: [a] method LoD (based on sample volume/mass); [b] assay LoD (based on reaction or L); [a,b] both method and assay LoD.

3.2.1. Nucleic Acid Targeting Methods

Nucleic acid targeting methods are designed to detect the specific DNA/RNA of pathogens. It is achieved by the hybridization between target nucleic acid sequences and synthetic oligonucleotides. Thus, the species-specific gene of pathogens and virulence genes can be detected through nucleic acid targeting methods. They are usually fast, efficient, and do not require the culture of the pathogens. These methods include polymerase chain reaction (PCR)-based methods such as conventional PCR, real-time/quantitative PCR (qPCR), droplet digital PCR (ddPCR), multiplex PCR (mPCR), and other methods such as microarrays, loop-mediated isothermal amplification (LAMP), sequencing, and fluorescence in situ hybridization (FISH).

PCR-Based Method

PCR is the most common molecular-based technique for the detection and quantification of pathogens. PCR enables the detection of a single pathogenic bacteria by targeting specific DNA sequences [57]. Through this method, a small sample of a DNA sequence could be rapidly amplified into a large amount. This advantage enables the detection and quantification of a low amount of the target DNA sequence. It is thus widely used in the diagnosis of human pathogens. It significantly increases the sensitivity of detection of microorganisms at low concentrations in environmental samples [103,104]. PCR has already been utilized to the detection of a series of pathogenic bacteria such as *E. coli* and spores of *C. perfringens* [104,105].

Conventional PCR needs gel electrophoresis to detect the formation of PCR products. Real-time polymerase chain reaction, also called quantitative PCR, is the real-time detection of the PCR process during the amplification of the target DNA sequence. qPCR determines the PCR amplification by measuring specific dual-labeled probes or fluorescent signals emitted by inserting dyes. The fluorescence intensity reflects the amount of the template DNA. There is a linear relationship between the cycle threshold (Ct or Cq) value and the initial concentration of the template gene during the exponential period of PCR amplification. Thus, the concentration of target sequences could be calculated from a well-established standard curve to achieve an absolute quantification. Real-time quantitative PCR is widely used in the real-time detective and quantitative analysis of target DNA sequences with higher specificity and sensitivity than conventional PCR [106].

Two main fluorescence systems have been developed for qPCR, i.e., the SYBR green method and the TaqMan probes method. SYBR green is a fluorescent pigment that can bind double-stranded DNA (dsDNA). This non-sequence-specific pigment enhances the fluorescence signal when it binds to DNA double helix minor grooves, thus enabling the quantification of the targeting sequence. In contrast, TaqMan probe does not require the addition of fluorescent pigment. The template-specific TaqMan probe further improves the specificity of qPCR by increasing primer specificity. For each amplification of a specific target, one molecule of fluorescent dye is released. The instrument detects the fluorescence produced by specific amplification, which is not impacted by non-specific amplification. This ensures the high specificity of the qPCR detection. There are many reporter–quencher sets with different wavelengths, which can be labeled with the TaqMan probe. This enables the TaqMan method to be able to detect multiple PCR reactions in the same tube, leading to reduced cost and improved efficiency and accuracy. It can also avoid the influence of different fluorescent dyes on the PCR reaction.

The mPCR is a faster detection methodology than simplex PCR, which can detect multiple gene targets simultaneously. Fan et al. (2008) reported one mPCR assay to achieve the simultaneous detection of various human pathogens in a single tube, with the detection sensitivities between 10 to 10^2 CFU/100 mL in seawater. To differentiate the pathogenic and commensal *E. coli* in clinical and water samples, an mPCR assay was developed to detect the occurrence of 11 virulence genes in *E. coli* [107]. Recently, the presence of enteropathogens in sewage was investigated by using the commercially available FilmArray® mPCR system [88].

Compared with simplex PCR, mPCR provides faster detection by simultaneously amplifying multiple gene targets. It can also differentiate closely related pathogenic bacteria.

Digital PCR is a biotechnology improvement on conventional PCR and can be used to directly amplify and quantify DNA, cDNA, or RNA. Droplet digital polymerase chain reaction (ddPCR) is a kind of dPCR technique that is emerging as a powerful analytical tool for absolute quantification. Similar to qPCR, ddPCR also utilizes Taq polymerase to amplify a target DNA sequence in a standard qPCR assay. The differences are that ddPCR separates the whole qPCR reaction into thousands of individual reactions before amplification, and ddPCR collects data at the reaction end point. These differences provide ddPCR many advantages, such as the direct and independent quantification of target DNA without standard curves and more precise and reproducible data than conventional qPCR, especially when PCR inhibition is present [108–110]. In comparison to qPCR, ddPCR shows better performance in detecting low concentrations of target genes in environmental samples. Moreover, it has the potential to reduce the effect of qPCR inhibitors, although its application to complex environmental samples needs further optimization [111]. However, qPCR is more reliable in detecting higher concentrations (2×10^5 or 2×10^4 gene copies/PCR), since ddPCR displays higher variability and less precision in these concentration ranges [112]. A recent study shows that, in addition to being faster, the ddPCR method exhibited higher sensitivity with a limitation of 10^{-5} ng/μL for genomic DNA templates and 10^{-1} CFU/mL for *Shigella* bacteria culture, when compared to PCR and qPCR [113].

Nowadays, with the increasing availability of sequencing data, it is theoretically possible to design qPCR assays for every microorganism [106]. The qPCR method has many benefits over other techniques. Firstly, the quantitative data produced by qPCR method could reach an accurate dynamic range of 7–8 log orders of magnitude without requiring post-amplification manipulation. Secondly, although the sensitivity of qPCR is varied towards different samples and might be inhibited by inhibitors, it has been reported to have higher sensitivity than many other molecular methods [114]. Theoretically, it is high enough to detect a single copy of a transcript. Studies have shown that qPCR is reliable and sufficient for the quantitative detection of various pathogens such as *E. coli* O157:H7 [115] and *Campylobacter* spp. [116] The qPCR method has been applied to detect and monitor the occurrence and concentrations of pathogens in drinking water sources [117], water and wastewater treatment plants [118,119], and recreational beaches [120]. The use of qPCR in water analysis enables quick microbial risk assessment, which may lead to immediate remedial actions.

However, for the detection and quantification of waterborne bacteria with low abundance, high requirements are demanded for qPCR, such as high accuracy, low limit of detection (LoD) and quantification (LoQ), and the ability to distinguish dead and viable cells. Studies have reported that it is possible to detect viable cells by detecting messenger RNA (mRNA), since it only exists in viable organisms [7,121]. However, not all mRNAs are present in all life phases of an organism, thus the target mRNA should be carefully chosen for viable organism detection. In addition, rRNA based RT qPCR assays were also confirmed to have a better association with the active bacterial populations in surface water samples than rDNA-based assays [122–124]. Furthermore, multiplex real-time PCR has been reported to be a valuable technique for the identification of viruses [125], bacteria [126], and parasites [127]. However, due to the limitation of instruments and the fluorescence groups, only four targets could be detected at the same time in TaqMan methods. This limitation prohibits its application in profiling microbiome communities in complex samples.

DNA Microarrays

DNA microarrays, also known as DNA arrays, are commonly known as gene chips. It is a special piece of glass or silicon chip with a DNA microarray, which places thousands or tens of thousands of nucleic acid probes on an area of several square centimeters [128]. DNA, complementary DNA (cDNA), and RNA in the sample are detected by fluorescence or electric signal after being combined with the probes. DNA microarrays enable the hybridization-based detection of multiple targets in a single experiment, which makes it suitable for the analysis of massive targets. Using a high-throughput DNA microarray assay, a study investigated the prevalence of 941 pathogenic bacteria in groundwater and differentiated their sources of origin [129]. In general, DNA microarray allows for the simultaneous detection of multiple pathogenic bacteria. It is thus a fast and reliable diagnostic method in analyzing large numbers of clinical/environmental samples. However, the complicated probe design work, the reliability of the microarray data, and the clinical applicability of the early results have been criticized [130]. The criticism and intensified competition from other technologies, such as next-generation sequencing (NGS), have hampered the growth of microarray-based testing in the molecular diagnostics market [131].

Loop-Mediated Isothermal Amplification (LAMP)

LAMP is an isothermal nucleic acid amplification technique. It has been utilized for the alternative detection of certain diseases because of its low cost. At present, LAMP has been applied to the identification and quantification of pathogenic bacteria with significant advantages in sensitivity, specificity, and rapidity [95,132]. Since LAMP requires four primers specifically designed for six different regions of the target, any incomplete matching of the primers will theoretically lead to the phenomenon of non-reaction and non-specific amplification. In addition, the LAMP method was confirmed to be 10–100 times more sensitive than PCR detection [133], with a detection limit of 10 copies or less in the template for one reaction. Furthermore, it can directly detect pathogenic microorganisms in diseased tissue, thus avoiding the tedious cultivation and nucleic acid extraction step [134]. Finally, and most importantly, the result of the reaction can be judged with naked eyes by demonstrating the absence of the target gene with the production of white precipitate of magnesium pyrophosphate. It is more difficult to design specific primers for LAMP than PCR (because LAMP requires 4–6 primers and PCR requires only two). A software tool named PrimerExplorer is available to help the primer design for LAMP (http://primerexplorer.jp/e/). Therefore, as a rapid detection method without the need of any equipment, LAMP shows great potential in the rapid diagnosis of human pathogens in various samples.

Fluorescent in Situ Hybridization (FISH)

FISH is a cytogenetic technique used to detect and locate nucleic acids in cells or sample matrices. Fluorescently labeled nucleic acid probes hybridize only with highly similar nucleic acids and can be used to locate genes on chromosomes or to label ribosomal RNA in different taxonomic bacteria or archaea in molecular ecology. FISH could be employed in the enumeration of particular microbial populations [135]. Compared to PCR, FISH is more suitable for complex matrices because of its lesser sensitivity to inhibitory substances. However, a major limitation of FISH is the small number of phylogenetically distinct targets that can be detected at the same time. A recent study developed a multi-FISH method that uses eight fluorophores, which is highly suitable for investigating the structure and function of microbial communities in different samples [96]. Furthermore, FISH has been used to discover emerging human pathogens in water, wastewater, and sludge, to produce quantitative descriptions of the microbial community in wastewater and activated sludge [136,137] and to investigate survival and infection mechanisms at the cellular level. However, this method is still partly based on cell culture.

Sequencing

Sequencing is the process of determining the sequence of nucleotides in a section of DNA. It includes any method or technique used to determine the order of the four bases: adenine, guanine, cytosine, and thymine (or uracil for RNA). In 1977, DNA sequencing technology was firstly developed by Frederick Sanger based on the chain-termination method (also known as Sanger sequencing). In the early stage, DNA sequencing was employed for small genomes such as viruses and organelles. Complete sequencing of a bacterium genome was not feasible because of the economic and technical limitations. Later, with the emergence of the shotgun method developed by Sanger et al., whole genome sequencing of bacteria was achieved. The shotgun method is considered the gold standard, and whole genome sequencing of many bacteria has been carried out using this method over years [138].

Next-generation sequencing (NGS), also known as high-throughput sequencing, is the overall term used to describe several different modern sequencing pathways. These technologies allow DNA and RNA to be sequenced faster and at lower cost than Sanger sequencing, which was previously used, thus revolutionizing genomics and molecular biology research [139]. NGS technologies include Illumina (Solexa) sequencing, Roche 454 sequencing, and proton/PGM sequencing. The NGS technologies achieve high throughput and reduced cost by using massively parallel analysis, which allows 300 Gb of DNA to be read in a single run on a single chip. The four main advantages of NGS over classical Sanger sequencing are: (i) NGS needs significantly less DNA, as it can obtain a sequence from a single strand; (ii) NGS is significantly quicker than Sanger sequencing by combining the two separate processes of Sanger sequencing, i.e., chemical reaction and signal detection, in some versions of NGS; (iii) NGS is more cost-effective due to reduced time, manpower, and reagents; (iv) repeats in NGS caused by many short overlapping reads lead to a more accurate and reliable sequence, even though individual reads are less accurate. These advantages enable a great potential of NGS in the application of environmental research. NGS is capable of producing large numbers of reads at exceptionally high coverages throughout the genome with dramatically reduced cost through the massively paralleled approach. However, NGS requires the amplification of DNA molecules, which introduces random errors in the DNA synthesis. The amplified DNA strands would become progressively out-of-sync, which means the signal quality deteriorates as the read length grows. Therefore, long DNA molecules must be broken up into smaller pieces to maintain the quality of the reading, leading to a critical limitation of second-generation sequencing [140].

To solve the limitation, third-generation sequencing (TGS) technologies were developed to produce substantially longer reads than NGS by the direct sequencing of single DNA molecules. Nanopore sequencing (Oxford Nanopore Technologies, Oxford, UK) is a representative TGS approach for the sequencing of biopolymers, specifically polynucleotides in the form of DNA/RNA. Through nanopore sequencing, individual molecules of a DNA/RNA can be sequenced without PCR amplification or chemical labeling of the sample. Nanopore sequencing has a great potential in providing relatively low-cost genotyping, high mobility for testing, and the ability to rapidly process samples and display results in real time [141]. Applications of this method in the rapid identification of viral pathogens [142], plant genome sequencing [143], monitoring of antibiotic resistance [144], and haplotyping [145] has been reported. One major limitation of nanopore sequencing is its high raw read error rate, which remains between 5% and 15% despite recent improvements in nanopore chemistry and computational tools [138]. However, according to the latest updates from Nanopore technologies (accessed on 15 October 2021), an accuracy of 98.3% could be achieved through the production software MinKNOW 4.3 ("Super-accuracy" basecalling model) and Guppy 5 (https://nanoporetech.com/accuracy). In addition, the quality of the sequencing result is affected by library quality and the presence of sequencing inhibitors. Although more efforts are needed to improve the quality of Nanopore sequencing results, studies have confirmed that it has better bacterial identification performance in complex samples than traditional Illumina platforms [146]. Winand et al. compared the bacterial identification performance of second (Illumina) and third-generation sequencing technologies (Nanopore sequencing, Oxford Nanopore Technologies, Oxford, UK) by targeting the 16S rRNA gene. The results revealed that both techniques provide reliable identification of bacterial genera but may mislead the identification of bacterial species and constitute viable alternatives to Sanger sequencing for rapid analysis of mixed samples without any culture steps [98].

3.2.2. Immunology-Based Methods

Immunological methods are based on the specific interaction between antibodies and antigens. These methods include enzyme-linked immunosorbent assays (ELISA), immunofluorescence assays (IFA), and serum neutralization tests (SNTs) [6]. For immunology-based methods, specific fluorochrome labeled antibodies are used to capture targeted antigens, which serves as the enumeration of fluorescently labeled cells by detecting the fluorescence signal using microscopy or flow cytometry. However, the biomarkers of these methods should be carefully chosen to achieve specific detection at different classification levels including genus, species, and serotypes. The detection of *S. typhimurium* on an immunochromatographic strip was reported by Park et al. (2010). This study achieved the quantitative detection of *S. typhimurium* in the range of 9.2×10^3 to 9.2×10^6 CFU/mL in river samples within 20 min [100].

Although these methods can specifically detect targeted bacteria and their toxins and can be multiplexed for multiple samples, they are still limited by false-negative results and cross-reactions with similar antigens. False-negative results are a serious problem and often happen to classical methods. They can be induced by various inhibitory compounds and matrices of different types of samples, which vary largely and thus might cause different effects on the analytical performance of different detection methods. In addition, cross-reaction is another big problem for immunology-based methods. In one study, a monoclonal antibody was used for specifically detecting *E. coli* O157:H7 lipopolysaccharide (LPS1) [147]. The cross-reactivity with other bacteria happened due to the presence of a constituent sugar of LPS. One recent method comparison study for *C. difficile* surveillance in Switzerland showed that, compared to the PCR method, enzyme immunoassay led to more false-negative results of human stool samples [148]. Immunological methods usually require pre-enrichment to expose surface antigens, which leads to extended detection time. Moreover, due to their lower sensitivity than other molecular methods, immunology-based methods were less employed in the direct detection of pathogenic bacteria in wastewater samples [9].

3.2.3. Biosensor-Based Methods

A biosensor is an analytical platform composed of two elements: a bio-receptor and a transducer. Bio-receptors are responsible for recognizing the targets such as enzymes, proteins, nucleic acids, and cell receptors. After recognition, the transducer converts the biological interactions into electrical signals that can be measured (e.g., optical, electrochemical, or magnetic). Biosensors provide a rapid, real-time, on-site, and multiple detection of bacteria. Optical biosensors are selective, sensitive, and can be used for real-time monitoring of toxins, drugs, and pathogens in wastewater [149]. For example, by applying a fluorescently labeled specific aptamer, Yildirim et al. developed a portable optical biosensor for the indirect sensing of an *E. coli* O157:H7 strain in wastewater [150]. Surface-enhanced Raman scattering (SERS) pathogen biosensors, with noble metal nanoparticles (e.g., silver and gold) as an impressive substrate, have become an attractive research field. The colorimetric changes induced by the hybridization between single-stranded DNA probes modified by gold nanoparticles and their complementary DNA can avoid the requirement of expensive and complex fluorescent labeling [151]. Gold nanoparticles are widely used in biosensor instruments, especially for dark water samples [152]. Another notable biosensor for *E. coli* O157:H7 used carboxyl functionalized graphene quantum dots (cf-GQDs) to label a specific antibody [101]. This sensor can specifically recognize *E. coli* O157:H7 from different sources, such as water and food, with the minimum detection limit of 100 CFU/mL. However, the sensitivity to changes in pH, mass, and temperature are some of the challenges that must be addressed in using biosensors for bacterial pathogens in wastewater [153].

3.2.4. Paper-Based Device

A paper-based device is a small analytical tool that is printed by a wax printer and has different functional areas. It can integrate all the processes required for nucleic acid detection (enrichment, extraction, amplification, and visual detection) into a cheap paper material [154]. The whole detection process can be completed by folding paper-based devices in different ways and in different sections, which overcomes the limitation of PCR tests. Paper-based device can achieve multichannel, sensitive detection, comparable to PCR detection, and provide a high-quality, rapid, and accurate diagnosis of pathogens [155]. Moreover, paper-based devices are easy to stack, store, and transport because they are thin, lightweight, and of different thicknesses [156]. Similar to biosensors, paper-based devices can also be used to target a variety of biomarkers, including nucleic acids, proteins, antigens, and chemicals [9,102,155,157]. By integrating various molecular detection methods, paper-based devices have emerged as a powerful platform for the fast diagnosis of pathogens and the determination of infection transmission [9,13,102]. However, the shelf life of paper-based device limits its further applications. Some paper-based devices contain reagents with a short shelf life, such as enzymes, and thus need to be stored in a refrigerator or freezer to maintain the activity of the reagents [158,159]. Studies about the shelf life of enzymes on paper-based devices have yielded some promising results and proven techniques, although further research is still needed. Furthermore, the analytical performance of paper-based devices is also limited by features of paper, including the paper fibers and pattern. Moreover, the coffee ring effect of paper-based devices can cause non-uniform distributions of detection reagents and samples, thus affecting the detection accuracy [158]. To overcome these limitations, future efforts should be put into developing more uniform papers and modifying the size and shape of the test zone.

4. Recent Advances of Molecular Methods for Pathogenic Bacteria in Wastewater

Current estimates of the burden of infectious diseases are often based on severe cases requiring hospitalization, which fails to cover asymptomatic patients. The emerging wastewater-based epidemiology (WBE) is based on the analysis of biomarkers in raw wastewater, which is then used to back-estimate the status of public health. Wastewater is a complex mixture of chemicals and microorganisms in water. It contains chemical and biological information directly discharged from our bodies. From a surveillance point of view, urban wastewater is an attractive resource, since it provides sampling material within a large and mostly healthy population.

The WBE approach was first outlined as a potential tool to evaluate the use of illicit drugs and misused therapeutic drugs within a community [160]. To date, WBE has become an important tool for estimating illicit and licit drug consumption by detecting and quantifying unchanged drugs and their human-specific metabolites in wastewater [12]. WBE studies also showed that wastewater can reveal not only illicit drug use and diet, but also infectious disease risk within a community [11,161,162]. Many studies have validated the feasibility of various molecular methods in wastewater. Sensitive and specific methods such as PCR, real-time PCR, and DNA sequencing have been employed in the analysis of wastewater to achieve the fast detection and accurate quantification of human pathogens [90,163,164] or the evaluation of community structure and antimicrobial resistance level [165–167].

4.1. Sample Processing and DNA/RNA Extraction Methods

Wastewater components, including fats, proteins, humic acids, and fulvic acids, can lead to problems in the downstream analysis (molecular detection). Wastewater sample processing is a key step for the detection of pathogenic bacteria by separating, concentrating, extracting, and purifying biomarkers for further analysis. The availability of different commercial DNA/RNA extraction kits showed variable efficiency when extracting samples such as wastewater and sediment. Table 3 lists several comparison studies of different sample storage, pre-treatment, and DNA/RNA extraction methods for various downstream analyses.

Table 3. Comparison studies of DNA/RNA extraction methods for various downstream molecular methods of wastewater and sediment samples.

Downstream Analysis	Targets	Best/Limited Fragment Length	Suggested Extraction Kits/Methods	Storage and Pretreatment of Samples	Recovery Efficiency	Sample Type	Reference
PCR-based method	Lambda DNA	-	FastDNA Spin Kit for Soil	Stored at −70 °C	15.5% to 43.3%	Sediment	[168]
	Ancylostoma caninum ova	-	MO Bio Power Max® Soil DNA Extraction Kit (MO BIO Laboratories Inc, Carlsbad, CA USA); Filtration	Stored at 4 °C in the dark	Treated wastewater: 39–50% Raw wastewater: 7.1–12%	Wastewater	[169]
Microarray	16S Rdna, cpn60, and wecE	Detection sensitivity is optimal when DNA targets > 500 bp	Bead beating separation and ammonium acetate purification	Centrifuged at 3000× *g* for 16 min at room temperature; stored at −20 °C	81 µg DNA/mL	Wastewater	[170]
NGS	ARG	150 bp (Limitation of the sequencing length)	FastDNA Spin Kit for Soil	Ethanol fixation (50%); filter-concentrated using 0.22-µm mixed cellulose ester filters; stored at -20 °C	10.3 ± 3.6 µg/sample	Wastewater	[171]
	16S rRNA amplicons		Qiagen Mini Kit and MO Bio PowerSoil Kit	Centrifuged at 10,000× *g* for 5 min to pellet; filtered through 0.22 µm cellulose nitrate membrane filters	-	Wastewater	[172]

Mumy and Findlay developed an external DNA recovery standard for sediments by comparing the performance of three commercial kits (UltraClean™Soil DNA, FastDNA®SPIN®, and Soil Master™DNA Extraction) [168]. The results indicated that the FastDNA®SPIN® kit has the highest recovery rate and makes it possible to collect additional DNA by cleaning beaded sediments. Gyawali et al. investigated six rapid DNA extraction methods for recovering *Ancylostoma caninum ova* DNA from wastewater and reported that the filtration method recovered higher DNA concentrations in both treated and raw wastewater than centrifugation, hollow fiber ultrafiltration (HFUF), and flotation [169]. A comparative study about the relative effectiveness of 10 different bacterial DNA extraction methods for wastewater samples showed that only a few could achieve satisfactory results when applied to bacterial pathogens [170]. The method of combined bead beating separation and ammonium acetate purification was suggested as the most suitable approach for bacterial DNA extraction from wastewater prior to specific microbial detection using microarray hybridization technology. Li et al. compared the ARG sequencing analysis results of three DNA extraction methods [171]. It was found that ARGs captured by the FastDNA SPIN Kit for Soil had the highest DNA yield, purity, and diversity. Moreover, no discernable effects were found on ARG profiles with fixation in ethanol, deep-freezing, and overseas transportation of samples compared with fresh samples. Another comparative study indicated that the DNA Mini kit and PowerSoil kit produce the most consistent sequencing results in water and wastewater [172]. Collectively speaking, the performance of DNA/RNA extraction methods of wastewater varies, and it is essential to develop standard DNA/RNA extraction methods for different downstream analysis methods to achieve high recovery and quality of the nucleic acids of various pathogens in wastewater.

4.2. Detection and Quantification of Pathogenic Bacteria

The low concentration of targeted pathogenic bacteria in wastewater samples brings difficulties to their detection, which entails high sensitivity and repeatability. As a complex matrix, wastewater contains various inhibitors for a number of molecular methods, which is a significant challenge for their application in wastewater analysis. In addition, from a disease surveillance perspective, wastewater samples should be analyzed quickly enough to provide an early warning. Several molecular approaches have showed great potential in rapid analysis. Table 4 critically compares various detection methods used for wastewater samples in the last five years.

Table 4. Applications of molecular methods in the analysis of pathogenic bacteria in wastewater.

Detection Method	Cultivation	DNA/RNA Extraction	Target Pathogen	Biomarker	Sample Type	Limit of Detection (LoD)	References
PCR	Yes	Yes	*Campylobacter* spp., *C. jejuni*, *C. coli*	16S rRNA, mapA, ceuE	Wastewater	2 CFU/100 mL	[173]
			E. coli O157:H7	stx2			
			S. typhimurium	stx1			
Real-time PCR	No	Yes	*E. coli*	uidA gene	Wastewater	10 gc/reaction (standard curve)	[174]
	No		*Simkania negevensis*	16S rRNA gene	Wastewater	5 gc/reaction (standard curve)	[175]
	No	Yes	*S. enterica* serovar *Typhi*	stgA	Wastewater	0.05–0.005 CFU/mL of seeded wastewater	[176]
			S. enterica serovar *Paratyphi A*	SSPAI			
Droplet digital PCR (ddPCR)	Yes	Yes	Shiga toxin-producing *E. coli*	stx2	River water, wastewater	6 gc/reaction of standard curve; 32 copies/100 mL in river water	[177]
qPCR array	No	Yes	All	285 ARGs and nine transposase genes	Wastewater	-	[166]
	No	Yes	All	229 ARGs and 25 mobile genetic elements	Wastewater	-	[178]
Microfluidic qPCR	No	Yes	All	ARGs, heavy metal resistance genes, genes encoding the integrase, and 16S rRNA genes	Wastewater, drinking water	-	[92]
LAMP	Yes	Yes	*Listeria monocytogenes*	lmo0753 gene	Wastewater	65 fg/µL of DNA and 38 CFU/mL in cell culture	[179]
FISH	Yes	No	*Salmonella* spp.	23S rRNA	Wastewater	10^2, 10, and 1 CFU/mL for 0 h, 6 h, and 24 h of enrichment in Rappaport-Vassiliadis broth, respectively.	[137]
Biosensor-based device	No	No	ARG	mecA gene	ARG-spiked wastewater effluent	70 pM (4×10^7 gc/µL) by bootstrapping	[180]
Paper-based device	No	No	*Salmonella typhimurium*	fimA	Spiked wastewater	10^2 CFU/mL	[181]
	Yes	No	β-lactamase-expressing bacteria	β-lactamase	Wastewater	3.8×10^6 CFU/mL	[157]
Sequencing	No	Yes	All	ARGs	Sludge	-	[182]
	No	Yes	Shotgun metagenomic for microbial community analysis and pathogen detection		Wastewater	Detected 87 pathogenic/opportunistic Bacteria, with most having <1% abundance.	[183]
	Yes	Yes	Nanopore and Illumina metagenomics analysis for mobile antibiotic resistome		Wastewater	-	[25]
	No	Yes	Full-length 16S rRNA		Wastewater	-	[184]
	No	Yes	16S-rRNA		Wastewater	-	[185]

So far, the PCR-based method is the most popular molecular approach for specific pathogenic bacteria detection and quantification (relative and absolute quantification) in wastewater because of its high sensitivity, specificity, and the low cost compared to the sequencing method. Plenty of primer-probe sets targeting various human pathogenic bacteria have been developed. Those primer-probe sets showed high specificity, sensitivity, and efficiency for pathogenic bacteria detection at genus and species levels. A series of PCR-based methods with or without bacteria isolation and cultivation procedures has been reported for the fast detection and quantification of pathogenic bacteria in wastewater samples [173–175]. Recently developed primer-probe sets for the bacterial pathogen detection of various PCR-based methods are listed in Table 5. Some of them have been confirmed to have good performance in wastewater samples and thus could be potentially used in wastewater analysis, although further feasibility studies should be conducted. Among the PCR-based methods, ddPCR has been increasingly reported as having better analysis performance than traditional PCR-based methods in wastewater [186]. However, some studies have still claimed that the sensitivity of qPCR is superior to ddPCR in some circumstances (e.g., high targeted gene concentration), and their performance might vary with different assays [187].

Table 5. PCR primer-probe sets available for the detection and quantification of typical pathogenic bacteria in wastewater and other samples.

Pathogenic Bacteria	Available PCR Primers and Probes (5′—3′)	Sensitivity	Reference
E. coli O157:H7	RFBEO157-F GGATGACAAATATCTGCGCTGC RFBEO157-R GGTGATTCCTTAATTCCTCTCTTTCC RFBEO157-P HEX-TACAAGTCCACAAGGAAAG-BHQ1	1 CFU/g of seeded meat products after 4 h enrichment period at 37 °C	[188]
	Rfb-F GTGTCCATTTATACGGACATCCATG Rfb-R CCTATAACGTCATGCCAATATTGCC	2 CFU/100 mL of raw sewage	[173]
Campylobacter spp.	16S-F CCTGAMGCAGCAACGCC 16S-R CGGAGTTAGCCGGTGCTTATT 16S-P FAM-CTCCGAAAAGTGTCATCCT –MGB	3.2 gene copies/reaction	[19,40]
C. jejuni	hipO-F CTTGCGGTCATGCTGGACATAC hipO-R AGCACCACCCAAACCCTCTTCA hipO-P VIC-ATTGCTTGCTGCAAAGT- MGB	2.0×10^2 CFU/g of feces	[31]
C. coli	glyA-F AAACCAAAGCTTATCGTGTGC glyA-R AGTGCAGCAATGTGTGCAATG glyA-P FAM-CAACTTCATCCGCAAT- MGB	2.5×10^2 CFU/g of feces	
C. lari	glyA-F CAGGCTTGGTTGTAGCAGGTG glyA-R ACCCCTTGGACCTCTTAAAGTTTT glyA-P TET-CATCCTAGTCCATTCCCTTATGCTC ATGTT-TAMRA	2.1 gene copies/reaction	[19]
Shigella spp.	ipaH-F CGCAATACCTCCGGATTCC ipaH-R TCCGCAGAGGCACTGAGTT ipaH-P FAM- AACAGGTCGCTGCATGGCTGGAA-BHQ1	10^{-5} ng/μL for genomic DNA templates, 10^{-1} CFU/mL for Shigella bacteria culture	[113]
Salmonella spp.	invA-F AACGTGTTTCCGTGCGTAAT invA-R TCCATCAAATTAGCGGAGGC invA-P TGGAAGCGCTCGCATTGTGG	9–15 CFU/25 g food sample	[189]
S. Typhi	stgA-F TATCGGCAACCCTGCTAATG stgA-R TATCCGCGCGG TTGTAAAT stgA-P FAM-CCATTACAG CATCTGGCGTAGCGA-BHQ1	0.05–0.005 CFU/mL of wastewater	[176]
S. enterica serovar *Paratyphi A*	SSPAI-F ACCATCCGCAGGACAAATC SSPAI-R GGGAGATTACTGATGGAGAGATTAC SSPAI-P Cy5-AGAGTGCAAGTGGAGTGCCTCAAA-BHQ2		
C. difficile	tpi-F AAAGAAGCTACTAAGGGTACAAA tpi-R CATAATATTGGGTCTATTCCTAC	For simultaneous identification and toxigenic type characterization (fecal and urban water samples)	[59,66]
	tcdB-F GGAAAAGAGAATGGTTTTATTAA tcdB-R ATCTTTAGTTATAACTTTGACATCTTT		
	tcdA-F AGATTCCTATATTTACATGACAATAT tcdA-R GTATCAGGCATAAAGTAATATACTTT		
Legionella spp.	PanLeg-F GGCGACCTGGCTTC PanLeg-R1 GGTCATCGTTTGCATTTATATTTA PanLeg-P1 FAM-ACGTGGGTTGCAA-MGBNFQ	5 genome units (GU)/reaction with water sample	[190]
L. pneumophila	Lp-F TTGTCTTATAGCATTGGTGCCG Lp-R CCAATTGAGCGCCACTCATAG Lp-P Quasar670-CGGAAGCAATGGCTAAAGGCATGCA-BHQ3		
L. pneumophila sg1	Lp1-F TGCCTCTGGCTTTGCAGTTA Lp1-R CACACAGGCACAGCAGAAACA Lp1-P VIC-TTTATTACTCCACTCCAGCGAT-MGBNFQ		
Mycobacterium spp.	16S rRNA-F: ATGCACCACCTGCACACAGG 16S rRNA-R: GGTGGTTTGTCGCGTTGTTC	10–100 copies of template plasmid/reaction (raw wastewater)	[191]

The LAMP method showed great potential in the application to wastewater samples because of its advantages such as inhibitor resistance, short reaction time (<1 h), and no need for advanced thermal cycling instruments [192]. By targeting the lmo0753 gene,

Nathaniel et al. developed a loop-mediated isothermal amplification assay for the detection of *L. monocytogenes* in wastewater. The LoD was 65 fg/μL of DNA and 38 CFU/mL, which was 10 times more sensitive than conventional PCR with primers targeting the HlyA gene. However, in the application to wastewater, a pre-culture procedure at 37 °C for 48 h was required [179]. An SA23 probe targeting *Salmonella* specifically by FISH has been developed by Santiago et al. (2008). The SA23 probe was shown to be capable of rapid and specific identification and visualization of *Salmonella* cells directly in the sample. By combining with the pre-enrichment, it could achieve the detection of 1 CFU/mL in seeded meat products [137]. This study also demonstrated the resistance of FISH to inhibitory substances in wastewater and the ability to differentiate viable but non-culturable (VBNC) cells. The advantage of the FISH method is that it is not inactivated by inhibitors and does not depend on the type of sample, even when dealing with large numbers of samples. Sequencing is also a powerful analysis tool for not only detection and relative quantification but also absolute quantification of bacteria in environmental samples, similar to the PCR-based method [193]. By spiking the samples with internal microorganism markers, the absolute bacterial number of targeting microbiomes could be calculated through the abundance of the internal markers with a known number. Different types of spiking markers have been used previously, including indigenous microorganism, synthetic, and heterogeneous markers [193–195]. The reference markers and spiking strategy should be optimized because only validated markers can be used to achieve reliable results [196].

In the application of biosensor-based methods, surface-enhanced Raman scattering/spectroscopy (SERS) has a high sensitivity, although its stability is unsatisfactory, and that limits its application in wastewater analysis [149]. Furthermore, colorimetric and fluorescent sensors generally have poorer LoDs than electrochemical, and they are easily disrupted by colored or turbid samples. Thus, it seems that the electrochemical aptasensor is more reliable in wastewater matrices [197]. In addition, the biosensor-based method showed good stability in wastewater samples. For example, Riquelme et al., developed a stable oligonucleotide-functionalized gold nanosensor for mecA ARG monitoring in 2017 [180]. This mecA-specific nanosensor can keep stable under environmental conditions and at high ionic strength, and it can demonstrate high selectivity even in the presence of target interference. This study supports the environmental suitability of a new, low-cost, field-deployable, and large-scale ARG analysis tool.

Most detection and quantification methods for pathogenic bacteria in wastewater involve sampling, which is followed by transportation to a central analytical laboratory for further analysis. In comparison, paper-based devices could achieve multiplexed, sensitive assays that rival PCR-based methods and provide high-quality, fast precision on-site diagnostics for pathogens. Although wastewater is a complex substrate, paper-based devices have shown the potential to detect pathogens in wastewater. Due to the outbreaks of COVID-19, paper-based devices are being quickly developed and employed in SARS-CoV-2 detection in sewage. In another previous study, it was successfully employed in the detection of various genes and microbiomes. A fully disposable and integrated paper-based sample-in-answer-out device was developed for nucleic acid testing, which can sensitively detect *S. typhimurium* with a detection limit of as low as 10^2 CFU/mL in wastewater [181]. The presence of β-lactamase-mediated resistance was also detected using paper-based analytical devices (PADs). It was shown that, compared to traditional methods including culture methods, antibiotic susceptibility testing, and PCR gene analysis, their method can still reduce the laboratory processing time by 14–20 h, although a laboratory is still required to concentrate the wastewater samples [157].

4.3. Profiling Potential Pathogens

Sequencing, as a powerful analysis tool, has been widely used for profiling bacterial diversity and potential pathogens in wastewater. The DNA sequencing-based method can perform large-scale parallel analysis of PCR products and environmental nucleic acids. This provides a new dimension for the analysis of pathogenic bacteria in wastewater. The application methodology of NGS technologies in wastewater study can be divided into four subcategories: whole genome sequencing (WGS), metagenomic sequencing, metatranscriptomic sequencing, and sequencing of an amplified targeted gene (e.g., 16S rRNA and 18S rRNA genes) [198]. WGS is a powerful approach for microorganism identification in wastewater, while it relies on bacteria isolation and culture, extraction of long DNA, and the development of long read sequencing platforms. Future development on these aspects could advance and simplify its application in wastewater analysis [199].

To date, sequencing assays based on the amplified gene regions take the dominant place in wastewater analysis. Microbial communities of waterborne pathogens were often studied by targeting high-variation region sequences of small subunit (SSU) rRNA genes (e.g., V1, V3, V4, V6) and large subunit (LSU) rRNA genes [182]. By using the full-length 16S rRNA gene sequence, Numberger et al. characterized and compared bacterial communities of the influent and effluent of a WWTP in Berlin, Germany [184]. The study found that during sewage treatment, the relative abundance of most pathogenic bacteria was effectively reduced, while *Legionella* and *Leptospirosis* showed an increase in relative proportion from inflow to effluent. This indicated that WWTPs may enrich and release certain potential pathogens into the environment, although they are effective in removing enteric bacteria. Oluseyi et al. studied the presence of pathogenic bacteria in three WWTPs in South Africa. Their study also confirmed the presence of bacterial pathogens in treated effluent, which may pose a potential contamination risk by transmission through soil, agriculture, water, or sediments [185]. Using Illumina MiSeq sequencing, Xue et al. (2019) investigated the spatial and temporal variability of bacterial structure and the presence of a human-associated Bacteroidale (HF183) marker in two WWTPs. Their findings illustrated how changes in bacterial communities can serve as a reliable means of monitoring the quality and performance of wastewater treatment plants for public and environmental health purposes [200].

Metagenomic study is an emerging methodology based on the sequencing data of genetic material recovered directly from environmental samples. This method has developed rapidly in the detection of microbial communities and their functional capabilities in wastewater. Currently, the application of metagenomics in wastewater is commonly employed for the identification of ARGs and genes associated with pathogens [183,201]. It is also increasingly utilized to support the assembly of whole or partial genomes from short-read sequencing data acquired from uncultured microbial communities [202]. Metatranscriptomic sequencing has similar principles with metagenomic sequencing, but it targets RNA rather than DNA, which is essential for identifying RNA viruses. Sequencing the transcribed mRNA could provide the information about which microbes are functionally active, as mRNA degradation rate varies among different species, thus challenging the preservation and analysis of mRNA [203]. In addition, the excess of ribosomal RNA (rRNA) in transcriptomes also interferes with the identification of mRNA in environmental samples. All these challenges limit its application in wastewater systems [198]. Comparison studies have also been conducted to access the performance of various methods to identify pathogens and associated virulence genes. For example, Yergeau et al. compared pre- and post-treatment biosolids from two WWTPs by using enumeration methods combined with molecular techniques including quantitative PCR, 16S rRNA and cpn60 gene amplicon sequencing, and shotgun metagenomic sequencing [204]. Their study showed that shotgun metagenomics indicted the widest range of pathogen DNA and was the only method that can obtain functional gene information in treated biosolids among all approaches.

4.4. *Antimicrobial Resistance Analysis*

With the growing concern about antimicrobial resistance, the WHO established the Global Antimicrobial Resistance Surveillance System (GLASS) in 2015 for sharing information on a global level to strengthen data on national and international actions and to aid decision making [205]. WWTPs treating wastewater from different sources provide a suitable circumstance for the emergence and spread of antibiotic resistance genes (ARGs) and antibiotic resistant bacteria (ARB) [79,206]. Analysis of ARGs in influent wastewater can provide a broader perspective for the study of ARGs in the population. Various ARGs have been investigated and reported in wastewater based on qPCR technology [207,208]. A quantitative analysis of ARGs and horizontal gene transfer (HGT) potential was conducted over four seasons at a WWTP using a high-throughput qPCR array [166]. In this study, 285 primer sets targeting ARGs and nine transposase genes related to HGT were successfully used with wastewater samples. A microfluidic quantitative polymerase chain reaction (MF-qPCR) method was developed and optimized for simultaneously quantifying 16S rRNA genes, ARGs, heavy metal resistance genes, and an integrase gene that encodes three different types of integrons. This MF-qPCR method has better detection limits than shotgun metagenomics, which has also been used to detect large amounts of ARGs in other studies [92,209].

Sequencing-based methods also contributed to the analysis of ARGs and ARB (e.g., prevalence, relative abundance, and persistence) in wastewater. Using metagenomic analysis, a pipeline covered the analysis of gene transfer potential and the potential, pathway, and phylogenetic origin of ARGs was developed for identifying antibiotic resistance determinants in wastewater samples [99,182,183]. Meanwhile, discrepancies in ARG quantification by using different sequencing approaches have been reported by several previous studies where some ARGs were only identified by Nanopore sequencing and some others were only detected via Illumina sequencing [210]. Sequencing platform biases on the ARG quantification outputs were due to different ARG identification or prediction algorithms. The Illumina algorithm is based on the similarity search using BLAST, whereas Nanopore sequencing is based on workflows designed for the alignment of long high-error-rate sequences [211].

4.5. Prospect of Molecular Methods for Pathogenic Bacteria in Wastewater Analysis

Based on the above recent advances of various molecular methods, their advantages and limitations in wastewater analysis are summarized in Table 6. All molecular methods are able to achieve the detection of pathogenic bacteria in wastewater with an LoD range from 1 to 100 CFU or gene copies per 100 mL. However, the LoD varies according to different targeted pathogenic bacteria and sample pre-treatment procedures. Overall, the sensitivity of the current molecular methods is adequate for WBE purposes.

Table 6. Advantages, limitations, and prospects of various molecular methods in analyzing pathogenic bacteria for WBE applications.

Molecular Method	Biomarkers	Advantages	Limitations	Reference
Nucleic acid targeting methods	DNA/RNA	- High sensitivity - High specificity - Multiple targets detection and quantification - Fast community profiling	- Require sample storage and processing - Require DNA/RNA extraction, which can cause DNA/RNA loss - Sensitive to inhibitors - High cost for large number of samples - Usually need specialized instruments	[11,212]
Immunology-based methods	Proteins	- Low cost - Can be automated - Can detect bacterial toxins	- Require pre-enrichment - Low sensitivity - Require labeling of antibodies and antigens	[9]
Biosensor-based methods	DNA/RNA, proteins, chemicals	- High sensitivity - Real-time detection - Label free	- High cost - Require specialized instruments - Low specificity - Not suitable for simultaneous detection of various organisms - Low reproducibility and insufficient stability	[197]
Paper-based device	DNA/RNA, proteins, chemicals	- Cost effective - Instrument free	- Detection limit (LoD) is usually high due to the traditional colorimetry - Limitations of the structure and material of paper device	[13,154,156]

Nucleic acid targeting methods have the potential to become the most suitable molecular method for wastewater analysis because one extracted nucleic acid sample could be analyzed by various methods through different downstream workflows to achieve comprehensive analysis, including target gene detection and quantification (relative/absolute), microbiome community profiling, and ARG/functional gene analysis. Moreover, nucleic acid targeting methods could realize the direct identification and quantification of specific genes rather than gene expression products and thus could reduce the mistakes induced by some sample pre-treatment procedures, including enrichment and cultivation. The methods need to be improved to overcome limitations including sensitivity to inhibitors in wastewater, DNA/RNA loss caused by sample processing, and high cost. In addition, more specific gene biomarkers including species-specific genes, host-specific genes, and reliable cell quantification genes (one DNA/RNA fragments per cell) should be carefully selected. Standardization for sample storage, pre-processing, inner-extraction control, recovery methods, and certain analytical methods will highly expand their prospects in WBE study.

Immunology-based methods are powerful methods for detecting and capturing gene expression products such as cell surface proteins and bacterial toxins. For wastewater analysis, they were widely used in specific bacteria detection, isolation, and enrichment. They are more suitable for specific pathogen and microbial antigen detection and isolation from wastewater. They can also combine with other molecular methods to achieve deep analysis of a target pathogen's genome. By combining with a biosensor platform or paper-based device, they are expected to achieve fast, on-site, real-time, and low-cost identification and enumeration of pathogens in wastewater.

Biosensor-based methods have showed great potential in pathogenic bacteria analysis in wastewater. However, in the real wastewater analysis, they are not cost-effective compared to other molecular methods and cannot process large numbers of environmental samples, such as wastewater, from a long-term surveillance point-of-view. Paper-based devices are a good platform to be combined with many other molecular detection methods to achieve fast analysis of pathogenic bacteria in wastewater. Their application prospect towards wastewater has been well confirmed as a fast, on-site, cost-effective, portable, and disposal device. However, their sensitivity and specificity should be further improved towards detecting various targets in wastewater.

According to different research objectives, appropriate molecular methods can be selected and combined to achieve satisfactory analytical performance. The sensitivity and specificity of different molecular methods should be evaluated on various wastewater samples to improve their analytical performance. Infectious disease surveillance based on WBE could be divided into three main phases including pathogen monitoring in raw sewage, disinfection evaluation and ARG variation in WWTPs, and risk evaluation for further spreading of effluent in the environment. Among those three phases, nucleic acid targeting methods are more suitable for the pathogen monitoring of raw sewage since this phase requires the fast analysis of deep and comprehensive genomic information of pathogenic microbiomes delivered from communities. Especially for the sequencing-based method, with the decreasing cost and by combining with different workflows, it is expected to realize all kinds of analysis, including identification, relative and absolute quantification, bacterial community profiling, and ARG and functional gene analysis. This sets it apart from other nucleic acid targeting methods. For disinfection and ARG evaluation in WWTPs, except for nucleic acid targeting methods, immunology-based methods and biosensor-based methods are powerful analysis tools for the assessment of the activity and infectivity of pathogenic bacteria. For the risk evaluation of effluent and the environment, since the concentration of pathogenic bacteria in effluent and in the surrounding environment is usually low, and this evaluation does not require accurate detection and quantification results, the paper-based device was considered as the best choice since it can achieve fast, cost-effective, and on-site screening of the concentration level of target genes and pathogens. In addition, immunology-based methods could also be used in this phase to evaluate the infectivity and activity of pathogenic bacteria for risk assessment. However, the gold standard approach of sampling, transport and storage, pre-treatment, and the enrichment of biomarkers in wastewater for various downstream molecular detections would dominate in WBE applications. The efficient recovery of high-quality biomarkers from wastewater samples should thus be the priority of future method development.

5. Conclusions

With the ever-increasing concerns over infectious diseases caused by pathogenic bacteria and their antimicrobial resistance, it is widely recognized that effective surveillance is key to the rapid intervention and control of outbreaks of infectious diseases. Wastewater-based epidemiology has become a popular tool due to its great potential as a population prevalence surveillance system and an early warning tool for disease outbreaks. With the development of molecular techniques for the detection of pathogenic bacteria and associated biomarkers, WBE applications are expanding to cover a wide range of pathogens using different advanced molecular methods tailored for wastewater analysis.

PCR-based methods have high sensitivity, and they are broadly used for the rapid analysis of pathogenic bacteria in wastewater, following DNA/RNA extraction procedures. Methods such as DNA microarray and sequencing-based methods are suitable for the in-depth study of bacterial communities and the presence of pathogenic bacteria and antimicrobial resistance due to their capability of large-scale parallel analysis of the whole microbiome. Alternative nucleic acid targeting methods such as FISH and LAMP are relatively sensitive, specific, and cost-efficient. However, nucleic acid targeting methods are not able to provide information about the activity and infectivity of various pathogens in wastewater. In addition, the need for sample pretreatment and research into multiplexing microorganism detection in a single sample are still challenges. Biosensors are easy to operate and do not need trained personnel for the detection of pathogenic bacteria in wastewater. Moreover, paper-based devices have recently emerged and have been widely used in pathogenic bacteria detection in wastewater because of their rapidness and cost effectiveness.

Molecular methods have found their diverse applications in WBE. Nucleic acid-based methods enable the direct and comprehensive analysis of the DNA/RNA of wastewater samples, including target gene (e.g., species-specific genes, ARGs, and functional genes) detection and quantification (both direct and relative quantification), profiling of the whole microbiome in the sample, genome sequencing, and analyzing. Thus, nucleic acid-based methods have the broadest prospect in wastewater analysis. Biosensor-based methods and paper-based devices exhibited great potential in fast and on-site detection of chemicals and microbiomes, which is suitable for the early warning of infectious disease outbreaks. Although, it seems that the application of immunology-based methods in wastewater is limited by a lot of disadvantages, it is the only method that targets the gene expression products of microbiomes. A lot of biosensors and paper-based devices are developed based on immunology-based methods. It is essential to improve immunology-based methods to suit WBE applications.

Although molecular methods have shown great potential in the analysis of pathogenic bacteria in wastewater, there are still several key challenges for their application in WBE. The low concentration of pathogenic bacteria in wastewater and inhibition of the complex wastewater matrix are additional concerns for these methods in comparison to the analysis of other types of samples. The low DNA/RNA recovery efficiency of pathogenic bacteria from wastewater needs to be improved and reported quantitatively with the results. The accuracy and reliability of WBE would be significantly enhanced with well-established molecular detection methods.

Author Contributions: Conceptualization, S.Z. and G.J.; writing—original draft preparation, S.Z.; writing—review, S.Z.; X.L.; J.W.; L.C.; J.O.; and F.H.; supervision, funding acquisition, and writing—review and editing, G.J. All authors have read and agreed to the published version of the manuscript.

Funding: This research was supported by the ARC Discovery project (DP190100385). Shuxin Zhang receives the support from a University of Wollongong PhD scholarship.

Institutional Review Board Statement: Not applicable.

Informed Consent Statement: Not applicable.

Conflicts of Interest: The authors declare no conflict of interest.

References

1. Hoffmann, V.; Moser, C.; Saak, A. Food safety in low and middle-income countries: The evidence through an economic lens. *World Dev.* **2019**, *123*, 104611. [CrossRef]
2. O'Brien, S.J. Foodborne Diseases: Prevalence of Foodborne Diseases in Europe. *Encycl. Food Saf.* **2014**, *13*, 302–311. [CrossRef]
3. Benedict, K.M.; Reses, H.; Vigar, M.; Roth, D.M.; Roberts, V.A.; Mattioli, M.; Cooley, L.A.; Hilborn, E.D.; Wade, T.J.; Fullerton, K.E.; et al. Surveillance for Waterborne Disease Outbreaks Associated with Drinking Water—United States, 2013–2014. *MMWR Morb. Mortal. Wkly. Rep.* **2017**, *66*, 1216–1221. [CrossRef]
4. Tiwari, A.; Oliver, D.M.; Bivins, A.; Sherchan, S.P.; Pitkänen, T. Bathing Water Quality Monitoring Practices in Europe and the United States. *Int. J. Environ. Res. Public Health* **2021**, *18*, 5513. [CrossRef] [PubMed]
5. Sohier, D.; Pavan, S.; Riou, A.; Combrisson, J.; Postollec, F. Evolution of microbiological analytical methods for dairy industry needs. *Front. Microbiol.* **2014**, *5*, 16. [CrossRef]
6. Deshmukh, R.A.; Joshi, K.; Bhand, S.; Roy, U. Recent developments in detection and enumeration of waterborne bacteria: A retrospective minireview. *Microbiologyopen* **2016**, *5*, 901–922. [CrossRef]
7. Gilbride, K. Molecular methods for the detection of waterborne pathogens. In *Waterborne Pathogens*; Elsevier: Amsterdam, The Netherlands, 2014; pp. 231–290.
8. Kudłak, B.; Wieczerzak, M. Aptamer based tools for environmental and therapeutic monitoring: A review of developments, applications, future perspectives. *Crit. Rev. Environ. Sci. Technol.* **2020**, *50*, 816–867. [CrossRef]
9. Pang, B.; Zhao, C.; Li, L.; Song, X.; Xu, K.; Wang, J.; Liu, Y.; Fu, K.; Bao, H.; Song, D.; et al. Development of a low-cost paper-based ELISA method for rapid *Escherichia coli* O157:H7 detection. *Anal. Biochem.* **2018**, *542*, 58–62. [CrossRef] [PubMed]
10. Jayan, H.; Pu, H.; Sun, D.-W. Recent development in rapid detection techniques for microorganism activities in food matrices using bio-recognition: A review. *Trends Food Sci. Technol.* **2020**, *95*, 233–246. [CrossRef]
11. Sims, N.; Kasprzyk-Hordern, B. Future Perspectives of Wastewater-Based Epidemiology: Monitoring Infectious Disease Spread and Resistance to the Community Level. *Environ. Int.* **2020**, *139*, 105689. [CrossRef] [PubMed]
12. Feng, L.; Zhang, W.; Li, X. Monitoring of regional drug abuse through wastewater-based epidemiology—A critical review. *Sci. China Earth Sci.* **2018**, *61*, 239–255. [CrossRef]
13. Hui, Q.; Pan, Y.; Yang, Z. Paper-based devices for rapid diagnostics and testing sewage for early warning of COVID-19 outbreak. *Case Stud. Chem. Environ. Eng.* **2020**, *2*, 100064. [CrossRef]
14. Choi, P.M.; Tscharke, B.J.; Donner, E.; O'Brien, J.W.; Grant, S.C.; Kaserzon, S.L.; Mackie, R.; O'Malley, E.; Crosbie, N.D.; Thomas, K.V. Wastewater-based epidemiology biomarkers: Past, present and future. *TrAC Trends Anal. Chem.* **2018**, *105*, 453–469. [CrossRef]
15. Ahmed, W.; Bibby, K.; D'Aoust, P.M.; Delatolla, R.; Gerba, C.P.; Haas, C.N.; Hamilton, K.A.; Hewitt, J.; Julian, T.R.; Kaya, D.; et al. Differentiating between the possibility and probability of SARS-CoV-2 transmission associated with wastewater: Empirical evidence is needed to substantiate risk. *FEMS Microbes* **2021**, *2*, xtab007. [CrossRef]
16. Tiwari, A.; Lipponen, A.; Hokajärvi, A.-M.; Luomala, O.; Sarekoski, A.; Rytkönen, A.; Österlund, P.; Al-Hello, H.; Juutinen, A.; Miettinen, I.T.; et al. Detection and quantification of SARS-CoV-2 RNA in wastewater influent in relation to reported COVID-19 incidence in Finland. *medRxiv* **2021**, *770*, 145274. [CrossRef]
17. Ramirez-Castillo, F.Y.; Loera-Muro, A.; Jacques, M.; Garneau, P.; Avelar-Gonzalez, F.J.; Harel, J.; Guerrero-Barrera, A.L. Waterborne pathogens: Detection methods and challenges. *Pathogens* **2015**, *4*, 307–334. [CrossRef] [PubMed]
18. Fijalkowski, K.L.; Kacprzak, M.J.; Rorat, A. Occurrence changes of *Escherichia coli* (including O157:H7 serotype) in wastewater and sewage sludge by quantitation method of (EMA) real time—PCR. *Desalination Water Treat.* **2014**, *52*, 3965–3972. [CrossRef]
19. Banting, G.S.; Braithwaite, S.; Scott, C.; Kim, J.; Jeon, B.; Ashbolt, N.; Ruecker, N.; Tymensen, L.; Charest, J.; Pintar, K. Evaluation of various *Campylobacter*-specific quantitative PCR (qPCR) assays for detection and enumeration of *Campylobacteraceae* in irrigation water and wastewater via a miniaturized most-probable-number–qPCR assay. *Appl. Environ. Microbiol.* **2016**, *82*, 4743–4756. [CrossRef] [PubMed]
20. Teklehaimanot, G.Z.; Genthe, B.; Kamika, I.; Momba, M.N.B. Prevalence of enteropathogenic bacteria in treated effluents and receiving water bodies and their potential health risks. *Sci. Total Environ.* **2015**, *518–519*, 441–449. [CrossRef] [PubMed]
21. Teklehaimanot, G.Z.; Coetzee, M.A.A.; Momba, M.N.B. Faecal pollution loads in the wastewater effluents and receiving water bodies: A potential threat to the health of Sedibeng and Soshanguve communities, South Africa. *Environ. Sci. Pollut. Res.* **2014**, *21*, 9589–9603. [CrossRef]
22. Koivunen, J.; Siitonen, A.; Heinonen-Tanski, H. Elimination of enteric bacteria in biological–chemical wastewater treatment and tertiary filtration units. *Water Res.* **2003**, *37*, 690–698. [CrossRef]
23. Sandhu, B.K.; McBride, S.M. Clostridioides difficile. *Trends Microbiol.* **2018**, *26*, 1049–1050. [CrossRef]
24. Xu, C.; Salsali, H.; Weese, S.; Warriner, K. Inactivation of *Clostridium difficile* in sewage sludge by anaerobic thermophilic digestion. *Can. J. Microbiol.* **2015**, *62*, 16–23. [CrossRef] [PubMed]

25. Caicedo, C.; Rosenwinkel, K.H.; Exner, M.; Verstraete, W.; Suchenwirth, R.; Hartemann, P.; Nogueira, R. *Legionella* occurrence in municipal and industrial wastewater treatment plants and risks of reclaimed wastewater reuse: Review. *Water Res.* **2019**, *149*, 21–34. [CrossRef] [PubMed]
26. Vermeulen, L.C.; Brandsema, P.S.; van de Kassteele, J.; Bom, B.C.J.; Sterk, H.A.M.; Sauter, F.J.; van den Berg, H.H.J.L.; de Roda Husman, A.M. Atmospheric dispersion and transmission of *Legionella* from wastewater treatment plants: A 6-year case-control study. *Int. J. Hyg. Environ. Health* **2021**, *237*, 113811. [CrossRef] [PubMed]
27. Radomski, N.; Betelli, L.; Moilleron, R.; Haenn, S.; Moulin, L.; Cambau, E.; Rocher, V.; Gonçalves, A.; Lucas, F.S. Mycobacterium Behavior in Wastewater Treatment Plant, A Bacterial Model Distinct From *Escherichia coli* and Enterococci. *Environ. Sci. Technol.* **2011**, *45*, 5380–5386. [CrossRef]
28. Manning, S.D.; Babcock, H. *Escherichia Coli Infections*; Chelsea House Publishers: New York, NY, USA, 2010.
29. Levin, R.E. *Rapid Detection and Characterization of Foodborne Pathogens by Molecular Techniques*; CRC Press: Boca Raton, FL, USA, 2009.
30. Singleton, P.; Sainsbury, D. *Dictionary of Microbiology and Molecular Biology*; John Wiley & Sons: Hoboken, NJ, USA, 2006.
31. Leblanc-Maridor, M.; Beaudeau, F.; Seegers, H.; Denis, M.; Belloc, C. Rapid identification and quantification of *Campylobacter coli* and *Campylobacter jejuni* by real-time PCR in pure cultures and in complex samples. *BMC Microbiol.* **2011**, *11*, 113. [CrossRef]
32. Rinsoz, T.; Hilfiker, S.; Oppliger, A. Quantification of Thermotolerant *Campylobacter* in Swiss Water Treatment Plants, by Real-Time Quantitative Polymerase Chain Reaction. *Water Environ. Res.* **2009**, *81*, 929–933. [CrossRef]
33. Moreno, Y.; Botella, S.; Alonso, J.L.; Ferrús, M.A.; Hernández, M.; Hernández, J. Specific detection of Arcobacter and *Campylobacter* strains in water and sewage by PCR and fluorescent in situ hybridization. *Appl. Environ. Microbiol.* **2003**, *69*, 1181–1186. [CrossRef]
34. Igwaran, A.; Okoh, A.I. Human campylobacteriosis: A public health concern of global importance. *Heliyon* **2019**, *5*, e02814. [CrossRef]
35. Hansson, I.; Sandberg, M.; Habib, I.; Lowman, R.; Engvall, E.O. Knowledge gaps in control of *Campylobacter* for prevention of campylobacteriosis. *Transbound. Emerg. Dis.* **2018**, *65*, 30–48. [CrossRef]
36. Bolton, D.J. *Campylobacter* virulence and survival factors. *Food Microbiol.* **2015**, *48*, 99–108. [CrossRef]
37. Eucker, T.P.; Konkel, M.E. The cooperative action of bacterial fibronectin-binding proteins and secreted proteins promote maximal *Campylobacter jejuni* invasion of host cells by stimulating membrane ruffling. *Cell. Microbiol.* **2012**, *14*, 226–238. [CrossRef] [PubMed]
38. Carvalho Alexandro, C.T.; Ruiz-Palacios Guillermo, M.; Ramos-Cervantes, P.; Cervantes, L.-E.; Jiang, X.; Pickering Larry, K. Molecular Characterization of Invasive and Noninvasive *Campylobacter jejuni* and *Campylobacter coli* Isolates. *J. Clin. Microbiol.* **2001**, *39*, 1353–1359. [CrossRef] [PubMed]
39. Mendez-Arancibia, E.; Vargas, M.; Soto, S.; Ruiz, J.; Kahigwa, E.; Schellenberg, D.; Urassa, H.; Gascón, J.; Vila, J. Prevalence of different virulence factors and biofilm production in enteroaggregative *Escherichia coli* isolates causing diarrhea in children in Ifakara (Tanzania). *Am. J. Trop Med. Hyg.* **2008**, *78*, 985–989. [CrossRef] [PubMed]
40. De Boer, R.F.; Ott, A.; Güren, P.; van Zanten, E.; van Belkum, A.; Kooistra-Smid, A.M. Detection of *Campylobacter* species and Arcobacter butzleri in stool samples by use of real-time multiplex PCR. *J. Clin. Microbiol.* **2013**, *51*, 253–259. [CrossRef] [PubMed]
41. Henry, R.; Schang, C.; Chandrasena, G.I.; Deletic, A.; Edmunds, M.; Jovanovic, D.; Kolotelo, P.; Schmidt, J.; Williamson, R.; McCarthy, D. Environmental monitoring of waterborne *Campylobacter*: Evaluation of the Australian standard and a hybrid extraction-free MPN-PCR method. *Front. Microbiol.* **2015**, *6*, 74. [CrossRef] [PubMed]
42. Dar, M.A.; Ahmad, S.M.; Bhat, S.A.; Ahmed, R.; Urwat, U.; Mumtaz, P.T.; Bhat, S.A.; Dar, T.A.; Shah, R.A.; Ganai, N.A. *Salmonella typhimurium* in poultry: A review. *Worlds Poult. Sci. J.* **2017**, *73*, 345–354. [CrossRef]
43. Liu, H.; Whitehouse, C.A.; Li, B. Presence and Persistence of *Salmonella* in Water: The Impact on Microbial Quality of Water and Food Safety. *Front. Public Health* **2018**, *6*, 159. [CrossRef]
44. House, D.; Bishop, A.; Parry, C.; Dougan, G.; Wain, J. Typhoid fever: Pathogenesis and disease. *Curr. Opin. Infect. Dis.* **2001**, *14*, 573–578. [CrossRef]
45. Cheng, C.-M.; Doran, T.; Lin, W.; Chen, K.-S.; Williams-Hill, D.; Pamboukian, R. Interlaboratory validation for a real-time PCR *Salmonella* detection method using the ABI 7500 FAST real-time PCR system. *J. Food Prot.* **2015**, *78*, 1119–1124. [CrossRef] [PubMed]
46. González-Escalona, N.; Brown, E.W.; Zhang, G. Development and evaluation of a multiplex real-time PCR (qPCR) assay targeting ttrRSBCA locus and invA gene for accurate detection of *Salmonella* spp. in fresh produce and eggs. *Food Res. Int.* **2012**, *48*, 202–208. [CrossRef]
47. Postollec, F.; Falentin, H.; Pavan, S.; Combrisson, J.; Sohier, D. Recent advances in quantitative PCR (qPCR) applications in food microbiology. *Food Microbiol.* **2011**, *28*, 848–861. [CrossRef] [PubMed]
48. Chaudhry, R.; Chandel, D.; Verma, N.; Singh, N.; Singh, P.; Dey, A. Rapid diagnosis of typhoid fever by an in-house flagellin PCR. *J. Med. Microbiol.* **2010**, *59*, 1391–1393. [CrossRef]
49. Andrews, J.R.; Ryan, E.T. Diagnostics for invasive *Salmonella* infections: Current challenges and future directions. *Vaccine* **2015**, *33*, C8–C15. [CrossRef] [PubMed]
50. Xu, J.; Suita, K.; Okuno, K.; Takaya, A.; Yamamoto, T.; Isogai, E. Membrane vesicle protein PagC as a novel biomarker for detecting pathogenic *Salmonella* in the viable but not culturable state. *J. Vet. Med. Sci* **2018**, *80*, 133–137. [CrossRef]

51. The, H.C.; Thanh, D.P.; Holt, K.E.; Thomson, N.R.; Baker, S. The genomic signatures of *Shigella* evolution, adaptation and geographical spread. *Nat. Rev. Microbiol.* **2016**, *14*, 235–250. [CrossRef] [PubMed]
52. Murugesan, P.; Revathi, K.; Elayaraja, S.; Vijayalakshmi, S.; Balasubramanian, T. Distribution of enteric bacteria in the sediments of Parangipettai and Cuddalore coast of India. *J. Environ. Biol.* **2012**, *33*, 705–711. [PubMed]
53. Binet, R.; Deer, D.M.; Uhlfelder, S.J. Rapid detection of *Shigella* and enteroinvasive *Escherichia coli* in produce enrichments by a conventional multiplex PCR assay. *Food Microbiol.* **2014**, *40*, 48–54. [CrossRef]
54. Thong, K.L.; Hoe, S.L.; Puthucheary, S.D.; Yasin, R. Detection of virulence genes in Malaysian *Shigella* species by multiplex PCR assay. *BMC Infect. Dis.* **2005**, *5*, 8. [CrossRef] [PubMed]
55. Vantarakis, A.; Komninou, G.; Venieri, D.; Papapetropoulou, M. Development of a multiplex PCR detection of *Salmonella* spp. and *Shigella* spp. in mussels. *Lett. Appl. Microbiol.* **2000**, *31*, 105–109. [CrossRef] [PubMed]
56. Farfán, M.J.; Garay, T.A.; Prado, C.A.; Filliol, I.; Ulloa, M.T.; Toro, C.S. A new multiplex PCR for differential identification of *Shigella flexneri* and *Shigella sonnei* and detection of *Shigella* virulence determinants. *Epidemiol. Infect.* **2010**, *138*, 525–533. [CrossRef] [PubMed]
57. Maheux, A.F.; Bissonnette, L.; Boissinot, M.; Bernier, J.-L.T.; Huppé, V.; Picard, F.J.; Bérubé, È.; Bergeron, M.G. Rapid concentration and molecular enrichment approach for sensitive detection of *Escherichia coli* and *Shigella* species in potable water samples. *Appl. Environ. Microbiol.* **2011**, *77*, 6199–6207. [CrossRef]
58. Romano, V.; Pasquale, V.; Lemee, L.; El Meouche, I.; Pestel-Caron, M.; Capuano, F.; Buono, P.; Dumontet, S. Clostridioides difficile in the environment, food, animals and humans in southern Italy: Occurrence and genetic relatedness. *Comp. Immunol. Microbiol. Infect. Dis.* **2018**, *59*, 41–46. [CrossRef] [PubMed]
59. Numberger, D.; Riedel, T.; McEwen, G.; Nübel, U.; Frentrup, M.; Schober, I.; Bunk, B.; Spröer, C.; Overmann, J.; Grossart, H.-P.; et al. Genomic analysis of three Clostridioides difficile isolates from urban water sources. *Anaerobe* **2019**, *56*, 22–26. [CrossRef]
60. Peniche, A.G.; Savidge, T.C.; Dann, S.M. Recent insights into Clostridium difficile pathogenesis. *Curr. Opin. Infect. Dis.* **2013**, *26*, 447–453. [CrossRef] [PubMed]
61. Dubberke, E.R.; Haslam, D.B.; Lanzas, C.; Bobo, L.D.; Burnham, C.A.D.; Gröhn, Y.T.; Tarr, P.I. The ecology and pathobiology of Clostridium difficile infections: An interdisciplinary challenge. *Zoonoses Public Health* **2011**, *58*, 4–20. [CrossRef] [PubMed]
62. Freeman, J.; Bauer, M.P.; Baines, S.D.; Corver, J.; Fawley, W.N.; Goorhuis, B.; Kuijper, E.J.; Wilcox, M.H. The changing epidemiology of Clostridium difficile infections. *Clin. Microbiol. Rev.* **2010**, *23*, 529–549. [CrossRef] [PubMed]
63. Hensgens, M.P.M.; Keessen, E.C.; Squire, M.M.; Riley, T.V.; Koene, M.G.J.; de Boer, E.; Lipman, L.J.A.; Kuijper, E.J. Clostridium difficile infection in the community: A zoonotic disease? *Clin. Microbiol. Infect.* **2012**, *18*, 635–645. [CrossRef] [PubMed]
64. O'Grady, K.; Knight, D.R.; Riley, T.V. Antimicrobial resistance in Clostridioides difficile. *Eur. J. Clin. Microbiol. Infect. Dis.* **2021**, *40*, 2459–2478. [CrossRef] [PubMed]
65. Nikaeen, M.; Aghili Dehnavi, H.; Hssanzadeh, A.; Jalali, M. Occurrence of Clostridium difficile in two types of wastewater treatment plants. *J. Formos. Med. Assoc.* **2015**, *114*, 663–665. [CrossRef]
66. Blasi, F.; Lovito, C.; Albini, E.; Bano, L.; Dalmonte, G.; Drigo, I.; Maresca, C.; Massacci, F.R.; Orsini, S.; Primavilla, S.; et al. Clostridioides difficile in Calves in Central Italy: Prevalence, Molecular Typing, Antimicrobial Susceptibility and Association with Antibiotic Administration. *Animals* **2021**, *11*, 515. [CrossRef]
67. Caicedo, C.; Beutel, S.; Scheper, T.; Rosenwinkel, K.H.; Nogueira, R. Occurrence of *Legionella* in wastewater treatment plants linked to wastewater characteristics. *Environ. Sci. Pollut. Res.* **2016**, *23*, 16873–16881. [CrossRef] [PubMed]
68. Guo, F.; Zhang, T.; Li, B.; Wang, Z.; Ju, F.; Liang, Y.-T. Mycobacterial species and their contribution to cholesterol degradation in wastewater treatment plants. *Sci. Rep.* **2019**, *9*, 836. [CrossRef]
69. Bierque, E.; Thibeaux, R.; Girault, D.; Soupé-Gilbert, M.-E.; Goarant, C. A systematic review of Leptospira in water and soil environments. *PLoS ONE* **2020**, *15*, e0227055. [CrossRef] [PubMed]
70. Leslie, E.; Hinds, J.; Hai, F.I. Causes, Factors, and Control Measures of Opportunistic Premise Plumbing Pathogens—A Critical Review. *Appl. Sci.* **2021**, *11*, 4474. [CrossRef]
71. Muder, R.R.; Victor, L.Y. Infection due to *Legionella* species other than *L. pneumophila*. *Clin. Infect. Dis.* **2002**, *35*, 990–998. [CrossRef]
72. Chauhan, D.; Shames, S.R. Pathogenicity and Virulence of *Legionella*: Intracellular replication and host response. *Virulence* **2021**, *12*, 1122–1144. [CrossRef]
73. Sabrià, M.; Pedro-Botet, M.L.; Gómez, J.; Roig, J.; Vilaseca, B.; Sopena, N.; Baños, V. Fluoroquinolones vs macrolides in the treatment of Legionnaires disease. *Chest* **2005**, *128*, 1401–1405. [CrossRef] [PubMed]
74. Postma, D.F.; Van Werkhoven, C.H.; Van Elden, L.J.; Thijsen, S.F.; Hoepelman, A.I.; Kluytmans, J.A.; Boersma, W.G.; Compaijen, C.J.; Van Der Wall, E.; Prins, J.M. Antibiotic treatment strategies for community acquired pneumonia in adults. *N. Engl. J. Med.* **2015**, *372*, 1312–1323. [CrossRef]
75. Garin, N.; Genné, D.; Carballo, S.; Chuard, C.; Eich, G.; Hugli, O.; Lamy, O.; Nendaz, M.; Petignat, P.-A.; Perneger, T. β-Lactam monotherapy vs β-lactam–macrolide combination treatment in moderately severe community-acquired pneumonia: A randomized noninferiority trial. *JAMA Intern. Med.* **2014**, *174*, 1894–1901. [CrossRef]
76. Johnson, M.M.; Odell, J.A. Nontuberculous mycobacterial pulmonary infections. *J. Thorac. Dis* **2014**, *6*, 210–220. [CrossRef] [PubMed]

77. Pontiroli, A.; Khera, T.T.; Oakley, B.B.; Mason, S.; Dowd, S.E.; Travis, E.R.; Erenso, G.; Aseffa, A.; Courtenay, O.; Wellington, E.M.H. Prospecting Environmental Mycobacteria: Combined Molecular Approaches Reveal Unprecedented Diversity. *PLoS ONE* **2013**, *8*, e68648. [CrossRef]
78. Chakaya, J.; Khan, M.; Ntoumi, F.; Aklillu, E.; Fatima, R.; Mwaba, P.; Kapata, N.; Mfinanga, S.; Hasnain, S.E.; Katoto, P.; et al. Global Tuberculosis Report 2020—Reflections on the Global TB burden, treatment and prevention efforts. *Int. J. Infect. Dis.* **2021**, *6*, 1201–9712. [CrossRef] [PubMed]
79. Rizzo, L.; Manaia, C.; Merlin, C.; Schwartz, T.; Dagot, C.; Ploy, M.C.; Michael, I.; Fatta-Kassinos, D. Urban wastewater treatment plants as hotspots for antibiotic resistant bacteria and genes spread into the environment: A review. *Sci. Total Environ.* **2013**, *447*, 345–360. [CrossRef] [PubMed]
80. Hyun, B.; Cha, H.-G.; Lee, N.; Yum, S.; Baek, S.H.; Shin, K. Development of an ATP assay for rapid onboard testing to detect living microorganisms in ballast water. *J. Sea Res.* **2018**, *133*, 73–80. [CrossRef]
81. Zhang, Y.; Liu, W.-T. The application of molecular tools to study the drinking water microbiome—Current understanding and future needs. *Crit. Rev. Environ. Sci. Technol.* **2019**, *49*, 1188–1235. [CrossRef]
82. Liu, Q.; Wu, J.e.; Lim, Z.Y.; Aggarwal, A.; Yang, H.; Wang, S. Evaluation of the metabolic response of *Escherichia coli* to electrolysed water by 1H NMR spectroscopy. *LWT-Food Sci. Technol.* **2017**, *79*, 428–436. [CrossRef]
83. Slavov, S.; Alusta, P.; Buzatu, D.A.; Wilkes, J.G. Feature selection from mass spectra of bacteria for serotyping *Salmonella*. *J. Anal. Appl. Pyrolysis* **2017**, *124*, 393–402. [CrossRef]
84. Zhang, Y.; Lai, B.S.; Juhas, M. Recent Advances in Aptamer Discovery and Applications. *Molecules* **2019**, *24*, 941. [CrossRef] [PubMed]
85. Zou, Y.; Duan, N.; Wu, S.; Shen, M.; Wang, Z. Selection, Identification, and Binding Mechanism Studies of an ssDNA Aptamer Targeted to Different Stages of *E. coli* O157:H7. *J. Agric. Food Chem.* **2018**, *66*, 5677–5682. [CrossRef] [PubMed]
86. Yu, X.; Chen, F.; Wang, R.; Li, Y. Whole-bacterium SELEX of DNA aptamers for rapid detection of E.coli O157:H7 using a QCM sensor. *J. Biotechnol.* **2018**, *266*, 39–49. [CrossRef] [PubMed]
87. Buss, J.E.; Cresse, M.; Doyle, S.; Buchan, B.W.; Craft, D.W.; Young, S. *Campylobacter* culture fails to correctly detect *Campylobacter* in 30% of positive patient stool specimens compared to non-cultural methods. *Eur. J. Clin. Microbiol. Infect. Dis.* **2019**, *38*, 1087–1093. [CrossRef] [PubMed]
88. Ormen, O.; Aalberg, K.; Madslien, E.H. Multiplex polymerase chain reaction detection of enteropathogens in sewage in Norway. *Acta Vet. Scand.* **2019**, *61*, 11. [CrossRef] [PubMed]
89. Yanagimoto, K.; Yamagami, T.; Uematsu, K.; Haramoto, E. Characterization of *Salmonella* Isolates from Wastewater Treatment Plant Influents to Estimate Unreported Cases and Infection Sources of Salmonellosis. *Pathogens* **2020**, *9*, 52. [CrossRef]
90. Kasturi, K.N.; Drgon, T. Real-Time PCR Method for Detection of *Salmonella* spp. in Environmental Samples. *Appl. Environ. Microbiol.* **2017**, *83*, e00644-17. [CrossRef]
91. Lekunberri, I.; Balcázar, J.L.; Borrego, C.M. Detection and quantification of the plasmid-mediated mcr-1 gene conferring colistin resistance in wastewater. *Int. J. Antimicrob. Agents* **2017**, *50*, 734–736. [CrossRef] [PubMed]
92. Sandberg, K.D.; Ishii, S.; LaPara, T.M. A Microfluidic Quantitative Polymerase Chain Reaction Method for the Simultaneous Analysis of Dozens of Antibiotic Resistance and Heavy Metal Resistance Genes. *Environ. Sci. Technol. Lett.* **2018**, *5*, 20–25. [CrossRef]
93. Pan, H.; Dong, K.; Rao, L.; Zhao, L.; Wu, X.; Wang, Y.; Liao, X. Quantitative detection of viable but nonculturable state *Escherichia coli* O157:H7 by ddPCR combined with propidium monoazide. *Food Control.* **2020**, *112*, 107140. [CrossRef]
94. Li, Y. Establishment and Application of a Visual DNA Microarray for the Detection of Food-borne Pathogens. *Anal. Sci.* **2016**, *32*, 215–218. [CrossRef]
95. Lee, J.-E.; Mun, H.; Kim, S.-R.; Kim, M.-G.; Chang, J.-Y.; Shim, W.-B. A colorimetric Loop-mediated isothermal amplification (LAMP) assay based on HRP-mimicking molecular beacon for the rapid detection of Vibrio parahaemolyticus. *Biosens. Bioelectron.* **2020**, *151*, 111968. [CrossRef] [PubMed]
96. Salimi, G.; Mousavi, Z.E.; Kiani, H. Efficiency of fluorescence in situ hybridization (FISH) method for the rapid detection of *Salmonella* in minced lamb meat: Method analysis and optimization. *J. Microbiol. Methods* **2020**, *175*, 105989. [CrossRef] [PubMed]
97. Lyu, J.; Zhang, J.; Ren, X. Detection and identification of bacterial pathogens directly from sputum samples by pyrosequencing. *J. Med. Microbiol.* **2019**, *68*, 368–373. [CrossRef] [PubMed]
98. Winand, R.; Bogaerts, B.; Hoffman, S.; Lefevre, L.; Delvoye, M.; Van Braekel, J.; Fu, Q.; Roosens, N.H.; De Keersmaecker, S.C.; Vanneste, K. Targeting the 16s rRNA gene for bacterial identification in complex mixed samples: Comparative evaluation of second (illumina) and third (oxford nanopore technologies) generation sequencing technologies. *Int. J. Mol. Sci.* **2020**, *21*, 298. [CrossRef] [PubMed]
99. Su, J.Q.; An, X.L.; Li, B.; Chen, Q.L.; Gillings, M.R.; Chen, H.; Zhang, T.; Zhu, Y.G. Metagenomics of urban sewage identifies an extensively shared antibiotic resistome in China. *Microbiome* **2017**, *5*, 84. [CrossRef]
100. Park, S.; Kim, Y.T.; Kim, Y.-K. Optical enzyme-linked immunosorbent assay on a strip for detection of *Salmonella typhimurium*. *BioChip J.* **2010**, *4*, 110–116. [CrossRef]
101. Yang, X.; Feng, L.; Qin, X. Preparation of the Cf-GQDs-*Escherichia coli* O157: H7 bioprobe and its application in optical imaging and sensing of *Escherichia coli* O157: H7. *Food Anal. Methods* **2018**, *11*, 2280–2286. [CrossRef]

102. Sun, Y.; Chang, Y.; Zhang, Q.; Liu, M. An Origami Paper-Based Device Printed with DNAzyme-Containing DNA Superstructures for *Escherichia coli* Detection. *Micromachines* **2019**, *10*, 531. [CrossRef] [PubMed]
103. Khan, I.U.; Gannon, V.; Kent, R.; Koning, W.; Lapen, D.R.; Miller, J.; Neumann, N.; Phillips, R.; Robertson, W.; Topp, E. Development of a rapid quantitative PCR assay for direct detection and quantification of culturable and non-culturable *Escherichia coli* from agriculture watersheds. *J. Microbiol. Methods* **2007**, *69*, 480–488. [CrossRef]
104. Maheux, A.F.; Bérubé, È.; Boudreau, D.K.; Villéger, R.; Cantin, P.; Boissinot, M.; Bissonnette, L.; Bergeron, M.G. Abilities of the mCP agar method and CRENAME alpha toxin-specific real-time PCR assay to detect *Clostridium perfringens* spores in drinking water. *Appl. Environ. Microbiol.* **2013**, *79*, 7654–7661. [CrossRef]
105. Ram, S.; Vajpayee, P.; Shanker, R. Rapid culture-independent quantitative detection of enterotoxigenic *Escherichia coli* in surface waters by real-time PCR with molecular beacon. *Environ. Sci. Technol.* **2008**, *42*, 4577–4582. [CrossRef]
106. Kralik, P.; Ricchi, M. A Basic Guide to Real Time PCR in Microbial Diagnostics: Definitions, Parameters, and Everything. *Front. Microbiol* **2017**, *8*, 108. [CrossRef] [PubMed]
107. Omar, K.; Barnard, T. Detection of diarrhoeagenic *Escherichia coli* in clinical and environmental water sources in South Africa using single-step 11-gene m-PCR. *World J. Microbiol. Biotechnol.* **2014**, *30*, 2663–2671. [CrossRef] [PubMed]
108. Racki, N.; Dreo, T.; Gutierrez-Aguirre, I.; Blejec, A.; Ravnikar, M. Reverse transcriptase droplet digital PCR shows high resilience to PCR inhibitors from plant, soil and water samples. *Plant. Methods* **2014**, *10*, 42. [CrossRef] [PubMed]
109. Taylor, S.C.; Carbonneau, J.; Shelton, D.N.; Boivin, G. Optimization of Droplet Digital PCR from RNA and DNA extracts with direct comparison to RT-qPCR: Clinical implications for quantification of Oseltamivir-resistant subpopulations. *J. Virol. Methods* **2015**, *224*, 58–66. [CrossRef] [PubMed]
110. Devonshire, A.S.; Sanders, R.; Whale, A.S.; Nixon, G.J.; Cowen, S.; Ellison, S.L.; Parkes, H.; Pine, P.S.; Salit, M.; McDaniel, J.; et al. An international comparability study on quantification of mRNA gene expression ratios: CCQM-P103.1. *Biomol. Detect. Quantif.* **2016**, *8*, 15–28. [CrossRef] [PubMed]
111. Kokkoris, V.; Vukicevich, E.; Richards, A.; Thomsen, C.; Hart, M.M. Challenges Using Droplet Digital PCR for Environmental Samples. *Appl. Microbiol.* **2021**, *1*, 74–88. [CrossRef]
112. Singh, G.; Sithebe, A.; Enitan, A.M.; Kumari, S.; Bux, F.; Stenstrom, T.A. Comparison of droplet digital PCR and quantitative PCR for the detection of *Salmonella* and its application for river sediments. *J. Water Health* **2017**, *15*, 505–508. [CrossRef]
113. Yang, J.; Zhang, N.; Lv, J.; Zhu, P.; Pan, X.; Hu, J.; Wu, W.; Li, S.; Li, H. Comparing the performance of conventional PCR, RTQ-PCR, and droplet digital PCR assays in detection of *Shigella*. *Mol. Cell. Probes* **2020**, *51*, 101531. [CrossRef] [PubMed]
114. Gosiewski, T.; Flis, A.; Sroka, A.; Kędzierska, A.; Pietrzyk, A.; Kędzierska, J.; Drwiła, R.; Bulanda, M. Comparison of nested, multiplex, qPCR.; FISH.; SeptiFast and blood culture methods in detection and identification of bacteria and fungi in blood of patients with sepsis. *BMC Microbiol.* **2014**, *14*, 313. [CrossRef]
115. Ibekwe, A.M.; Watt, P.M.; Grieve, C.M.; Sharma, V.K.; Lyons, S.R. Multiplex fluorogenic real-time PCR for detection and quantification of *Escherichia coli* O157:H7 in dairy wastewater wetlands. *Appl Env. Microbiol* **2002**, *68*, 4853–4862. [CrossRef]
116. LaGier, M.J.; Joseph, L.A.; Passaretti, T.V.; Musser, K.A.; Cirino, N.M. A real-time multiplexed PCR assay for rapid detection and differentiation of *Campylobacter jejuni* and *Campylobacter coli*. *Mol. Cell. Probes* **2004**, *18*, 275–282. [CrossRef]
117. Reischer, G.H.; Kasper, D.C.; Steinborn, R.; Farnleitner, A.H.; Mach, R.L. A quantitative real-time PCR assay for the highly sensitive and specific detection of human faecal influence in spring water from a large alpine catchment area. *Lett. Appl. Microbiol.* **2007**, *44*, 351–356. [CrossRef]
118. Nayak, A.; Rose, J. Detection of *Helicobacter pylori* in sewage and water using a new quantitative PCR method with SYBR®green. *J. Appl. Microbiol.* **2007**, *103*, 1931–1941. [CrossRef] [PubMed]
119. Wery, N.; Lhoutellier, C.; Ducray, F.; Delgenes, J.-P.; Godon, J.-J. Behaviour of pathogenic and indicator bacteria during urban wastewater treatment and sludge composting, as revealed by quantitative PCR. *Water Res.* **2008**, *42*, 53–62. [CrossRef] [PubMed]
120. Haugland, R.A.; Siefring, S.C.; Wymer, L.J.; Brenner, K.P.; Dufour, A.P. Comparison of Enterococcus measurements in freshwater at two recreational beaches by quantitative polymerase chain reaction and membrane filter culture analysis. *Water Res.* **2005**, *39*, 559–568. [CrossRef] [PubMed]
121. Baeumner, A.J.; Humiston, M.C.; Montagna, R.A.; Durst, R.A. Detection of viable oocysts of *Cryptosporidium parvum* following nucleic acid sequence based amplification. *Anal. Chem.* **2001**, *73*, 1176–1180. [CrossRef] [PubMed]
122. Pitkänen, T.; Ryu, H.; Elk, M.; Hokajärvi, A.-M.; Siponen, S.; Vepsäläinen, A.; Räsänen, P.; Santo Domingo, J.W. Detection of Fecal Bacteria and Source Tracking Identifiers in Environmental Waters Using rRNA-Based RT-qPCR and rDNA-Based qPCR Assays. *Environ. Sci. Technol.* **2013**, *47*, 13611–13620. [CrossRef] [PubMed]
123. Tiwari, A.; Niemelä, S.I.; Vepsäläinen, A.; Rapala, J.; Kalso, S.; Pitkänen, T. Comparison of Colilert-18 with miniaturised most probable number method for monitoring of *Escherichia coli* in bathing water. *J. Water Health* **2015**, *14*, 121–131. [CrossRef] [PubMed]
124. Inkinen, J.; Siponen, S.; Jayaprakash, B.; Tiwari, A.; Hokajärvi, A.-M.; Pursiainen, A.; Ikonen, J.; Kauppinen, A.; Miettinen, I.T.; Paananen, J.; et al. Diverse and active archaea communities occur in non-disinfected drinking water systems-Less activity revealed in disinfected and hot water systems. *Water Res. X* **2021**, *12*, 100101. [CrossRef]
125. Fumian, T.M.; Leite, J.P.G.; Castello, A.A.; Gaggero, A.; Caillou, M.S.L.d.; Miagostovich, M.P. Detection of rotavirus A in sewage samples using multiplex qPCR and an evaluation of the ultracentrifugation and adsorption-elution methods for virus concentration. *J. Virol. Methods* **2010**, *170*, 42–46. [CrossRef]

126. Al-Tebrineh, J.; Pearson, L.A.; Yasar, S.A.; Neilan, B.A. A multiplex qPCR targeting hepato- and neurotoxigenic cyanobacteria of global significance. *Harmful Algae* **2012**, *15*, 19–25. [CrossRef]
127. Kamau, E.; Alemayehu, S.; Feghali, K.C.; Saunders, D.; Ockenhouse, C.F. Multiplex qPCR for detection and absolute quantification of malaria. *PLoS ONE* **2013**, *8*, e71539. [CrossRef]
128. Bumgarner, R. Overview of DNA microarrays: Types, applications, and their future. *Curr. Protoc. Mol. Biol.* **2013**, *101*, 22-1. [CrossRef]
129. Inoue, D.; Hinoura, T.; Suzuki, N.; Pang, J.; Malla, R.; Shrestha, S.; Chapagain, S.K.; Matsuzawa, H.; Nakamura, T.; Tanaka, Y. High-throughput DNA microarray detection of pathogenic bacteria in shallow well groundwater in the Kathmandu Valley, Nepal. *Curr. Microbiol.* **2015**, *70*, 43–50. [CrossRef] [PubMed]
130. Lemuth, K.; Rupp, S. Microarrays as Research Tools and Diagnostic Devices. In *RNA and DNA Diagnostics*; Springer: Cham, Switzerland, 2015; pp. 259–280. [CrossRef]
131. Cristescu, M.E.; Hebert, P.D.N. Uses and Misuses of Environmental DNA in Biodiversity Science and Conservation. *Annu. Rev. Ecol. Evol. Syst.* **2018**, *49*, 209–230. [CrossRef]
132. Sheet, O.; Grabowski, N.; Klein, G.; Abdulmawjood, A. Development and validation of a loop mediated isothermal amplification (LAMP) assay for the detection of *Staphylococcus aureus* in bovine mastitis milk samples. *Mol. Cell. Probes* **2016**, *30*, 320–325. [CrossRef] [PubMed]
133. Niu, J.H.; Jian, H.; Guo, Q.X.; Chen, C.L.; Wang, X.Y.; Liu, Q.; Guo, Y.D. Evaluation of loop-mediated isothermal amplification (LAMP) assays based on 5S rDNA-IGS2 regions for detecting *Meloidogyne enterolobii. Plant. Pathol.* **2012**, *61*, 809–819. [CrossRef]
134. Nzelu, C.O.; Cáceres, A.G.; Guerrero-Quincho, S.; Tineo-Villafuerte, E.; Rodriquez-Delfin, L.; Mimori, T.; Uezato, H.; Katakura, K.; Gomez, E.A.; Guevara, A.G. A rapid molecular diagnosis of cutaneous leishmaniasis by colorimetric malachite green-loop-mediated isothermal amplification (LAMP) combined with an FTA card as a direct sampling tool. *Acta Trop.* **2016**, *153*, 116–119. [CrossRef]
135. Amann, R.; Fuchs, B.M. Single-cell identification in microbial communities by improved fluorescence in situ hybridization techniques. *Nat. Rev. Microbiol.* **2008**, *6*, 339–348. [CrossRef]
136. Lukumbuzya, M.; Schmid, M.; Pjevac, P.; Daims, H. A Multicolor Fluorescence in situ Hybridization Approach Using an Extended Set of Fluorophores to Visualize Microorganisms. *Front. Microbiol.* **2019**, *10*, 1383. [CrossRef]
137. Santiago, P.; Jiménez-Belenguer, A.; García-Hernández, J.; Estellés, R.M.; Hernández Pérez, M.; Castillo López, M.A.; Ferrús, M.A.; Moreno, Y. High prevalence of *Salmonella* spp. in wastewater reused for irrigation assessed by molecular methods. *Int. J. Hyg. Environ. Health* **2018**, *221*, 95–101. [CrossRef] [PubMed]
138. Gwinn, M.; MacCannell, D.; Armstrong, G.L. Next-Generation Sequencing of Infectious Pathogens. *JAMA* **2019**, *321*, 893–894. [CrossRef]
139. Behjati, S.; Tarpey, P.S. What is next generation sequencing? *Arch. Dis. Child. Educ. Pract.* **2013**, *98*, 236–238. [CrossRef]
140. Treangen, T.J.; Salzberg, S.L. Repetitive DNA and next-generation sequencing: Computational challenges and solutions. *Nat. Rev. Genet.* **2012**, *13*, 36–46. [CrossRef] [PubMed]
141. Zhu, X.; Yan, S.; Yuan, F.; Wan, S. The Applications of Nanopore Sequencing Technology in Pathogenic Microorganism Detection. *Can. J. Infect. Dis. Med. Microbiol.* **2020**, *2020*, 6675206. [CrossRef]
142. Greninger, A.L.; Naccache, S.N.; Federman, S.; Yu, G.; Mbala, P.; Bres, V.; Stryke, D.; Bouquet, J.; Somasekar, S.; Linnen, J.M.; et al. Rapid metagenomic identification of viral pathogens in clinical samples by real-time nanopore sequencing analysis. *Genome Med.* **2015**, *7*, 99. [CrossRef]
143. Schmidt, M.H.W.; Vogel, A.; Denton, A.K.; Istace, B.; Wormit, A.; van de Geest, H.; Bolger, M.E.; Alseekh, S.; Maß, J.; Pfaff, C.; et al. De Novo Assembly of a New Solanum pennellii Accession Using Nanopore Sequencing. *Plant. Cell* **2017**, *29*, 2336–2348. [CrossRef] [PubMed]
144. Cao, M.D.; Ganesamoorthy, D.; Elliott, A.G.; Zhang, H.; Cooper, M.A.; Coin, L.J.M. Streaming algorithms for identification of pathogens and antibiotic resistance potential from real-time MinION(TM) sequencing. *Gigascience* **2016**, *5*, 32. [CrossRef]
145. Ammar, R.; Paton, T.A.; Torti, D.; Shlien, A.; Bader, G.D. Long read nanopore sequencing for detection of HLA and CYP2D6 variants and haplotypes. *F1000Res* **2015**, *4*, 17. [CrossRef]
146. De Siqueira, G.M.V.; Pereira-Dos-Santos, F.M.; Silva-Rocha, R.; Guazzaroni, M.-E. Nanopore Sequencing Provides Rapid and Reliable Insight Into Microbial Profiles of Intensive Care Units. *Front. Public Health* **2021**, *9*, 710985. [CrossRef]
147. Wattam, A.R.; Inzana, T.J.; Williams, K.P.; Mane, S.P.; Shukla, M.; Almeida, N.F.; Dickerman, A.W.; Mason, S.; Moriyón, I.; O'Callaghan, D.; et al. Comparative genomics of early-diverging Brucella strains reveals a novel lipopolysaccharide biosynthesis pathway. *mBio* **2012**, *3*, e00246-12. [CrossRef] [PubMed]
148. Widmer, A.F.; Frei, R.; Kuijper, J.; Wilcox, M.H.; Schindler, R.; Spaniol, V.; Goldenberger, D.; Egli, A.; Tschudin-Sutter, S. Multicenter Prevalence Study Comparing Molecular and Toxin Assays for *Clostridioides difficile* Surveillance, Switzerland. *Emerg. Infect. Dis.* **2020**, *26*, 2370. [CrossRef] [PubMed]
149. Mao, K.; Zhang, H.; Pan, Y.; Yang, Z. Biosensors for wastewater-based epidemiology for monitoring public health. *Water Res.* **2021**, *191*, 116787. [CrossRef] [PubMed]
150. Yildirim, N.; Long, F.; Gu, A.Z. Aptamer based *E. coli* detection in waste waters by portable optical biosensor system. In Proceedings of the 40th Annual Northeast Bioengineering Conference (NEBEC), Boston, MA, USA, 25–27 April 2014; pp. 1–3.

151. Doria, G.; Conde, J.; Veigas, B.; Giestas, L.; Almeida, C.; Assunção, M.; Rosa, J.; Baptista, P.V. Noble metal nanoparticles for biosensing applications. *Sensors* **2012**, *12*, 1657–1687. [CrossRef] [PubMed]
152. Jyoti, A.; Tomar, R.S.; Shanker, R. Nanosensors for the detection of pathogenic bacteria. In *Nanoscience in Food and Agriculture 1*; Springer: Cham, Switzerland, 2016; pp. 129–150.
153. Ahmed, A.; Rushworth, J.V.; Hirst, N.A.; Millner, P.A. Biosensors for whole-cell bacterial detection. *Clin. Microbiol. Rev.* **2014**, *27*, 631–646. [CrossRef] [PubMed]
154. Choi, J.R.; Yong, K.W.; Tang, R.; Gong, Y.; Wen, T.; Li, F.; Pingguan-Murphy, B.; Bai, D.; Xu, F. Advances and challenges of fully integrated paper-based point-of-care nucleic acid testing. *TrAC Trends Anal. Chem.* **2017**, *93*, 37–50. [CrossRef]
155. Suea-Ngam, A.; Choopara, I.; Li, S.; Schmelcher, M.; Somboonna, N.; Howes, P.D.; deMello, A.J. In Situ Nucleic Acid Amplification and Ultrasensitive Colorimetric Readout in a Paper-Based Analytical Device Using Silver Nanoplates. *Adv. Healthc. Mater.* **2021**, *10*, 2001755. [CrossRef]
156. Mao, K.; Zhang, H.; Yang, Z. Can a Paper-Based Device Trace COVID-19 Sources with Wastewater-Based Epidemiology? *Environ. Sci. Technol.* **2020**, *54*, 3733–3735. [CrossRef]
157. Boehle, K.E.; Gilliand, J.; Wheeldon, C.R.; Holder, A.; Adkins, J.A.; Geiss, B.J.; Ryan, E.P.; Henry, C.S. Utilizing Paper-Based Devices for Antimicrobial-Resistant Bacteria Detection. *Angew. Chem. Int. Ed.* **2017**, *56*, 6886–6890. [CrossRef]
158. Nishat, S.; Jafry, A.T.; Martinez, A.W.; Awan, F.R. Paper-based microfluidics: Simplified fabrication and assay methods. *Sens. Actuators B Chem.* **2021**, *336*, 129681. [CrossRef]
159. Ganaja, K.A.; Chaplan, C.A.; Zhang, J.; Martinez, N.W.; Martinez, A.W. Paper Microzone Plates as Analytical Tools for Studying Enzyme Stability: A Case Study on the Stabilization of Horseradish Peroxidase Using Trehalose and SU-8 Epoxy Novolac Resin. *Anal. Chem.* **2017**, *89*, 5333–5341. [CrossRef]
160. Daughton, C.G. Illicit Drugs in Municipal Sewage. In *Pharmaceuticals and Care Products in the Environment*; American Chemical Society: Washington, DC, USA, 2001; pp. 348–364.
161. Aarestrup, F.M.; Woolhouse, M.E.J. Using sewage for surveillance of antimicrobial resistance. *Science* **2020**, *367*, 630–632. [CrossRef] [PubMed]
162. Hellmer, M.; Paxeus, N.; Magnius, L.; Enache, L.; Arnholm, B.; Johansson, A.; Bergstrom, T.; Norder, H. Detection of pathogenic viruses in sewage provided early warnings of hepatitis A virus and norovirus outbreaks. *Appl. Env. Microbiol.* **2014**, *80*, 6771–6781. [CrossRef] [PubMed]
163. Park, S.H.; Hanning, I.; Jarquin, R.; Moore, P.; Donoghue, D.J.; Donoghue, A.M.; Ricke, S.C. Multiplex PCR assay for the detection and quantification of *Campylobacter* spp., *Escherichia coli* O157:H7, and *Salmonella* serotypes in water samples. *FEMS Microbiol. Lett.* **2011**, *316*, 7–15. [CrossRef] [PubMed]
164. Marangi, M.; Giangaspero, A.; Lacasella, V.; Lonigro, A.; Gasser, R.B. Multiplex PCR for the detection and quantification of zoonotic taxa of Giardia, Cryptosporidium and Toxoplasma in wastewater and mussels. *Mol. Cell. Probes* **2015**, *29*, 122–125. [CrossRef] [PubMed]
165. An, X.L.; Su, J.Q.; Li, B.; Ouyang, W.Y.; Zhao, Y.; Chen, Q.L.; Cui, L.; Chen, H.; Gillings, M.R.; Zhang, T.; et al. Tracking antibiotic resistome during wastewater treatment using high throughput quantitative PCR. *Environ. Int.* **2018**, *117*, 146–153. [CrossRef] [PubMed]
166. Karkman, A.; Johnson, T.A.; Lyra, C.; Stedtfeld, R.D.; Tamminen, M.; Tiedje, J.M.; Virta, M. High-throughput quantification of antibiotic resistance genes from an urban wastewater treatment plant. *FEMS Microbiol. Ecol.* **2016**, *92*, fiw01. [CrossRef] [PubMed]
167. Urbaniak, C.; Sielaff, A.C.; Frey, K.G.; Allen, J.E.; Singh, N.; Jaing, C.; Wheeler, K.; Venkateswaran, K. Detection of antimicrobial resistance genes associated with the International Space Station environmental surfaces. *Sci. Rep.* **2018**, *8*, 814. [CrossRef] [PubMed]
168. Mumy, K.L.; Findlay, R.H. Convenient determination of DNA extraction efficiency using an external DNA recovery standard and quantitative-competitive PCR. *J. Microbiol. Methods* **2004**, *57*, 259–268. [CrossRef] [PubMed]
169. Gyawali, P.; Ahmed, W.; Jagals, P.; Sidhu, J.P.S.; Toze, S. Comparison of concentration methods for rapid detection of hookworm ova in wastewater matrices using quantitative PCR. *Exp. Parasitol.* **2015**, *159*, 160–167. [CrossRef]
170. Lemarchand, K.; Berthiaume, F.; Maynard, C.; Harel, J.; Payment, P.; Bayardelle, P.; Masson, L.; Brousseau, R. Optimization of microbial DNA extraction and purification from raw wastewater samples for downstream pathogen detection by microarrays. *J. Microbiol. Methods* **2005**, *63*, 115–126. [CrossRef]
171. Li, A.-D.; Metch, J.W.; Wang, Y.; Garner, E.; Zhang, A.N., Riquelme, M.V., Vikesland, P.J.; Pruden, A.; Zhang, T. Effects of sample preservation and DNA extraction on enumeration of antibiotic resistance genes in wastewater. *FEMS Microbiol. Ecol.* **2017**, *94*, fix189. [CrossRef] [PubMed]
172. Walden, C.; Carbonero, F.; Zhang, W. Assessing impacts of DNA extraction methods on next generation sequencing of water and wastewater samples. *J. Microbiol. Methods* **2017**, *141*, 10–16. [CrossRef] [PubMed]
173. Bonetta, S.; Pignata, C.; Lorenzi, E.; De Ceglia, M.; Meucci, L.; Bonetta, S.; Gilli, G.; Carraro, E. Detection of pathogenic *Campylobacter*, *E. coli* O157:H7 and *Salmonella* spp. in wastewater by PCR assay. *Environ. Sci. Pollut. Res.* **2016**, *23*, 15302–15309. [CrossRef] [PubMed]
174. Kibbee, R.J.; Örmeci, B. Development of a sensitive and false-positive free PMA-qPCR viability assay to quantify VBNC *Escherichia coli* and evaluate disinfection performance in wastewater effluent. *J. Microbiol. Methods* **2017**, *132*, 139–147. [CrossRef] [PubMed]

175. Baccari, O.; Elleuch, J.; Barkallah, M.; Boukedi, H.; Ayed, N.B.; Hammami, A.; Fendri, I.; Abdelkafi, S. Development of a new TaqMan-based PCR assay for the specific detection and quantification of Simkania negevensis. *Mol. Cell. Probes* **2020**, *53*, 101645. [CrossRef]
176. Liu, P.; Ibaraki, M.; Kapoor, R.; Amin, N.; Das, A.; Miah, R.; Mukhopadhyay, A.K.; Rahman, M.; Dutta, S.; Moe, C.L. Development of Moore Swab and Ultrafiltration Concentration and Detection Methods for *Salmonella Typhi* and *Salmonella Paratyphi* A in Wastewater and Application in Kolkata, India and Dhaka, Bangladesh. *Front. Microbiol.* **2021**, *12*, 684094. [CrossRef]
177. Jikumaru, A.; Ishii, S.; Fukudome, T.; Kawahara, Y.; Iguchi, A.; Masago, Y.; Nukazawa, K.; Suzuki, Y. Fast, sensitive, and reliable detection of waterborne pathogens by digital PCR after coagulation and foam concentration. *J. Biosci. Bioeng.* **2020**, *130*, 76–81. [CrossRef] [PubMed]
178. Pärnänen, K.M.M.; Narciso-da-Rocha, C.; Kneis, D.; Berendonk, T.U.; Cacace, D.; Do, T.T.; Elpers, C.; Fatta-Kassinos, D.; Henriques, I.; Jaeger, T.; et al. Antibiotic resistance in European wastewater treatment plants mirrors the pattern of clinical antibiotic resistance prevalence. *Sci. Adv.* **2019**, *5*, eaau9124. [CrossRef]
179. Nathaniel, B.R.; Ghai, M.; Druce, M.; Maharaj, I.; Olaniran, A.O. Development of a loop-mediated isothermal amplification assay targeting lmo0753 gene for detection of Listeria monocytogenes in wastewater. *Lett. Appl. Microbiol.* **2019**, *69*, 264–270. [CrossRef]
180. Riquelme, M.V.; Leng, W.; Carzolio, M.; Pruden, A.; Vikesland, P. Stable oligonucleotide-functionalized gold nanosensors for environmental biocontaminant monitoring. *J. Environ. Sci.* **2017**, *62*, 49–59. [CrossRef] [PubMed]
181. Tang, R.; Yang, H.; Gong, Y.; You, M.; Liu, Z.; Choi, J.R.; Wen, T.; Qu, Z.; Mei, Q.; Xu, F. A fully disposable and integrated paper-based device for nucleic acid extraction, amplification and detection. *Lab. A Chip* **2017**, *17*, 1270–1279. [CrossRef] [PubMed]
182. Guo, J.; Li, J.; Chen, H.; Bond, P.L.; Yuan, Z. Metagenomic analysis reveals wastewater treatment plants as hotspots of antibiotic resistance genes and mobile genetic elements. *Water Res.* **2017**, *123*, 468–478. [CrossRef] [PubMed]
183. Yasir, M. Analysis of Microbial Communities and Pathogen Detection in Domestic Sewage Using Metagenomic Sequencing. *Diversity* **2021**, *13*, 6. [CrossRef]
184. Numberger, D.; Ganzert, L.; Zoccarato, L.; Mühldorfer, K.; Sauer, S.; Grossart, H.-P.; Greenwood, A.D. Characterization of bacterial communities in wastewater with enhanced taxonomic resolution by full-length 16S rRNA sequencing. *Sci. Rep.* **2019**, *9*, 9673. [CrossRef]
185. Oluseyi Osunmakinde, C.; Selvarajan, R.; Mamba, B.B.; Msagati, T.A.M. Profiling Bacterial Diversity and Potential Pathogens in Wastewater Treatment Plants Using High-Throughput Sequencing Analysis. *Microorganisms* **2019**, *7*, 506. [CrossRef] [PubMed]
186. Ciesielski, M.; Blackwood, D.; Clerkin, T.; Gonzalez, R.; Thompson, H.; Larson, A.; Noble, R. Assessing sensitivity and reproducibility of RT-ddPCR and RT-qPCR for the quantification of SARS-CoV-2 in wastewater. *J. Virol. Methods* **2021**, *297*, 114230. [CrossRef]
187. Nshimyimana, J.P.; Cruz, M.C.; Wuertz, S.; Thompson, J.R. Variably improved microbial source tracking with digital droplet PCR. *Water Res.* **2019**, *159*, 192–202. [CrossRef] [PubMed]
188. Gordillo, R.; Rodríguez, A.; Werning, M.L.; Bermúdez, E.; Rodríguez, M. Quantification of viable *Escherichia coli* O157: H7 in meat products by duplex real-time PCR assays. *Meat Sci.* **2014**, *96*, 964–970. [CrossRef]
189. Garrido, A.; Chapela, M.-J.; Román, B.; Ferreira, M.; Lago, J.; Vieites, J.M.; Cabado, A.G. Development of a multiplex real-time PCR method for simultaneous detection of *Salmonella enterica*, *Shigella flexneri* and *Listeria monocytogenes* in processed food samples. *Eur. Food Res. Technol.* **2012**, *234*, 571–580. [CrossRef]
190. Collins, S.; Jorgensen, F.; Willis, C.; Walker, J. Real-time PCR to supplement gold-standard culture-based detection of *Legionella* in environmental samples. *J. Appl. Microbiol.* **2015**, *119*, 1158–1169. [CrossRef] [PubMed]
191. Cui, B.; Liang, S. Monitoring Opportunistic Pathogens in Domestic Wastewater from a Pilot-Scale Anaerobic Biofilm Reactor to Reuse in Agricultural Irrigation. *Water* **2019**, *11*, 1283. [CrossRef]
192. Wong, Y.P.; Othman, S.; Lau, Y.L.; Radu, S.; Chee, H.Y. Loop-mediated isothermal amplification (LAMP): A versatile technique for detection of micro-organisms. *J. Appl. Microbiol.* **2018**, *124*, 626–643. [CrossRef] [PubMed]
193. Yang, L.; Lou, J.; Wang, H.; Wu, L.; Xu, J. Use of an improved high-throughput absolute abundance quantification method to characterize soil bacterial community and dynamics. *Sci. Total Environ.* **2018**, *633*, 360–371. [CrossRef]
194. Lou, J.; Yang, L.; Wang, H.; Wu, L.; Xu, J. Assessing soil bacterial community and dynamics by integrated high-throughput absolute abundance quantification. *PeerJ* **2018**, *6*, e4514. [CrossRef]
195. Tourlousse, D.M.; Yoshiike, S.; Ohashi, A.; Matsukura, S.; Noda, N.; Sekiguchi, Y. Synthetic spike-in standards for high-throughput 16S rRNA gene amplicon sequencing. *Nucleic Acids Res.* **2017**, *45*, e23. [CrossRef]
196. Wang, X.; Howe, S.; Deng, F.; Zhao, J. Current Applications of Absolute Bacterial Quantification in Microbiome Studies and Decision-Making Regarding Different Biological Questions. *Microorganisms* **2021**, *9*, 1797. [CrossRef]
197. Ejeian, F.; Etedali, P.; Mansouri-Tehrani, H.-A.; Soozanipour, A.; Low, Z.-X.; Asadnia, M.; Taheri-Kafrani, A.; Razmjou, A. Biosensors for wastewater monitoring: A review. *Biosens. Bioelectron.* **2018**, *118*, 66–79. [CrossRef]
198. Garner, E.; Davis, B.C.; Milligan, E.; Blair, M.F.; Keenum, I.; Maile-Moskowitz, A.; Pan, J.; Gnegy, M.; Liguori, K.; Gupta, S.; et al. Next generation sequencing approaches to evaluate water and wastewater quality. *Water Res.* **2021**, *194*, 116907. [CrossRef]
199. Shapiro, E.; Biezuner, T.; Linnarsson, S. Single-cell sequencing-based technologies will revolutionize whole-organism science. *Nat. Rev. Genet.* **2013**, *14*, 618–630. [CrossRef]

200. Xue, J.; Schmitz, B.W.; Caton, K.; Zhang, B.; Zabaleta, J.; Garai, J.; Taylor, C.M.; Romanchishina, T.; Gerba, C.P.; Pepper, I.L.; et al. Assessing the spatial and temporal variability of bacterial communities in two Bardenpho wastewater treatment systems via Illumina MiSeq sequencing. *Sci. Total Environ.* **2019**, *657*, 1543–1552. [CrossRef]
201. Stamps, B.W.; Spear, J.R. Identification of Metagenome-Assembled Genomes Containing Antimicrobial Resistance Genes, Isolated from an Advanced Water Treatment Facility. *Microbiol. Resour. Announc.* **2020**, *9*, e00003-20. [CrossRef]
202. Alneberg, J.; Karlsson, C.M.G.; Divne, A.-M.; Bergin, C.; Homa, F.; Lindh, M.V.; Hugerth, L.W.; Ettema, T.J.G.; Bertilsson, S.; Andersson, A.F.; et al. Genomes from uncultivated prokaryotes: A comparison of metagenome-assembled and single-amplified genomes. *Microbiome* **2018**, *6*, 173. [CrossRef] [PubMed]
203. Pascault, N.; Loux, V.; Derozier, S.; Martin, V.; Debroas, D.; Maloufi, S.; Humbert, J.-F.; Leloup, J. Technical challenges in metatranscriptomic studies applied to the bacterial communities of freshwater ecosystems. *Genetica* **2015**, *143*, 157–167. [CrossRef] [PubMed]
204. Yergeau, E.; Masson, L.; Elias, M.; Xiang, S.; Madey, E.; Huang, H.; Brooks, B.; Beaudette, L.A. Comparison of Methods to Identify Pathogens and Associated Virulence Functional Genes in Biosolids from Two Different Wastewater Treatment Facilities in Canada. *PLoS ONE* **2016**, *11*, e0153554. [CrossRef] [PubMed]
205. World Health Organization. *Global Antimicrobial Resistance Surveillance System: Manual for Early Implementation*; World Health Organization: Geneva, Switzerland, 2015.
206. Bitton, G. *Microbiology of Drinking Water Production and Distribution*; John Wiley & Sons: Hoboken, NJ, USA, 2014; p. 312.
207. Sun, Y.; Shen, Y.-X.; Liang, P.; Zhou, J.; Yang, Y.; Huang, X. Multiple antibiotic resistance genes distribution in ten large-scale membrane bioreactors for municipal wastewater treatment. *Bioresour. Technol.* **2016**, *222*, 100–106. [CrossRef] [PubMed]
208. Mao, D.; Yu, S.; Rysz, M.; Luo, Y.; Yang, F.; Li, F.; Hou, J.; Mu, Q.; Alvarez, P. Prevalence and proliferation of antibiotic resistance genes in two municipal wastewater treatment plants. *Water Res.* **2015**, *85*, 458–466. [CrossRef]
209. Nesme, J.; Cécillon, S.; Delmont, T.O.; Monier, J.-M.; Vogel, T.M.; Simonet, P. Large-scale metagenomic-based study of antibiotic resistance in the environment. *Curr. Biol.* **2014**, *24*, 1096–1100. [CrossRef]
210. Xia, Y.; Li, A.-D.; Deng, Y.; Jiang, X.-T.; Li, L.-G.; Zhang, T. MinION Nanopore Sequencing Enables Correlation between Resistome Phenotype and Genotype of Coliform Bacteria in Municipal Sewage. *Front. Microbiol.* **2017**, *8*, 2105. [CrossRef]
211. Che, Y.; Xia, Y.; Liu, L.; Li, A.-D.; Yang, Y.; Zhang, T. Mobile antibiotic resistome in wastewater treatment plants revealed by Nanopore metagenomic sequencing. *Microbiome* **2019**, *7*, 44. [CrossRef] [PubMed]
212. Law, J.W.-F.; Ab Mutalib, N.-S.; Chan, K.-G.; Lee, L.-H. Rapid methods for the detection of foodborne bacterial pathogens: Principles, applications, advantages and limitations. *Front. Microbiol.* **2015**, *5*, 770. [CrossRef] [PubMed]

water

MDPI

Review

Tracing COVID-19 Trails in Wastewater: A Systematic Review of SARS-CoV-2 Surveillance with Viral Variants

Ananda Tiwari [1,*,†], Sangeet Adhikari [2], Shuxin Zhang [3], Tamunobelema B. Solomon [4], Anssi Lipponen [1], Md. Aminul Islam [5,6], Ocean Thakali [7], Sarawut Sangkham [8], Mohamed N. F. Shaheen [9], Guangming Jiang [3,10], Eiji Haramoto [11], Payal Mazumder [12], Bikash Malla [11], Manish Kumar [12,13], Tarja Pitkänen [1,14] and Samendra P. Sherchan [4,15,*,†]

1 Expert Microbiology Unit, Department of Health Security, Finnish Institute for Health and Welfare, 70701 Kuopio, Finland
2 Thermo Fisher Scientific, San Jose, CA 95134, USA
3 School of Civil, Mining, Environmental and Architecture Engineering, University of Wollongong, Wollongong 2522, Australia
4 Department of Biology, Morgan State University, Baltimore, MD 11428, USA
5 COVID-19 Diagnostic Lab, Department of Microbiology, Noakhali Science and Technology University, Noakhali 3814, Bangladesh
6 Advanced Molecular Lab, Department of Microbiology, President Abdul Hamid Medical College, Karimganj 2310, Bangladesh
7 Department of Civil Engineering, University of Ottawa, Ottawa, ON K1N 6N5, Canada
8 Department of Environmental Health, School of Public Health, University of Phayao, Muang District, Phayao 56000, Thailand
9 Department of Water Pollution Research, Environment and Climate Change Research Institute, National Research Center, Giza 2310, Egypt
10 Illawarra Health and Medical Research Institute (IHMRI), University of Wollongong, Wollongong 2522, Australia
11 Interdisciplinary Center for River Basin Environment, University of Yamanashi, 4-3-11 Takeda, Kofu 400-8511, Yamanashi, Japan
12 Sustainability Cluster, School of Engineering, University of Petroleum & Energy Studies, Dehradun 248007, Uttarakhand, India
13 Escuela de Ingeniería y Ciencias, Tecnologico de Monterrey, Campus Monterey, Monterrey 64849, Nuevo Leon, Mexico
14 Department of Food Hygiene and Environmental Health, Faculty of Veterinary Medicine, University of Helsinki, 00014 Helsinki, Finland
15 Department of Environmental Health Sciences, School of Public Health and Tropical Medicine, Tulane University, New Orleans, LA 70118, USA
* Correspondence: ananda.tiwari@thl.fi (A.T.); sshercha@tulane.edu (S.P.S.); Tel.: +358-442535468 (A.T.); +1-5049887283 (S.P.S.)
† These authors contributed equally to this work.

check for updates

Citation: Tiwari, A.; Adhikari, S.; Zhang, S.; Solomon, T.B.; Lipponen, A.; Islam, M.A.; Thakali, O.; Sangkham, S.; Shaheen, M.N.F.; Jiang, G.; et al. Tracing COVID-19 Trails in Wastewater: A Systematic Review of SARS-CoV-2 Surveillance with Viral Variants. *Water* **2023**, *15*, 1018. https://doi.org/10.3390/w15061018

Academic Editor: Abasiofiok Mark Ibekwe

Received: 9 February 2023
Revised: 2 March 2023
Accepted: 4 March 2023
Published: 7 March 2023

Abstract: The emergence of new variants of SARS-CoV-2 associated with varying infectivity, pathogenicity, diagnosis, and effectiveness against treatments challenged the overall management of the COVID-19 pandemic. Wastewater surveillance (WWS), i.e., monitoring COVID-19 infections in communities through detecting viruses in wastewater, was applied to track the emergence and spread of SARS-CoV-2 variants globally. However, there is a lack of comprehensive understanding of the use and effectiveness of WWS for new SARS-CoV-2 variants. Here we systematically reviewed published articles reporting monitoring of different SARS-CoV-2 variants in wastewater by following the PRISMA guidelines and provided the current state of the art of this study area. A total of 80 WWS studies were found that reported different monitoring variants of SARS-CoV-2 until November 2022. Most of these studies (66 out of the total 80, 82.5%) were conducted in Europe and North America, i.e., resource-rich countries. There was a high variation in WWS sampling strategy around the world, with composite sampling (50/66 total studies, 76%) as the primary method in resource-rich countries. In contrast, grab sampling was more common (8/14 total studies, 57%) in resource-limited countries. Among detection methods, the reverse transcriptase polymerase chain reaction (RT-PCR)-based sequencing method and quantitative RT-PCR method were commonly used for monitoring SARS-CoV-2 variants in wastewater.

Among different variants, the B1.1.7 (Alpha) variant that appeared earlier in the pandemic was the most reported (48/80 total studies), followed by B.1.617.2 (Delta), B.1.351 (Beta), P.1 (Gamma), and others in wastewater. All variants reported in WWS studies followed the same pattern as the clinical reporting within the same timeline, demonstrating that WWS tracked all variants in a timely way when the variants emerged. Thus, wastewater monitoring may be utilized to identify the presence or absence of SARS-CoV-2 and follow the development and transmission of existing and emerging variants. Routine wastewater monitoring is a powerful infectious disease surveillance tool when implemented globally.

Keywords: COVID-19; SARS-CoV-2 variants; Alpha (B.1.1.7); Delta (B.1.617.2); Omicron (B.1.1.529); wastewater-based epidemiology

---

1. Introduction

Severe acute respiratory syndrome coronavirus 2 (SARS-CoV-2), the causative agent of coronavirus disease 2019 (COVID-19), continuously underwent mutations leading to the emergence of new variants [1]. These variants are of great concern [2–4], as they might be associated with increased infectivity [1,5], severity [1,6,7], could have higher shedding rates [8], the potential to escape natural or vaccine-induced immunity [9,10], and can also affect the performance of diagnostic methodologies [11,12]. Such changes in virus characteristics affected the overall management plan for the COVID-19 pandemic. For example, it led to travel restrictions both locally and internationally for people from infected areas [1,7], and many more consequences on the daily lives of individuals. Therefore, the emergence of SARS-CoV-2 variants increased the need for genomic surveillance and other innovative tools to protect public health.

Whole-genome sequencing (WGS) of clinical specimens is a primary approach for identifying new emerging variants [13], by comparing the sample genome with the reference genome [14]. However, using WGS for monitoring each clinical specimen is time-consuming, labor-intensive, and expensive, and is usually conducted for individuals with clinical symptoms. Many of the COVID-19-infected individuals can be asymptomatic, so only relying on a clinical monitoring approach in the surveillance can miss the mutant variants carried by asymptomatic individuals.

Wastewater surveillance (WWS), also known as wastewater-based epidemiology (WBE), of infectious diseases through analyzing municipal sewage proved to be a cost-effective approach for monitoring the circulation of SARS-CoV-2 at a population level, covering both symptomatic and asymptomatic individuals [15–20]. In contrast to the clinical approach, WWS is a comprehensive, rapid technique for regular monitoring and tracking of the possible emergence of new variants at a population level [19–23]. From a surveillance point of view, municipal raw sewage can be a good material for SARS-CoV-2 monitoring, as it comprises the entire population of a community, both healthy and infected individuals (symptomatic, asymptomatic, pre-symptomatic, and post-symptomatic), contributing through feces, nasal mucus, and sputum to sewage from households, hospitals, and nursing homes [16,17,24]. Globally, many studies reported monitoring different variants of SARS-CoV-2 in wastewater [11,15–17,20,24–28], thereby highlighting WWS as an alternative tool for detecting different variants in communities. However, a comprehensive evaluation of the state-of-art use of WWS for monitoring SARS-CoV-2 variants is lacking. Such data can help evaluate and optimize WWS for monitoring SARS-CoV-2 variants. Such information can also be useful in managing future infectious outbreaks, such as how the wild and mutated variants differ among geological locations. Here, this review provides a comprehensive evaluation of the use of WWS for monitoring emerging SARS-CoV-2 variants, considering the opportunities and limitations of different methods used to analyze variants and the corresponding results (Figure 1).

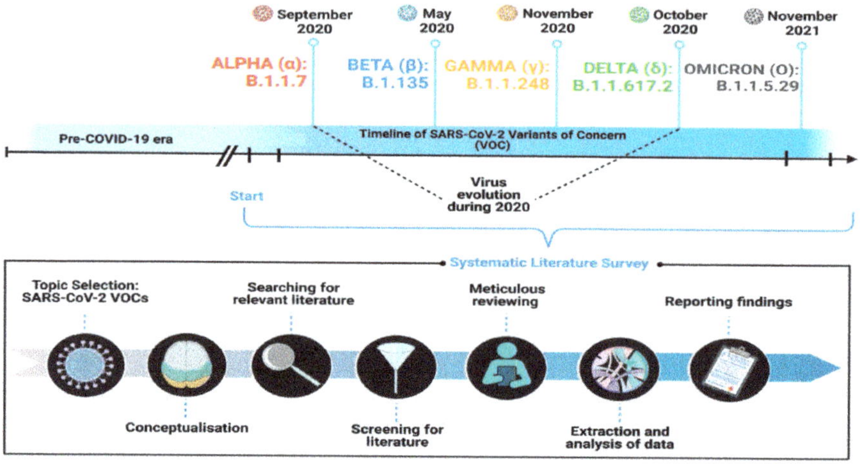

Figure 1. Conceptual framework of the study.

2. Theoretical Background: The Emergence of SARS-CoV-2 Variants

SARS-CoV-2 is an enveloped single-strand RNA (ssRNA) virus belonging to the *Coronaviridae* family and genus *Betacoronavirus* [9,29]. As with other ssRNA viruses, SARS-CoV-2 contains RNA-dependent RNA polymerase (RdRP), which is responsible for subgenomic mRNA synthesis for producing viral proteins, including the virus envelope and spike proteins [30]. RNA viruses are relatively prone to adapt more rapidly to a changed environment by changing their genome structure.

SARS-CoV-2 continuously evolves into new variants due to genetic mutation and viral recombination [1,2,13,31]. Mutation refers to at least a single change in a virus's genetic code. Genetic modifications can change the virus's characteristics [1]. A SARS-CoV-2 variant can have one or more mutations that differentiate its features from other variants. SARS-CoV-2 has a similar mutation mechanism to other ssRNA viruses that lack proofreading capability, giving rise to new variants [25]. Uncorrected mutations occur during genome replication, recombination, and RNA editing by the deaminase of the infected host [13]. A recombinant variant is created due to a combination of genetic material from two different variants, and a mutant variant is created due to a mutation in RNA. A lineage is a group of closely related viruses with a common ancestor [32]. The ancestral SARS-CoV-2 (wild variant) genome evolved into several lineages (https://cov-lineages.org/lineage_list.html, accessed on 28 November 2022), such as the Alpha (B.1.1.7), Delta (B. 1.617.2), and Omicron (B.1.1.529) [2,3,7,11,28,32–37], due to exposure to some selective pressure [38]. Most of these new variants were developed due to viral spike protein (S-protein) mutation [39].

2.1. *Alpha (B.1.1.7 and Q Lineages)*

The Alpha variant was first isolated in the United Kingdom in September 2020 and was followed by an upsurge in infection in December 2020 [40]. Soon after, it became the dominant variant until August 2021 in many countries, including the US, India, Sweden, and globally in at least 189 countries (Table 1). The World Health Organization (WHO) classified the Alpha variant as a variant of concern (VOC) on 29 December 2020 [10], after rising hospitalization cases and creating a strain on the public health system and facilities across countries [41]. The Alpha variant was reported to be about 100-fold more lethal than the original SARS-CoV-2 strain [6]. Further, mRNA vaccines were reported to be about 68% less effective against this variant [6]. On 21 September 2021, the WHO designated the Alpha variant as the "variant being monitored" [1,7]. After 2022, this variant's circulation drastically reduced worldwide, following the emergence of Delta variants, probably due to the impact on vaccine-induced immunity (Table 1).

Table 1. SARS-CoV-2 variants and lineages [1,7,33].

WHO Label/Pango Lineage	Country First Detected	Spike Mutations of Interest	Outbreak Countries	Major Outbreak Peaks	Classification (WHO) during November 2022	Outbreak Condition during November 2022
Alpha/B.1.1.7	United Kingdom, September 2020	N501Y, D614G, P681H	At least 189 countries, predominant in the US, India, Sweden, France, Spain, Australia, Nigeria, and so on (1.2 million cases globally reported).	November 2020 to August 2021	VOC: 29 December 2020 VBM: 21 September 2021 PVOC: 9 March 2022	Drastically reduced circulation globally, with almost no reporting at the time of writing the manuscript.
Delta/B.1.617.2	India, October 2020	L452R, T478K, D614G, P681R	At least 208 countries, predominant in the US, UK, Japan, Italy, India, Germany, Canada, Denmark, France, and so on (4.4 million cases globally).	May 2021–January 2022	VOC: 15 June 2021 VBM: 14 April 2022, PVOC: 7 June 2022	Abundance is very low at the time of writing the manuscript.
Beta/B.1.351	South Africa, May 2020	K417N, E484K, N501Y, D614G, A701V	At least 127 countries, predominant in South Africa, the US, India, Sweden, France, Spain, Australia, Nigeria, Iran, and so on (43,000 cases globally).	November 2020 to August 2021	VOC: 29 December 2020 VBM: 21 September 2021, PVOC: 9 March 2022	Drastically reduced circulation globally, with almost no reporting at the time of writing the manuscript.
Gamma/P.1	Brazil, November 2020	K417T, E484K, N501Y, D614G, H655Y	At least 93 countries, predominant in the US, Canada, Brazil, Argentina, Chile, Italy, Peru, Mexico, Sweden, South Korea, Venezuela, and so on (74,300 cases globally).	Feb 2021–November 2021	VOC: 29 December 2020 VBM: 21 September 2021 PVOC: 9 March 2022	No longer detected or detected at extremely low levels globally.
Epsilon/B.1.427, B.1429	California USA, July 2020	I4205V and D1183Y in the ORF1ab gene, and S13I, W152C, L452R in the spike protein's S-gene	At least 45 countries.	November 2020–March 2021	VOC, March 2021. VBM-September 2021. Previously circulating VOI: March 2022 (WHO),	After an initial increase, its prevalence rapidly decreased from February 2021 and was outcompeted by the more transmissible Alpha variant.
Lambda/C.37	Peru, August 2020	Virus's spike protein code: G75V, T76I, Δ246–252, L452Q, F490S, D614G and T859N	At least 45 countries (predominant in Peru, Chile, US). Total global cases of less than 10,000.	November 2020–November 2021	VOI June 2021	No longer reported.
Omicron/BA.2, BA.4, BA.5, BA.2.75, BQ.1, XBB	South Africa and Botswana November, 2021	BA.2 (y*), BA.4 (L452R, F486V, R493Q), BA.5 (L452R, F486V, R493Q). BA.2.75 (z**), BQ.1 (K444T, N460K), XBB (N460K, F490S)	At least 208 countries, predominant in the US, UK, Denmark, Canada, India, Japan, Germany, France, and so on (6.2 million cases globally, 14 December 2022).	November 2021-**Currently circulating lineages:** BA.2, BA.4, BA.5, and their recombinants and sub-lineages	BA.2, BA.4, and BA.5 are VOC, and BA.2.75, BQ.1, XBB are current VOI at the time of drafting the manuscript, December 2022.	Many lineages are currently going around the world (83,046 in the last four weeks of 14 December 2022).

Notes: Other variants monitored earlier: Eta/B.1.525 (VOI on 02/29/21, but VBM since 09/21/21), IOTA/B.1.526 (VOI on 02/29/21, but VBM since 09/21/21), Kappa/B.1.617.1 (VOI on 05/07/21 but VBM since 09/21/21), B.1.617.3 (VOI on 05/07/21, but VBM since 09/21/21), Zeta/P.2 (VOI on 02/26/21, but VBM since 09/21/21), and Mu/B.1.621 and B.1.621.1 (VBM on 09/21/21but VBM since 09/21/21) *Key: VOI—Variant of interest; VOC—Variants of concern; VBM—Variant being monitored; PVOC—Previous variant of concern.* y*: G142D, N211I, Δ212, V213G, G339D, S371F, S373P, S375F, T376A, D405N, R408S, K417N, N440K, S477N, T478K, E484A, Q493R, Q498R, N501Y, Y505H, D614G, H655Y, N679K, P681H, N764K, D796Y, Q954H, N969K, z**: W152R, F157L, I210V, G257S, D339H, G446S, N460K, and Q493 (reversion).

2.2. *Delta (B.1.617.2 and AY Lineages)*

The Delta variant was first detected in India in October 2020 [10,42], and it swept rapidly through India and then the United Kingdom by mid-April 2021 before spreading to the US and the rest of the world [42]. The Delta variant was reported to be 60% more infectious and lethal than the Alpha variant [1]. It was reported that a single and double dose of AstraZeneca vaccine was 33% and 60% effective in reducing the Delta lineage infection, respectively, compared to 60% and 66% on the Alpha lineage [43,44]. Similar results (i.e., less effective vaccine than with Alpha lineage) were observed in a study when Pfizer vaccines were administered [42]. The emergence of this variant caused a delay in the United Kingdom's reopening plans after several months of lockdown beyond June 2021 [9]. It became a VOC in the US on 15 June 2021 [1], after it became evident that people infected with the Delta variant were twice as likely to become hospitalized than those with the Alpha variant [41]. This implied that the Delta variant exhibited more infectivity than earlier variants [42]. Until October 2021, Delta was the most dominant variant in the world, with about 90% sequences in the Global Initiative on Sharing Avian Influenza Data (GISAID).

2.3. Omicron (B.1.1.529 and BA Lineages)

The Omicron variant was declared a VOC immediately after it was reported in South Africa in November 2021 [1]. The Omicron variant has the highest number of mutations, compared to the reference wild SARS-CoV-2 genome, with 37 mutations in the spike (S) protein, three mutations in the nucleocapsid (N) protein, one mutation in the envelope (E) protein, three mutations in the membrane (M) protein, and 10 synonymous mutations [45]. This variant is more contagious than the earlier variants, with a reported rise of cases from hundreds per day to thousands per day in South Africa over two weeks [9]. It soon began to spread to several other countries and became one of the most dominant variants after December 2021 [25,46,47]. A subvariant known as BA.2 was also discovered and monitored as it accounted for 23% of cases in the US as of March 2022 [48]. The Omicron BA.2 sub-variant has a mutation on the spike protein, which is responsible for infecting host cells, thereby increasing infectability and having the capacity to evade immunity [49], most especially those who recovered from previous COVID-19 variants infection but were yet to be vaccinated [50]. The Omicron variant and its sub-lineages were the most dominant variants circulating globally in 2022, for more than 98% of sequences shared on GISAID between February 2022 and November 2022 belonging to Omicron. Omicron is a complex variant that continues to evolve, leading to descendent lineages with different genetic constellations of mutations [7]. Despite its high transmissibility, it has lower severity than previous Delta and Alpha variants. In December 2022, the world was still passing through the pandemic of the Omicron variant. A total of 83,046 cases were reported in GISAID within four weeks (https://gisaid.org/hcov19-variants/, accessed on: 7 December 2022) (Table 1).

2.4. Other Variants

Aside from the variants mentioned above, many other variants were first declared as VOC but were later re-designated as variants of interest (VOI) [1,33] or declared as variants being monitored (VBM), previously circulating VOCs or VOIs or formerly monitored variants (FMVs). The designation of VOC, VOI, VBM, and FMVs are working definitions periodically updated by WHO and CDC (US and EU) (Table S1).

Among the emergence of different variants, the Beta (B.1.351) variant was detected in South Africa [51], and the Gamma (P1) variant was identified in Brazil in November 2020 [34,46]. Both lineages of Epsilon (B.1.427 and B.1.429) were identified in California [35] and were reported to have higher transmissibility, infectibality, and severity than the preceding variants and lineages [41]. The Lambda variant (lineage C.37) was first reported in Peru in August 2020 and was designated as VOI by WHO on 14 June 2021 [36]. This variant was reported to be more resistant to neutralizing antibodies than other variants [36]. The Lambda variant was suspected to be more resistant to vaccines than the Alpha and Gamma variants [37]. Table 1 shows details of SARS-CoV-2 variants and their WHO designation at the time of review. This highlights the importance of genetic surveillance of SARS-CoV-2 variants worldwide.

3. Methodology

A thorough literature search was conducted in November 2022 using ScienceDirect, Google Scholar, PubMed, Web of Science, Scopus, and NCBI databases. No publication date or language restrictions were applied during the search, and Booleans "AND" and

"OR" to combine keywords were used. Searches were directed toward the review objectives with pertinent keyword combinations (a) wastewater surveillance; (b) SARS-CoV-2 and emerging variants; (c) variants of concern in wastewater; (d) genomic tools for SARS-CoV-2 monitoring; (e) wastewater-based epidemiology for SARS-CoV-2 variants; (f) next-generation sequencing for SARS-CoV-2; (g) SARS-CoV-2 variants in wastewater; and (h) Alpha, Delta, and Omicron variants in wastewater. The literature search process is presented in Figure 2, following PRISMA guidelines [52]. The included literature was mostly peer-reviewed journal articles (except two pre-print). Sole clinical monitoring of SARS-CoV-2 variants not including WWS, methodology comparison studies without surveillance aim, review papers without original data, and numerical modeling papers were excluded. The database search identified 718 studies, as shown in Figure 2.

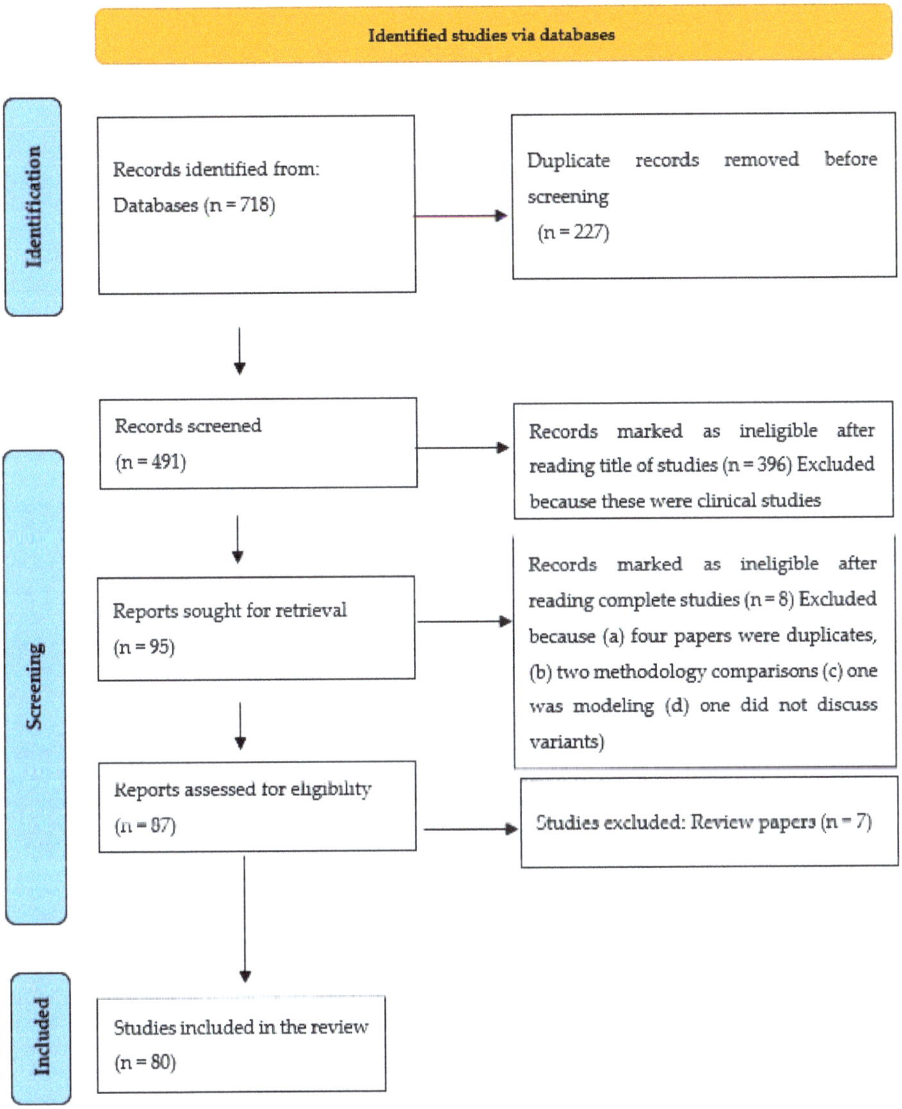

Figure 2. PRISMA flow diagram of literature search.

After the de-duplication steps, a total of 491 papers remained. Upon further screening of titles and abstracts, 396 research articles related to clinical testing were removed, leaving 95 articles for full-text screening. After reading 95 articles, an additional 15 articles were removed from the list as seven were found to be review papers: four were duplicates of earlier studies, two studies were only methodology comparisons, one was a modeling study, and one did not report variants. In total, 80 publications reporting SARS-CoV-2 variants in wastewater were included in this systematic review (Figure 2).

4. Results

4.1. Geospatial Distribution

The majority of the WWS studies (66/80, 82.5%) included in this review were conducted in resource-rich countries in Europe and North America (Figure 3). Out of the total of 80 studies, 38 (47.5%) were in Europe [4,9,18,23,28,53–83], 28 (35.0%) in North America (20 studies in the US [11,84–101], and 8 studies in Canada [102–109]), 10 studies (12.5%) in Asia [110–117], 2 studies in South America [118,119], and 1 study each were conducted in Africa [120] and Australia [121] (Table S1).

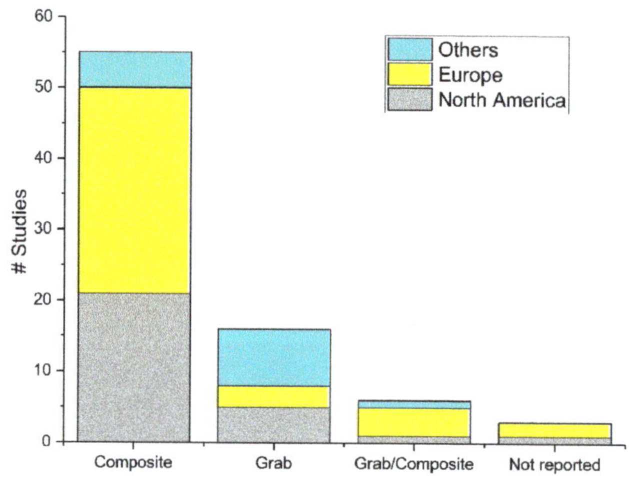

Figure 3. Number of studies based on sample collection type in different continents (others = countries from Asia, Africa, Australia, and South America).

4.2. Sampling Techniques

In general, autosamplers were used for taking composite samples that were either time proportional (equal aliquot is taken at a fixed time interval) or flow proportional (after a specific volume based on the flow rate of the source, an aliquot of sample is taken) [15]. Out of 80 studies, 55 (68.8%) used composite sampling, 16 (20%) used grab sampling, six (7.5%) used both composite and grab sampling, and three studies (3.8%) did not report the sampling technique used in their research. As shown in Figure 3, composite sampling was relatively more frequent in high-resource countries compared to other countries. Furthermore, the review found that resource-limited countries lack composite samplers due to the associated high costs and are most likely to rely on grab sampling (Figure 3). In resource-rich countries, grab samplers were primarily used in sub-catchments or facilities (such as hospital sewage or college campuses) or small regions of communities [18,57,101].

Out of 55 studies using composite sampling, 30 used time proportional, 14 studies used flow proportional, and the others used composite settled solids or did not report on sampling. In a previous report, proportional flow sampling was considered less biased than the time-proportional sampling mode [122]. These autosamplers have higher capital and operating (installation and maintenance) costs and may not be easily accessible in low-resource settings. In a recent study, grab samples and 24 h composite samples showed comparable results during relatively low COVID-19 incidence [123]. Grab sampling is relatively easier and faster and does not need any automated equipment. It provides the status of SARS-CoV-2 only during the sample collection, so it is less representative of fecal community contributions than composite samples. Composite sampling is the collection of multiple grab samples, so it is more representative and provides the average situation of SARS-CoV-2 for a certain time interval. Such discrepancy in the sampling mode among resource-rich and resource-limited countries explains the lack of resources against emerging diseases or conditions, such as the COVID-19 global pandemic. Therefore, resource-limited countries deserve more global attention and funding for fighting emerging and re-emerging diseases.

4.3. Concentration Methods

Virus concentration is a critical step to detecting the low concentration of SARS-CoV-2 and its variants in wastewater. Our review noted a high variation in virus concentration methods. The most used methods were polyethylene glycol (PEG) precipitation, ultrafiltration, electro-negative membrane filtration, ultracentrifugation, aluminum hydroxide adsorption–precipitation, $AlCl_3$ precipitation, and skimmed milk (SM) flocculation. Many studies combined more than one concentration method and optimized it by using positive control during the extraction process. Among various virus concentration methods, PEG precipitation, skimmed milk (SM) flocculation, and aluminum polychloride (PAC) flocculation are reported to be relatively more cost-effective methods than ultrafiltration and ultracentrifugation [124]. The SM method can handle a large volume of samples at a time, it does not require special equipment, and the number of processing steps is relatively smaller than many other methods [124]. An earlier study reported that PEG precipitation and PAC flocculation were the most effective virus concentration methods from wastewater (62.2% and 45.0%, respectively) among eleven compared concentration methods [124].

4.4. Analytical Methods to Detect Variants

Among variant detection methods, PCR amplification-based sequencing was the dominant (49/80 of total studies) for SARS-CoV-2 variants analysis in wastewater. Among the various PCR-based amplification methods, including RT-PCR, five studies used nested PCR (nPCR), and multiplex tiling PCR was employed to prepare the cDNA and library for the downstream sequencing analysis of SARS-CoV-2 variants [4,23,102]. The workflow of sequencing-based detection methods includes three main steps. The first started with pre-screening of the total SARS-CoV-2 RNA by using the classical reverse transcriptase quantitative polymerase chain reaction (RT-qPCR assays targeting N1, N2, RdRP, and E genes) suggested by US CDC to screen SARS-CoV-2-positive samples. Then, positive samples with a Ct value (<35) were amplified by RT-PCR to generate a cDNA and sequencing library. The RT-PCR generating cDNA can also be a two-step approach, where cDNA synthesis is conducted separately and only after that PCR amplification takes place. Finally, high-throughput sequencing methods were employed to acquire sequence information on the samples, and the alignment between sample sequences and reference genomes was used to detect single-nucleotide variants (SNVs) or specific mutations associated with the variant/lineage.

Wurtz et al. also confirmed that direct sequencing is more accurate than the RT-PCR method in detecting SARS-CoV-2 variants [83]. However, according to another report, it is an instrument of high cost and requires some technical skills [125]. The sample size must also be certain to give an acceptable result. Hence, it was advised by Vo et al. (2022) that other less costly methods, such as the TaqMan RT-PCR, characterize the strain before using the WGS to confirm [11].

The RT-qPCR/RT-digital PCR (dPCR)-based method was the next most widely used detection method (31/80 of total studies) after the sequencing-based method among all 80 studies included in this study. RT-PCR-based quantitative detection methods (RT-qPCR and RT-dPCR) could generate relative or absolute quantification results that can directly reveal the concentration of SARS-CoV-2 RNA in wastewater associated with clinical data to enhance the monitoring of SARS-CoV-2 prevalence in communities [126]. Thus, it became one of the most sensitive methods for monitoring SARS-CoV-2 RNA in wastewater [127]. However, compared with sequencing methods, the RT-qPCR method has a certain lag in discovering the emergence of new variants because it requires a specific primer–probe design according to the details of the genomic information of new variants reported by clinical tests [126]. In addition, the gene targets of RT-qPCR/RT-dPCR methods were limited by fluorophores and the detection instrument. Thus, it is mainly adopted for detecting and quantifying only the already known variants circulating in communities or elsewhere.

Multiplexed RT-qPCR /RT-dPCR methods exhibited great potential in monitoring SARS-CoV-2 variants in wastewater, since they can be regarded as a low-cost replacement of sequencing methods to increase the output of RT-qPCR methods and achieve the quantification of new variants [126]. Peterson et al. developed an RT-qPCR allelic discrimination assay that was sufficiently sensitive and specific to achieve the detection and relative quantitation of SARS-CoV-2 variants (B.1.1.7, B.1.351, and P.1) in wastewater [109]. Caduff et al. also developed a drop-off RT-dPCR assay to analyze temporal dynamics of SARS-CoV-2 signature mutations (spike Δ69–70 and ORF1a Δ3675–3677) in wastewater [70]. Furthermore, Boogaerts et al. successfully optimized, validated, and applied a multiplex digital polymerase chain reaction (dPCR) assay to measure the emerging SARS-CoV-2 variants of concern (VOC) in influent wastewater [76]. All these methods successfully distinguished between SARS-CoV-2 RNA originating from the wild-type and B.1.1.7, B.1.351, P.1, and B.1.617.2 variants, thus having great potential in the application of SARS-CoV-2 monitoring in wastewater. In addition to enabling the detection and absolute quantification of various SARS-CoV-2 variants simultaneously, the multiplexed RT-qPCR methods could also reveal the proportion of new variants out of all prevalent strains in wastewater [97]. Through the relative quantification results acquired by simultaneously detecting the highly conserved sequence of SARS-CoV-2 genomes and specific genes of various new variants, multiplexed RT-PCR methods might be further used to provide insightful information on the proportion of new variants in communities. Thus, the various methods used by different studies reviewed have both advantages and disadvantages.

4.5. Detected Variants

Out of the total of 80 studies, the B.1.1.7 (Alpha) variant was the most targeted and detected variant (48/80 of total studies), followed by B.1.617.2 (Delta, 24/80 total studies), B.1.351 (Beta, 19/80 total studies), P.1 (Gamma, 19/80 total studies), and others in wastewater (Figure 4). As all variants reported as WWS followed the same pattern as the clinical reporting of variants in the same timeline, this demonstrated that WWS tracked all variants timely when they were developed (Table S1). The reason why the B1.1.7 (Alpha) variant was the most reported is that it appeared earlier than other variants in the pandemic, and thus during the gathering of papers for the present review (November 2022), there was more time to report the Alpha variant findings than findings of the other variants. Although Omicron is the largest among the different variants and constituted an ongoing pandemic at the time of writing this review, its importance was not yet reflected in the number of publications available by November 2022, being only the fifth most reported (Figure 4). This may be due to reporting delays.

In 49 sequencing-based analyses, 28 studies detected the B.1.1.7 (Alpha) variant, 20 studies detected the B.1.617.2 (Delta) variant, 15 detected the P.1 (Gamma) variant, 15 detected B.1.351 (Beta) variant, 7 detected various lineages Omicron (e.g., BA.1 and BA.2), and 7 other studies reported only other uncommon variants (Table S1). Similarly, out of 31 RT-qPCR/RT-dPCR methods-based studies, 20 detected the B.1.1.7 (Alpha) variant by targeting the △HV69/70 deletion, the N501Y mutation, or/and the N-D3L mutation. Nine studies detected the B.1.617.2 (Delta) variant by targeting SΔ157 mutation [70,76,109].

Six studies detected the B.1.351 (Beta) variant by targeting the N501Y mutation, ORF1a Δ3675–3677 deletions, and SΔ241 mutation. Five studies detected the P.1 (Gamma) variant by targeting the N501Y mutation, ORF1a Δ3675–3677 deletions, and the insertion in the 28227–28286 region. In addition, ten studies reported the detection of different lineages of the Omicron (e.g., BA.1 and BA.2) variant by targeting a region of five adjacent SNPs common to BA.1 and BA.2 and distinguished these two variants by detecting the specific 143–145 deletion of BA.1 and the specific LPPA24S (a 9 bp deletion) mutation of BA.2 [100].

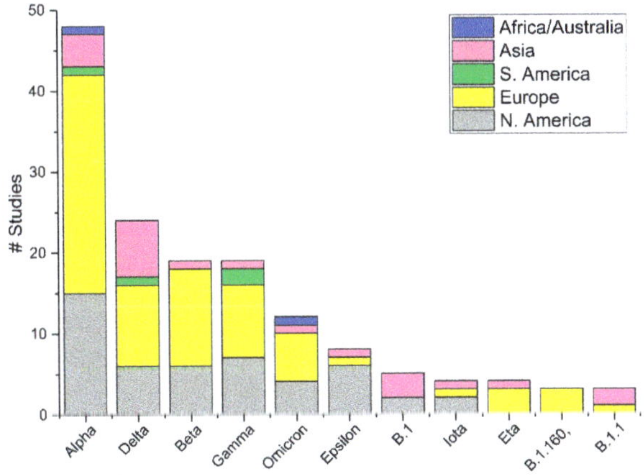

Figure 4. Number of studies targeting or detecting different SARS-CoV-2 variants. Variants are included in the figure only if they are detected by three or more studies.

5. Discussion

The monitoring of SARS-CoV-2 variants in wastewater was established as a promising tool for evaluating spatial and temporal trends of emerging variants at a population level. However, this review showed that the approach was not yet equally popular in resource-limited countries compared to resource-rich ones (Europe and North America). This can be due to a lack of WWS knowledge, awareness, and functional governmental public health institutions in resource-limited countries. Significantly low reporting of the WWS approach in resource-limited settings may indicate a major weakness of current global policy and institutions working on emerging pathogens. It may challenge strengthening the institutions and managing the resources (monetary, technology, and manpower) in resource-limited regions for fighting current and future pandemics. As SARS-CoV-2 demonstrated, the emergence of a communicable disease from one nation can easily sweep through all nations globally.

Indeed, WWS is an economic approach for monitoring emerging pathogens at a population level, and its promotion in developing countries can help fight future emerging pandemics. Many resource-limited countries successfully conducted a variant survey in clinical patients and reported the results in GISAID and WHO reporting systems. However, such studies and reporting are not conducted for wastewater, although the economic cost and analysis technology of one WWS sample analysis is roughly the same as that for the analysis of one single clinical patient sample. Countries with large populations, poor healthcare facilities, and the incapacity of extensive individual clinical testing could greatly upgrade their sentinel surveillance systems by using WWS. The WWS approach provides a near real-time snapshot of the ongoing pandemic [39]. For example, clinical surveillance needs a series of stages for understanding the status of the spread of new variants at a population level: virus colonization to individuals, maturation of symptoms, clinical testing, diagnosis, and reporting. Nonetheless, a variant is excreted to wastewater as soon as it is colonized in individuals and shed from feces, nasal secretion, and other bodily fluids [21]. WWS is an important complementary survey approach to the conventional clinical or hospital-based epidemiological survey of infectious diseases.

Two approaches, PCR-based (RT-qPCR and RT-dPCR) and next-generation sequencing (NGS) or high-throughput sequencing followed by bioinformatics analysis (Table 2), are widely used for monitoring SARS-CoV-2 variants in wastewater [39]. The PCR-based approaches are relatively fast, simple, sensitive, and cost-effective for monitoring specific variants. These methods target signature mutation regions of the SARS-CoV-2 genome and detect the variant. Targeting the ORF1a gene that exists in many variants and the HV69/70 deletion present in the Alpha variant helps to differentiate this variant from others. For discriminating between variants, European authorities recommend that targeting should cover at least the S gene, particularly in that it encodes the entire N-terminal region and the receptor-binding domain (RBD) corresponding to amino acids [128]. RT-qPCR/RT-dPCR methodologies are often designed as duplex or multiplex, allowing the simultaneous detection of many variants and estimating their percentages among other simultaneously occurring variants. The relatively low cost and low time and labor requirements make these methods (RT-qPCR, RT-dPCR) more suitable for monitoring known variants. The targeting of mutation assays applied to SARS-CoV-2 RNA extracted from wastewater can be a rapid, efficient, and reliable way of monitoring variants introduced and circulated in a community [88]. Compared to RT-qPCR, RT-dPCR was recently popularized and reported as more sensitive than RT-qPCR, as it measures the absolute count of gene copies and is less affected by PCR inhibitors than RT-qPCR [126,129].

Table 2. Comparing pros and cons of sequencing and RT-qPCR/RT-dPCR based methods for detecting SARS-CoV-2 variants.

	Sequencing	RT-qPCR/RT-dPCR
Pros	• Can detect all circulating variants, even silently circulating in a population, and enable showing the diversity of circulating variants; • Suitable for early warning of emerging variants; • Accurately provide information about the full mutation patterns specific to all variants; • Possible simultaneous detection of many variants and estimating of their percentages.	• Fast, sensitive, low labor required, and cost-effective for screening particular known variants; • Powerful screening tool; • Simple to use and interpret; • Possible duplexing or multiplexing allows the simultaneous detection of many variants and estimating of their percentages.
Cons	• Relatively expensive, as it demands high reagent and consumable costs, specific equipment, and technical analysis skills; • Labor- and time-consuming as extra bioinformatics knowledge is required; • Multiple lineages in raw sewage complicate the proper assembly of reads to determine the complex sequence circulating in communities.	• Earlier knowledge about the mutated sequence is needed for designing primers and probes; • Detect only the targeted signature mutations and cannot detect non-targeted variants; • Less suitable for tracking new emerging variants and for early warning.

The requirement that PCR-based methods (RT-qPCR/RT-dPCR) use specific primers is a limitation, as these methods cannot detect non-targeted variants. Such primers and probes need to be designed and validated for each new mutant/variant (Table 2). Therefore, PCR-based methods are less suitable for detecting novel variants compared to NGS. For designing a primer, previous information about the region of mutation in the genome is needed. Nevertheless, such information can be obtained for primer design from the GISAID database after clinical testing deposits [130]. Further, a variant-specific reverse transcription-nested PCR approach can be applied to determine the key regions of the viral spike protein.

The next-generation sequencing (NGS)-based approach enables the detection of signature mutations of different variants and provides a real picture of the variant circulation

in real-time [39]. In comparison to the PCR-based method, the NGS-based method can detect all circulating variants (even silently circulating) in a population and enables us to show the diversity of circulating variants. As the NGS-based method covers longer genome coverage and also the design and validation of new assay is not needed for each variant. Additionally, the NGS-based method enables the possibility of retrospective analysis and finding variants circulating earlier in a particular sample. These methods are the most efficient analysis method for monitoring the emergence of the new SARS-CoV-2 variants even earlier than clinical cases. However, NGS-based methods require high reagent and consumable costs, specific equipment, and technical analysis skills. In addition, multiple lineages in raw sewage do not allow the proper assembly of reads to determine the complex sequence circulating in communities [69,93].

Other methods used for monitoring SARS-CoV-2 variants in wastewater were reverse transcription-nested PCR (RT-nPCR) assays followed by Sanger sequencing or NGS analysis, and amplicon sequencing of the only selective gene of SARS-CoV-2 variants instead of targeting the whole or nearly complete genome of SARS-CoV-2 from environmental samples.

WWS played an important role in enabling decision-making among health officials and researchers. It has positive impacts on public health departments and governments to prepare for the possible spread of new variants, though they generally do not pay attention to other variants below the set detectable limits in wastewater due to dilution. The use of WWS also highlights how fast variants of concern transmit from one nation (where the first case is recorded) to others connected via direct or indirect transportation connections, such as land, sea, and air transport routes [26,121]. The detection of Alpha, Beta, Delta, and Omicron variants, amongst others, in multiple countries, is an indication of how global an impact the COVID-19 pandemic had. This makes the use of WWS tools more about understanding and providing adequate control measures in forestalling the spread of SARS-CoV-2. Wastewater surveillance tools can be further developed to track the spread of other infectious diseases [21,22,131–135], especially viruses that can be excreted through human feces and/or urine and saliva. Incorporating WWS as a reliable surveillance tool in a global policy about tackling new and emerging pathogens can be prominent. It may be prominent for strengthening institutions and managing resources (monetary, technology, and human) in resource-limited countries, as the emergence of a communicable disease from one nation can easily sweep through all countries globally, as SARS-CoV-2 is showing.

In addition to its many benefits, WWS of the SARS-CoV-2 variant has some limitations; the most prominent being that it does not reach up to an individual level. Further, wastewater can have many variants in different proportions and PCR inhibitors [126]. Unlike when isolated from an individual case, which consists of a single genome, wastewater samples are likely to contain material from multiple SARS-CoV-2 variants shed from different individuals. Further, each variant at low concentrations and in various stages of genomic integrity because of degradation makes accurate detection challenging [88]. In addition, multiple lineages in raw sewage do not allow for the proper assembly of reads to determine the complex sequence circulating in communities [69,93]. Thus, in the future, more efforts should be put into developing a comprehensive monitoring workflow of SARS-CoV-2 variants, including the sequencing-based screen of unknown variants and the RT-qPCR/RT-dPCR methods-based long-term detection and quantification of specific variants.

Finally, some limitations of this type of systematic review need to be considered while interpreting the findings [29,133]. For example, there is high heterogeneity in the reviewed studies in terms of sample size, target, and detection methods, and in many cases, there can be a lack of transparency in reporting results. For example, there could be reporting biases (most likely positive reports could be more frequently published than negative results [29,133]). Further, there could be target selection bias, particularly while using PCR-based methods, as primers could be targeted only toward some most common variants. Locally circulating variants could guide such interest in communities based on clinical reporting. However, there could already be new variants in communities, which could be detected with WWS.

6. Conclusions

This review revealed that several nations, mostly from resource-rich regions (Europe and North America) actively and regularly engage in WWS to track different variants of SARS-CoV-2. Among different variants, the B1.1.7 (Alpha) variant, which first appeared as an emerging variant during the pandemic, was the most reported (48/80 total studies), followed by B.1.617.2 (Delta), B.1.351 (Beta), P.1 (Gamma), and others in wastewater. All variants reported as WWS followed the same pattern as the clinical reporting of the variants in a timeline and demonstrated that WWS tracked all variants in a timely way when they were developing. Still, developing a highly sensitive, accurate, cost-effective, automated, and reliable tool, which can be practically feasible in both resource-rich and resource-limited settings, for monitoring the trends of current and future COVID-19 outbreaks is highly important for increasing the reliability of WWS.

Supplementary Materials: The following supporting information can be downloaded at: https://www.mdpi.com/article/10.3390/w15061018/s1; Table S1: Summary of literature data on the detection of SARS-CoV-2 variants by wastewater surveillance.

Author Contributions: Conceptualization, A.T. and S.P.S.; literature review, A.T., S.A., T.B.S., O.T., S.P.S., B.M., M.A.I. and P.M.; formal analysis, A.T.; writing—original draft preparation, A.T., S.A. and S.Z.; writing—review and editing, T.P., S.P.S., G.J. and M.K.; visualization, A.T., M.K., P.M. and M.A.I.; supervision, A.T., T.P., S.P.S., M.K., G.J., B.M., S.S., A.L., E.H. and M.N.F.S.; project administration, A.T. All authors have read and agreed to the published version of the manuscript.

Funding: This research received no external funding.

Conflicts of Interest: Sangeet Adhikari is an employee of Thermo Fisher Scientific. The findings and conclusions contained within are those of the author and do not reflect the positions or policies of Thermo Fisher Scientific. All authors declare no conflict of interest in this manuscript.

References

1. CDC. SARS-CoV-2 Variant Classifications and Definitions. Available online: https://www.cdc.gov/coronavirus/2019-ncov/variants/variant-classifications.html (accessed on 27 May 2022).
2. Cosar, B.; Karagulleoglu, Z.Y.; Unal, S.; Ince, A.T.; Uncuoglu, D.B.; Tuncer, G.; Kilinc, B.R.; Ozkan, Y.E.; Ozkoc, H.C.; Demir, I.N.; et al. SARS-CoV-2 Mutations and Their Viral Variants. *Cytokine Growth Factor Rev.* **2022**, *63*, 10–22. [CrossRef]
3. Callaway, E. Remember Beta? New Data Reveal Variant's Deadly Powers. *Nature* **2021**. [CrossRef]
4. Wilton, T.; Bujaki, E.; Klapsa, D.; Majumdar, M.; Zambon, M.; Fritzsche, M.; Mate, R.; Martin, J. Rapid Increase of SARS-CoV-2 Variant B.1.1.7 Detected in Sewage Samples from England between October 2020 and January 2021. *mSystems* **2021**, *6*, e0035321. [CrossRef] [PubMed]
5. Yurkovetskiy, L.; Wang, X.; Pascal, K.E.; Tomkins-Tinch, C.; Nyalile, T.P.; Wang, Y.; Baum, A.; Diehl, W.E.; Dauphin, A.; Carbone, C.; et al. Structural and Functional Analysis of the D614G SARS-CoV-2 Spike Protein Variant. *Cell* **2020**, *183*, 739–751.e8. [CrossRef] [PubMed]
6. Radvak, P.; Kwon, H.J.; Kosikova, M.; Ortega-Rodriguez, U.; Xiang, R.; Phue, J.N.; Shen, R.F.; Rozzelle, J.; Kapoor, N.; Rabara, T.; et al. SARS-CoV-2 B.1.1.7 (Alpha) and B.1.351 (Beta) Variants Induce Pathogenic Patterns in K18-HACE2 Transgenic Mice Distinct from Early Strains. *Nat. Commun.* **2021**, *12*, 6559. [CrossRef] [PubMed]
7. WHO. Tracking SARS-CoV-2 Variants. Available online: https://www.who.int/activities/tracking-SARS-CoV-2-variants (accessed on 15 December 2022).
8. Saguti, F.; Magnil, E.; Enache, L.; Patzi, M.; Johansson, A. Surveillance of Wastewater Revealed Peaks of SARS-CoV-2 Preceding Those of Hospitalized Patients with COVID-19. *Water Res.* **2020**, *189*, 116620. [CrossRef]
9. La Rosa, G.; Mancini, P.; Bonanno Ferraro, G.; Veneri, C.; Iaconelli, M.; Lucentini, L.; Bonadonna, L.; Brusaferro, S.; Brandtner, D.; Fasanella, A.; et al. Rapid Screening for SARS-CoV-2 Variants of Concern in Clinical and Environmental Samples Using Nested RT-PCR Assays Targeting Key Mutations of the Spike Protein. *Water Res.* **2020**, *197*, 117104. [CrossRef]
10. Wang, P.; Nair, M.S.; Liu, L.; Iketani, S.; Luo, Y.; Guo, Y.; Wang, M.; Yu, J.; Zhang, B.; Kwong, P.D.; et al. Antibody Resistance of SARS-CoV-2 Variants B.1.351 and B.1.1.7. *Nature* **2021**, *593*, 130–135. [CrossRef]
11. Vo, V.; Tillett, R.L.; Papp, K.; Shen, S.; Gu, R.; Gorzalski, A.; Siao, D.; Markland, R.; Chang, C.-L.; Baker, H.; et al. Use of Wastewater Surveillance for Early Detection of Alpha and Epsilon SARS-CoV-2 Variants of Concern and Estimation of Overall COVID-19 Infection Burden. *Sci. Total Environ.* **2022**, *835*, 155410. [CrossRef]
12. Bei, Y.; Pinet, K.; Vrtis, K.B.; Borgaro, J.G.; Sun, L.; Campbell, M.; Apone, L.; Langhorst, B.W.; Nichols, N.M. Overcoming Variant Mutation-Related Impacts on Viral Sequencing and Detection Methodologies. *Front. Med.* **2022**, *9*, 989913. [CrossRef]

13. Ugurel, O.M.; Ata, O.; Turgut-Balik, D. An Updated Analysis of Variations in SARS-CoV-2 Genome. *Turk. J. Biol.* **2020**, *44*, 157–167. [CrossRef] [PubMed]
14. Yin, R.; Kwoh, C.K.; Zheng, J. Whole Genome Sequencing Analysis. In *Encyclopedia of Bioinformatics and Computational Biology*; Elsevier: Amsterdam, The Netherlands, 2019; pp. 176–183.
15. Tiwari, A.; Lipponen, A.; Hokajärvi, A.-M.; Luomala, O.; Sarekoski, A.; Rytkönen, A.; Österlund, P.; Al-Hello, H.; Juutinen, A.; Miettinen, I.T.; et al. Detection and Quantification of SARS-CoV-2 RNA in Wastewater Influent in Relation to Reported COVID-19 Incidence in Finland. *Water Res.* **2022**, *215*, 118220. [CrossRef]
16. Ahmed, W.; Angel, N.; Edson, J.; Bibby, K.; Bivins, A.; O'Brien, J.W.; Choi, P.M.; Kitajima, M.; Simpson, S.L.; Li, J.; et al. First Confirmed Detection of SARS-CoV-2 in Untreated Wastewater in Australia: A Proof of Concept for the Wastewater Surveillance of COVID-19 in the Community. *Sci. Total Environ.* **2020**, *728*, 138764. [CrossRef] [PubMed]
17. Sherchan, S.P.; Shahin, S.; Ward, L.M.; Tandukar, S.; Aw, T.G.; Schmitz, B.; Ahmed, W.; Kitajima, M. First Detection of SARS-CoV-2 RNA in Wastewater in North America: A Study in Louisiana, USA. *Sci. Total Environ.* **2020**, *743*, 140621. [CrossRef] [PubMed]
18. Pérez-Cataluña, A.; Chiner-Oms, Á.; Cuevas-Ferrando, E.; Díaz-Reolid, A.; Falcó, I.; Randazzo, W.; Girón-Guzmán, I.; Allende, A.; Bracho, M.A.; Comas, I.; et al. Spatial and Temporal Distribution of SARS-CoV-2 Diversity Circulating in Wastewater. *Water Res.* **2022**, *211*, 118007. [CrossRef] [PubMed]
19. Fontenele, R.S.; Kraberger, S.; Hadfield, J.; Driver, E.M.; Bowes, D.; Holland, L.A.; Faleye, T.O.C.; Adhikari, S.; Kumar, R.; Inchausti, R.; et al. High-Throughput Sequencing of SARS-CoV-2 in Wastewater Provides Insights into Circulating Variants. *Water Res.* **2021**, *205*, 117710. [CrossRef]
20. Islam, M.A.; Rahman, M.A.; Jakariya, M.; Bahadur, N.M.; Hossen, F.; Mukharjee, S.K.; Hossain, M.S.; Tasneem, A.; Haque, M.A.; Sera, F.; et al. A 30-Day Follow-up Study on the Prevalence of SARS-CoV-2 Genetic Markers in Wastewater from the Residence of COVID-19 Patient and Comparison with Clinical Positivity. *Sci. Total Environ.* **2023**, *858*, 159350. [CrossRef] [PubMed]
21. Tiwari, A.; Adhikari, S.; Kaya, D.; Islam, M.A.; Malla, B.; Sherchan, S.P.; Al-Mustapha, A.I.; Kumar, M.; Aggarwal, S.; Bhattacharya, P.; et al. Monkeypox Outbreak: Wastewater and Environmental Surveillance Perspective. *Sci. Total Environ.* **2022**, *856*, 159166. [CrossRef]
22. Tiwari, A.; Paakkanen, J.; Österblad, M.; Kirveskari, J.; Hendriksen, R.S.; Heikinheimo, A. Wastewater Surveillance Detected Carbapenemase Enzymes in Clinically Relevant Gram-Negative Bacteria in Helsinki, Finland; 2011–2012. *Front. Microbiol.* **2022**, *13*, 887888. [CrossRef]
23. Martin, J.; Klapsa, D.; Wilton, T.; Zambon, M.; Bentley, E.; Bujaki, E.; Fritzsche, M.; Mate, R.; Majumdar, M. Tracking SARS-CoV-2 in Sewage: Evidence of Changes in Virus Variant Predominance during COVID-19 Pandemic. *Viruses* **2020**, *12*, 1144. [CrossRef]
24. Hokajärvi, A.M.; Rytkönen, A.; Tiwari, A.; Kauppinen, A.; Oikarinen, S.; Lehto, K.M.; Kankaanpää, A.; Gunnar, T.; Al-Hello, H.; Blomqvist, S.; et al. The Detection and Stability of the SARS-CoV-2 RNA Biomarkers in Wastewater Influent in Helsinki, Finland. *Sci. Total Environ.* **2021**, *770*, 145274. [CrossRef]
25. Wurtzer, S.; Waldman, P.; Levert, M.; Cluzel, N.; Almayrac, J.L.; Charpentier, C.; Masnada, S.; Gillon-Ritz, M.; Mouchel, J.M.; Maday, Y.; et al. SARS-CoV-2 Genome Quantification in Wastewaters at Regional and City Scale Allows Precise Monitoring of the Whole Outbreaks Dynamics and Variants Spreading in the Population. *Sci. Total Environ.* **2022**, *810*, 152213. [CrossRef]
26. Ahmed, W.; Bertsch, P.M.; Angel, N.; Bibby, K.; Bivins, A.; Dierens, L.; Edson, J.; Ehret, J.; Gyawali, P.; Hamilton, K.A.; et al. Detection of SARS-CoV-2 RNA in Commercial Passenger Aircraft and Cruise Ship Wastewater: A Surveillance Tool for Assessing the Presence of COVID-19 Infected Travellers. *J. Travel Med.* **2020**, *27*, taaa116. [CrossRef]
27. Gonzalez, R.; Curtis, K.; Bivins, A.; Bibby, K.; Weir, M.H.; Yetka, K.; Thompson, H.; Keeling, D.; Mitchell, J.; Gonzalez, D. COVID-19 Surveillance in Southeastern Virginia Using Wastewater-Based Epidemiology. *Water Res.* **2020**, *186*, 116296. [CrossRef]
28. La Rosa, G.; Brandtner, D.; Mancini, P.; Veneri, C.; Ferraro, G.B.; Bonadonna, L.; Lucentini, L.; Suffredini, E. Key Sars-Cov-2 Mutations of Alpha, Gamma, and Eta Variants Detected in Urban Wastewaters in Italy by Long-Read Amplicon Sequencing Based on Nanopore Technology. *Water* **2021**, *13*, 2503. [CrossRef]
29. Tiwari, A.; Phan, N.; Tandukar, S.; Ashoori, R.; Thakali, O.; Mousazadesh, M.; Dehghani, M.H.; Sherchan, S.P. Persistence and Occurrence of SARS-CoV-2 in Water and Wastewater Environments: A Review of the Current Literature. *Environ. Sci. Pollut. Res.* **2021**, *29*, 85658–85668. [CrossRef] [PubMed]
30. Kitajima, M.; Ahmed, W.; Bibby, K.; Carducci, A.; Gerba, C.P.; Hamilton, K.A.; Haramoto, E.; Rose, J.B. SARS-CoV-2 in Wastewater: State of the Knowledge and Research Needs. *Sci. Total Environ.* **2020**, *739*, 139076. [CrossRef] [PubMed]
31. Naqvi, A.A.T.; Fatima, K.; Mohammad, T.; Fatima, U.; Singh, I.K.; Singh, A.; Atif, S.M.; Hariprasad, G.; Hasan, G.M.; Hassan, M.I. Insights into SARS-CoV-2 Genome, Structure, Evolution, Pathogenesis and Therapies: Structural Genomics Approach. *Biochim. Biophys. Acta (BBA)-Mol. Basis Dis.* **2020**, *1866*, 165878. [CrossRef]
32. Rambaut, A.; Holmes, E.C.; O'Toole, Á.; Hill, V.; McCrone, J.T.; Ruis, C.; du Plessis, L.; Pybus, O.G. A Dynamic Nomenclature Proposal for SARS-CoV-2 Lineages to Assist Genomic Epidemiology. *Nat. Microbiol.* **2020**, *5*, 1403–1407. [CrossRef]
33. ECDC. SARS-CoV-2 Variants of Concern as of 8 December 2022. Available online: https://www.ecdc.europa.eu/en/covid-19/variants-concern (accessed on 15 December 2022).
34. da Silva, J.C.; Félix, V.B.; Leão, S.A.B.F.; Trindade-Filho, E.M.; Scorza, F.A. New Brazilian Variant of the SARS-CoV-2 (P1/Gamma) of COVID-19 in Alagoas State. *Braz. J. Infect. Dis.* **2021**, *25*, 101588. [CrossRef] [PubMed]
35. Wadman, M. California Coronavirus Strain May Be More Infectious—And Lethal. *Science* **2021**, *23*, abh2101. [CrossRef]

36. Kimura, I.; Kosugi, Y.; Wu, J.; Zahradnik, J.; Yamasoba, D.; Butlertanaka, E.P.; Tanaka, Y.L.; Uriu, K.; Liu, Y.; Morizako, N.; et al. The SARS-CoV-2 Lambda Variant Exhibits Enhanced Infectivity and Immune Resistance. *Cell Rep.* **2022**, *38*, 110218. [CrossRef]
37. Acevedo, M.L.; Gaete-Argel, A.; Alonso-Palomares, L.; de Oca, M.M.; Bustamante, A.; Gaggero, A.; Paredes, F.; Cortes, C.P.; Pantano, S.; Martínez-Valdebenito, C.; et al. Differential Neutralizing Antibody Responses Elicited by CoronaVac and BNT162b2 against SARS-CoV-2 Lambda in Chile. *Nat. Microbiol.* **2022**, *7*, 524–529. [CrossRef] [PubMed]
38. López-Cortés, G.I.; Palacios-Pérez, M.; Zamudio, G.S.; Veledíaz, H.F.; Ortega, E.; José, M.V. Neutral Evolution Test of the Spike Protein of SARS-CoV-2 and Its Implications in the Binding to ACE2. *Sci. Rep.* **2021**, *11*, 18847. [CrossRef] [PubMed]
39. Itarte, M.; Bofill-Mas, S.; Martínez-Puchol, S.; Torrell, H.; Ceretó, A.; Carrasco, M.; Forés, E.; Canela, N.; Girones, R.; Rusiñol, M. Looking for a Needle in a Haystack. SARS-CoV-2 Variant Characterization in Sewage. *Curr. Opin. Environ. Sci. Health* **2021**, *24*, 100308. [CrossRef] [PubMed]
40. Walker, A.S.; Vihta, K.-D.; Gethings, O.; Pritchard, E.; Jones, J.; House, T.; Bell, I.; Bell, J.I.; Newton, J.N.; Farrar, J.; et al. Tracking the Emergence of SARS-CoV-2 Alpha Variant in the United Kingdom. *N. Engl. J. Med.* **2021**, *385*, 2582–2585. [CrossRef]
41. Davies, N.G.; Abbott, S.; Barnard, R.C.; Jarvis, C.I.; Kucharski, A.J.; Munday, J.D.; Pearson, C.A.B.; Russell, T.W.; Tully, D.C.; Washburne, A.D.; et al. Estimated Transmissibility and Impact of SARS-CoV-2 Lineage B.1.1.7 in England. *Science* **2021**, *372*, eabg3055. [CrossRef] [PubMed]
42. Liu, C.; Ginn, H.M.; Dejnirattisai, W.; Supasa, P.; Wang, B.; Tuekprakhon, A.; Nutalai, R.; Zhou, D.; Mentzer, A.J.; Zhao, Y.; et al. Reduced Neutralization of SARS-CoV-2 B.1.617 by Vaccine and Convalescent Serum. *Cell* **2021**, *184*, 4220–4236.e13. [CrossRef]
43. Yaniv, K.; Ozer, E.; Lewis, Y.; Kushmaro, A. RT-QPCR Assays for SARS-CoV-2 Variants of Concern in Wastewater Reveals Compromised Vaccination-Induced Immunity. *Water Res.* **2021**, *207*, 117808. [CrossRef]
44. Yaniv, K.; Ozer, E.; Shagan, M.; Lakkakula, S.; Plotkin, N.; Bhandarkar, N.S.; Kushmaro, A. Direct RT-QPCR Assay for SARS-CoV-2 Variants of Concern (Alpha, B.1.1.7 and Beta, B.1.351) Detection and Quantification in Wastewater. *Environ. Res.* **2021**, *201*, 111653. [CrossRef]
45. Hodcroft, E.B.; Zuber, M.; Nadeau, S.; Vaughan, T.G.; Crawford, K.H.D.; Althaus, C.L.; Reichmuth, M.L.; Bowen, J.E.; Walls, A.C.; Corti, D.; et al. Spread of a SARS-CoV-2 Variant through Europe in the Summer of 2020. *Nature* **2021**, *595*, 707–712. [CrossRef]
46. Maida, C.M.; Amodio, E.; Mazzucco, W.; La Rosa, G.; Lucentini, L.; Suffredini, E.; Palermo, M.; Andolina, G.; Iaia, F.R.; Merlo, F.; et al. Wastewater-Based Epidemiology for Early Warning of SARS-CoV-2 Circulation: A Pilot Study Conducted in Sicily, Italy. *Int. J. Hyg. Environ. Health* **2022**, *242*, 113948. [CrossRef]
47. Sharif, S.; Ikram, A.; Khurshid, A.; Salman, M.; Mehmood, N.; Arshad, Y.; Ahmed, J.; Safdar, R.M.; Rehman, L.; Mujtaba, G.; et al. Detection of SARs-CoV-2 in Wastewater Using the Existing Environmental Surveillance Network: A Potential Supplementary System for Monitoring COVID-19 Transmission. *PLoS ONE* **2021**, *16*, e0249568. [CrossRef]
48. Sidik, S.M. Vaccines Protect against Infection from Omicron Subvariant—But Not for Long. *Nature* **2022**. [CrossRef]
49. Jensen, B.; Luebke, N.; Feldt, T.; Keitel, V.; Brandenburger, T.; Kindgen-Milles, D.; Lutterbeck, M.; Freise, N.F.; Schoeler, D.; Haas, R.; et al. Emergence of the E484K Mutation in SARS-CoV-2-Infected Immunocompromised Patients Treated with Bamlanivimab in Germany. *Lancet Reg. Health-Eur.* **2021**, *8*, 100164. [CrossRef] [PubMed]
50. Kreier, F. Long-COVID Symptoms Less Likely in Vaccinated People, Israeli Data Say. *Nature* **2022**. [CrossRef]
51. Abdool Karim, S.S.; de Oliveira, T. New SARS-CoV-2 Variants—Clinical, Public Health, and Vaccine Implications. *N. Engl. J. Med.* **2021**, *384*, 1866–1868. [CrossRef] [PubMed]
52. Moher, D.; Liberati, A.; Tetzlaff, J.; Altman, D.G.; PRISMA Group. Preferred Reporting Items for Systematic Reviews and Meta-Analyses: The PRISMA Statement. *BMJ* **2009**, *339*, b2535. [CrossRef] [PubMed]
53. Wurtzer, S.; Levert, M.; Dhenain, E.; Accrombessi, H.; Manco, S.; Fagour, N.; Goulet, M.; Boudaud, N.; Gaillard, L.; Bertrand, I.; et al. From Alpha to Omicron BA.2: New Digital RT-PCR Approach and Challenges for SARS-CoV-2 VOC Monitoring and Normalization of Variant Dynamics in Wastewater. *Sci. Total Environ.* **2022**, *848*, 157740. [CrossRef]
54. Barbé, L.; Schaeffer, J.; Besnard, A.; Jousse, S.; Wurtzer, S.; Moulin, L.; Le Guyader, F.S.; Desdouits, M. SARS-CoV-2 Whole-Genome Sequencing Using Oxford Nanopore Technology for Variant Monitoring in Wastewaters. *Front. Microbiol.* **2022**, *13*, 889811. [CrossRef]
55. Heijnen, L.; Elsinga, G.; de Graaf, M.; Molenkamp, R.; Koopmans, M.P.G.; Medema, G. Droplet Digital RT-PCR to Detect SARS-CoV-2 Signature Mutations of Variants of Concern in Wastewater. *Sci. Total Environ.* **2021**, *799*, 149456. [CrossRef]
56. Ho, J.; Stange, C.; Suhrborg, R.; Wurzbacher, C.; Drewes, J.E.; Tiehm, A. SARS-CoV-2 Wastewater Surveillance in Germany: Long-Term RT-Digital Droplet PCR Monitoring, Suitability of Primer/Probe Combinations and Biomarker Stability. *Water Res.* **2022**, *210*, 117977. [CrossRef]
57. Rubio-Acero, R.; Beyerl, J.; Muenchhoff, M.; Roth, M.S.; Castelletti, N.; Paunovic, I.; Radon, K.; Springer, B.; Nagel, C.; Boehm, B.; et al. Spatially Resolved Qualified Sewage Spot Sampling to Track SARS-CoV-2 Dynamics in Munich—One Year of Experience. *Sci. Total Environ.* **2021**, *797*, 149031. [CrossRef] [PubMed]
58. Schumann, V.-F.; de Castro Cuadrat, R.R.; Wyler, E.; Wurmus, R.; Deter, A.; Quedenau, C.; Dohmen, J.; Faxel, M.; Borodina, T.; Blume, A.; et al. SARS-CoV-2 Infection Dynamics Revealed by Wastewater Sequencing Analysis and Deconvolution. *Sci. Total Environ.* **2022**, *853*, 158931. [CrossRef]
59. Agrawal, S.; Orschler, L.; Tavazzi, S.; Greither, R.; Gawlik, B.M.; Lackner, S. Genome Sequencing of Wastewater Confirms the Arrival of the SARS-CoV-2 Omicron Variant at Frankfurt Airport but Limited Spread in the City of Frankfurt, Germany, in November 2021. *Microbiol. Resour. Announc.* **2022**, *11*, e01229-21. [CrossRef]

60. La Rosa, G.; Iaconelli, M.; Veneri, C.; Mancini, P.; Bonanno Ferraro, G.; Brandtner, D.; Lucentini, L.; Bonadonna, L.; Rossi, M.; Grigioni, M.; et al. The Rapid Spread of SARS-CoV-2 Omicron Variant in Italy Reflected Early through Wastewater Surveillance. *Sci. Total Environ.* **2022**, *837*, 155767. [CrossRef]
61. Lee, W.L.; Armas, F.; Guarneri, F.; Gu, X.; Formenti, N.; Wu, F.; Chandra, F.; Parisio, G.; Chen, H.; Xiao, A.; et al. Rapid Displacement of SARS-CoV-2 Variant Delta by Omicron Revealed by Allele-Specific PCR in Wastewater. *Water Res.* **2022**, *221*, 118809. [CrossRef]
62. D'Agostino, Y.; Rocco, T.; Ferravante, C.; Porta, A.; Tosco, A.; Cappa, V.M.; Lamberti, J.; Alexandrova, E.; Memoli, D.; Terenzi, I.; et al. Rapid and Sensitive Detection of SARS-CoV-2 Variants in Nasopharyngeal Swabs and Wastewaters. *Diagn. Microbiol. Infect. Dis.* **2022**, *102*, 115632. [CrossRef]
63. Cutrupi, F.; Cadonna, M.; Manara, S.; Postinghel, M.; La Rosa, G.; Suffredini, E.; Foladori, P. The Wave of the SARS-CoV-2 Omicron Variant Resulted in a Rapid Spike and Decline as Highlighted by Municipal Wastewater Surveillance. *Environ. Technol. Innov.* **2022**, *28*, 102667. [CrossRef] [PubMed]
64. Róka, E.; Déri, D.; Khayer, B.; Kis, Z.; Schuler, E.; Magyar, N.; Pályi, B.; Pándics, T.; Vargha, M. SARS-CoV-2 Variant Detection from Wastewater: Rapid Spread of B.1.1.7 Lineage in Hungary. *J. Water Health* **2022**, *20*, 277–286. [CrossRef] [PubMed]
65. Avgeris, M.; Adamopoulos, P.G.; Galani, A.; Xagorari, M.; Gourgiotis, D.; Trougakos, I.P.; Voulgaris, N.; Dimopoulos, M.-A.; Thomaidis, N.S.; Scorilas, A. Novel Nested-Seq Approach for SARS-CoV-2 Real-Time Epidemiology and In-Depth Mutational Profiling in Wastewater. *Int. J. Mol. Sci.* **2021**, *22*, 8498. [CrossRef]
66. Galani, A.; Aalizadeh, R.; Kostakis, M.; Markou, A.; Alygizakis, N.; Lytras, T.; Adamopoulos, P.G.; Peccia, J.; Thompson, D.C.; Kontou, A.; et al. SARS-CoV-2 Wastewater Surveillance Data Can Predict Hospitalizations and ICU Admissions. *Sci. Total Environ.* **2022**, *804*, 150151. [CrossRef]
67. Pechlivanis, N.; Tsagiopoulou, M.; Maniou, M.C.; Togkousidis, A.; Mouchtaropoulou, E.; Chassalevris, T.; Chaintoutis, S.C.; Petala, M.; Kostoglou, M.; Karapantsios, T.; et al. Detecting SARS-CoV-2 Lineages and Mutational Load in Municipal Wastewater and a Use-Case in the Metropolitan Area of Thessaloniki, Greece. *Sci. Rep.* **2022**, *12*, 2659. [CrossRef]
68. Chassalevris, T.; Chaintoutis, S.C.; Koureas, M.; Petala, M.; Moutou, E.; Beta, C.; Kyritsi, M.; Hadjichristodoulou, C.; Kostoglou, M.; Karapantsios, T.; et al. SARS-CoV-2 Wastewater Monitoring Using a Novel PCR-Based Method Rapidly Captured the Delta-to-Omicron BA.1 Transition Patterns in the Absence of Conventional Surveillance Evidence. *Sci. Total Environ.* **2022**, *844*, 156932. [CrossRef]
69. Jahn, K.; Dreifuss, D.; Topolsky, I.; Kull, A.; Ganesanandamoorthy, P.; Fernandez-Cassi, X.; Bänziger, C.; Devaux, A.J.; Stachler, E.; Caduff, L.; et al. Early Detection and Surveillance of SARS-CoV-2 Genomic Variants in Wastewater Using COJAC. *Nat. Microbiol.* **2022**, *7*, 1151–1160. [CrossRef] [PubMed]
70. Caduff, L.; Dreifuss, D.; Schindler, T.; Devaux, A.J.; Ganesanandamoorthy, P.; Kull, A.; Stachler, E.; Fernandez-Cassi, X.; Beerenwinkel, N.; Kohn, T.; et al. Inferring Transmission Fitness Advantage of SARS-CoV-2 Variants of Concern from Wastewater Samples Using Digital PCR, Switzerland, December 2020 through March 2021. *Eurosurveillance* **2022**, *27*, 2100806. [CrossRef]
71. Agrawal, S.; Orschler, L.; Schubert, S.; Zachmann, K.; Heijnen, L.; Tavazzi, S.; Gawlik, B.M.; de Graaf, M.; Medema, G.; Lackner, S. Prevalence and Circulation Patterns of SARS-CoV-2 Variants in European Sewage Mirror Clinical Data of 54 European Cities. *Water Res.* **2022**, *214*, 118162. [CrossRef]
72. Radu, E.; Masseron, A.; Amman, F.; Schedl, A.; Agerer, B.; Endler, L.; Penz, T.; Bock, C.; Bergthaler, A.; Vierheilig, J.; et al. Emergence of SARS-CoV-2 Alpha Lineage and Its Correlation with Quantitative Wastewater-Based Epidemiology Data. *Water Res.* **2022**, *215*, 118257. [CrossRef] [PubMed]
73. Amman, F.; Markt, R.; Endler, L.; Hupfauf, S.; Agerer, B.; Schedl, A.; Richter, L.; Zechmeister, M.; Bicher, M.; Heiler, G.; et al. Viral Variant-Resolved Wastewater Surveillance of SARS-CoV-2 at National Scale. *Nat. Biotechnol.* **2022**, *40*, 1814–1822. [CrossRef]
74. Markt, R.; Endler, L.; Amman, F.; Schedl, A.; Penz, T.; Büchel-Marxer, M.; Grünbacher, D.; Mayr, M.; Peer, E.; Pedrazzini, M.; et al. Detection and Abundance of SARS-CoV-2 in Wastewater in Liechtenstein, and the Estimation of Prevalence and Impact of the B.1.1.7 Variant. *J. Water Health* **2022**, *20*, 114–125. [CrossRef] [PubMed]
75. Herold, M.; D'Herouël, A.F.; May, P.; Delogu, F.; Wienecke-Baldacchino, A.; Tapp, J.; Walczak, C.; Wilmes, P.; Cauchie, H.-M.; Fournier, G.; et al. Genome Sequencing of SARS-CoV-2 Allows Monitoring of Variants of Concern through Wastewater. *Water* **2021**, *13*, 3018. [CrossRef]
76. Boogaerts, T.; Van den Bogaert, S.; Van Poelvoorde, L.A.E.; El Masri, D.; De Roeck, N.; Roosens, N.H.C.; Lesenfants, M.; Lahousse, L.; Van Hoorde, K.; van Nuijs, A.L.N.; et al. Optimization and Application of a Multiplex Digital PCR Assay for the Detection of SARS-CoV-2 Variants of Concern in Belgian Influent Wastewater. *Viruses* **2022**, *14*, 610. [CrossRef] [PubMed]
77. Reynolds, L.J.; Gonzalez, G.; Sala Comorera, L.; Martin, N.A.; Byrne, A.; Fennema, S.; Holohan, N.; Kuntamukkula, S.R.; Sarwar, N.; Nolan, T.M.; et al. SARS-CoV-2 Variant Trends in Ireland: Wastewater-Based Epidemiology and Clinical Surveillance. *Sci. Total Environ.* **2022**, *838*, 155828. [CrossRef] [PubMed]
78. Rasmussen, L.D.; Richter, S.R.; Midgley, S.E.; Franck, K.T. Detecting SARS-CoV-2 Omicron B.1.1.529 Variant in Wastewater Samples by Using Nanopore Sequencing. *Emerg. Infect. Dis.* **2022**, *28*, 1296. [CrossRef]
79. Mishra, S.; Mindermann, S.; Sharma, M.; Whittaker, C.; Mellan, T.A.; Wilton, T.; Klapsa, D.; Mate, R.; Fritzsche, M.; Zambon, M.; et al. Changing Composition of SARS-CoV-2 Lineages and Rise of Delta Variant in England. *EClinicalMedicine* **2021**, *39*, 101064. [CrossRef]

80. Carcereny, A.; Garcia-Pedemonte, D.; Martínez-Velázquez, A.; Quer, J.; Garcia-Cehic, D.; Gregori, J.; Antón, A.; Andrés, C.; Pumarola, T.; Chacón-Villanueva, C.; et al. Dynamics of SARS-CoV-2 Alpha (B.1.1.7) Variant Spread: The Wastewater Surveillance Approach. *Environ. Res.* **2022**, *208*, 112720. [CrossRef]
81. Novoa, B.; Ríos-Castro, R.; Otero-Muras, I.; Gouveia, S.; Cabo, A.; Saco, A.; Rey-Campos, M.; Pájaro, M.; Fajar, N.; Aranguren, R.; et al. Wastewater and Marine Bioindicators Surveillance to Anticipate COVID-19 Prevalence and to Explore SARS-CoV-2 Diversity by next Generation Sequencing: One-Year Study. *Sci. Total Environ.* **2022**, *833*, 155140. [CrossRef]
82. Rios, G.; Lacoux, C.; Leclercq, V.; Diamant, A.; Lebrigand, K.; Lazuka, A.; Soyeux, E.; Lacroix, S.; Fassy, J.; Couesnon, A.; et al. Monitoring SARS-CoV-2 Variants Alterations in Nice Neighborhoods by Wastewater Nanopore Sequencing. *Lancet Reg. Health-Eur.* **2021**, *10*, 100202. [CrossRef] [PubMed]
83. Wurtz, N.; Revol, O.; Jardot, P.; Giraud-Gatineau, A.; Houhamdi, L.; Soumagnac, C.; Annessi, A.; Lacoste, A.; Colson, P.; Aherfi, S.; et al. Monitoring the Circulation of SARS-CoV-2 Variants by Genomic Analysis of Wastewater in Marseille, South-East France. *Pathogens* **2021**, *10*, 1042. [CrossRef] [PubMed]
84. Rothman, J.A.; Loveless, T.B.; Kapcia, J.; Adams, E.D.; Steele, J.A.; Zimmer-Faust, A.G.; Langlois, K.; Wanless, D.; Griffith, M.; Mao, L.; et al. RNA Viromics of Southern California Wastewater and Detection of SARS-CoV-2 Single-Nucleotide Variants. *Appl. Environ. Microbiol.* **2021**, *87*, e01448-21. [CrossRef]
85. Ai, Y.; Davis, A.; Jones, D.; Lemeshow, S.; Tu, H.; He, F.; Ru, P.; Pan, X.; Bohrerova, Z.; Lee, J. Wastewater SARS-CoV-2 Monitoring as a Community-Level COVID-19 Trend Tracker and Variants in Ohio, United States. *Sci. Total Environ.* **2021**, *801*, 149757. [CrossRef]
86. Wolfe, M.; Hughes, B.; Duong, D.; Chan-Herur, V.; Wigginton, K.R.; White, B.J.; Boehm, A.B. Detection of SARS-CoV-2 Variants Mu, Beta, Gamma, Lambda, Delta, Alpha, and Omicron in Wastewater Settled Solids Using Mutation-Specific Assays Is Associated with Regional Detection of Variants in Clinical Samples. *Appl. Environ. Microbiol.* **2022**, *88*, e00045-22. [CrossRef]
87. Sutton, M.; Radniecki, T.S.; Kaya, D.; Alegre, D.; Geniza, M.; Girard, A.-M.; Carter, K.; Dasenko, M.; Sanders, J.L.; Cieslak, P.R.; et al. Detection of SARS-CoV-2 B.1.351 (Beta) Variant through Wastewater Surveillance before Case Detection in a Community, Oregon, USA. *Emerg. Infect. Dis.* **2022**, *28*, 1101. [CrossRef] [PubMed]
88. Yu, A.T.; Hughes, B.; Wolfe, M.K.; Leon, T.; Duong, D.; Rabe, A.; Kennedy, L.C.; Ravuri, S.; White, B.J.; Wigginton, K.R.; et al. Estimating Relative Abundance of 2 SARS-CoV-2 Variants through Wastewater Surveillance at 2 Large Metropolitan Sites, United States. *Emerg. Infect. Dis.* **2022**, *28*, 940–947. [CrossRef]
89. Oh, C.; Sashittal, P.; Zhou, A.; Wang, L.; El-Kebir, M.; Nguyen, T.H. Design of SARS-CoV-2 Variant-Specific PCR Assays Considering Regional and Temporal Characteristics. *Appl. Environ. Microbiol.* **2022**, *88*, e02289-21. [CrossRef] [PubMed]
90. Brumfield, K.D.; Leddy, M.; Usmani, M.; Cotruvo, J.A.; Tien, C.-T.; Dorsey, S.; Graubics, K.; Fanelli, B.; Zhou, I.; Registe, N.; et al. Microbiome Analysis for Wastewater Surveillance during COVID-19. *mBio* **2022**, *13*, e00591-22. [CrossRef]
91. Layton, B.A.; Kaya, D.; Kelly, C.; Williamson, K.J.; Alegre, D.; Bachhuber, S.M.; Banwarth, P.G.; Bethel, J.W.; Carter, K.; Dalziel, B.D.; et al. Evaluation of a Wastewater-Based Epidemiological Approach to Estimate the Prevalence of SARS-CoV-2 Infections and the Detection of Viral Variants in Disparate Oregon Communities at City and Neighborhood Scales. *Environ. Health Perspect.* **2022**, *130*, 067010. [CrossRef]
92. Silva, C.S.; Tryndyak, V.P.; Camacho, L.; Orloff, M.S.; Porter, A.; Garner, K.; Mullis, L.; Azevedo, M. Temporal Dynamics of SARS-CoV-2 Genome and Detection of Variants of Concern in Wastewater Influent from Two Metropolitan Areas in Arkansas. *Sci. Total Environ.* **2022**, *849*, 157546. [CrossRef] [PubMed]
93. Smyth, D.S.; Trujillo, M.; Gregory, D.A.; Cheung, K.; Gao, A.; Graham, M.; Guan, Y.; Guldenpfennig, C.; Hoxie, I.; Kannoly, S.; et al. Tracking Cryptic SARS-CoV-2 Lineages Detected in NYC Wastewater. *Nat. Commun.* **2022**, *13*, 635. [CrossRef] [PubMed]
94. Crits-Christoph, A.; Kantor, R.S.; Olm, M.R.; Whitney, O.N.; Al-Shayeb, B.; Lou, Y.C.; Flamholz, A.; Kennedy, L.C.; Greenwald, H.; Hinkle, A.; et al. Genome Sequencing of Sewage Detects Regionally Prevalent SARS-CoV-2 Variants. *mBio* **2021**, *12*, e02703-20. [CrossRef]
95. Swift, C.L.; Isanovic, M.; Correa Velez, K.E.; Norman, R.S. Community-Level SARS-CoV-2 Sequence Diversity Revealed by Wastewater Sampling. *Sci. Total Environ.* **2021**, *801*, 149691. [CrossRef]
96. Gregory, D.A.; Wieberg, C.G.; Wenzel, J.; Lin, C.-H.; Johnson, M.C. Monitoring SARS-CoV-2 Populations in Wastewater by Amplicon Sequencing and Using the Novel Program SAM Refiner. *Viruses* **2021**, *13*, 1647. [CrossRef]
97. Ash, K.T.; Alamilla, I.; Li, Y.; Joyner, D.C.; Williams, D.E.; McKay, P.J.; Green, B.M.; Iler, C.; DeBlander, S.E.; Kara-Murdoch, F.; et al. Coding-Complete Genome Sequence of a SARS-CoV-2 Variant Obtained from Raw Sewage at the University of Tennessee—Knoxville Campus. *Microbiol. Resour. Announc.* **2021**, *10*, e01049-21. [CrossRef]
98. Lee, W.L.; Imakaev, M.; Armas, F.; McElroy, K.A.; Gu, X.; Duvallet, C.; Chandra, F.; Chen, H.; Leifels, M.; Mendola, S.; et al. Quantitative SARS-CoV-2 Alpha Variant B.1.1.7 Tracking in Wastewater by Allele-Specific RT-QPCR. *Environ. Sci. Technol. Lett.* **2021**, *8*, 675–682. [CrossRef]
99. Rainey, A.L.; Loeb, J.C.; Robinson, S.E.; Lednicky, J.A.; McPherson, J.; Colson, S.; Allen, M.; Coker, E.S.; Sabo-Attwood, T.; Maurelli, A.T.; et al. Wastewater Surveillance for SARS-CoV-2 in a Small Coastal Community: Effects of Tourism on Viral Presence and Variant Identification among Low Prevalence Populations. *Environ. Res.* **2022**, *208*, 112496. [CrossRef] [PubMed]
100. Boehm, A.B.; Hughes, B.; Wolfe, M.K.; White, B.J.; Duong, D.; Chan-Herur, V. Regional Replacement of SARS-CoV-2 Variant Omicron BA.1 with BA.2 as Observed through Wastewater Surveillance. *Environ. Sci. Technol. Lett.* **2022**, *9*, 575–580. [CrossRef] [PubMed]

101. Li, L.; Uppal, T.; Hartley, P.; Gorzalski, A.; Pandori, M.; Picker, M.A.; Verma, S.; Pagilla, K. Detecting SARS-CoV-2 variants in wastewater and their correlation with circulating variants in the communities. *Sci Rep* **2022**, *12*, 16141. [CrossRef] [PubMed]
102. Lin, X.; Glier, M.; Kuchinski, K.; Ross-Van Mierlo, T.; McVea, D.; Tyson, J.R.; Prystajecky, N.; Ziels, R.M. Assessing Multiplex Tiling PCR Sequencing Approaches for Detecting Genomic Variants of SARS-CoV-2 in Municipal Wastewater. *mSystems* **2021**, *6*, e01068-21. [CrossRef]
103. Graber, T.E.; Mercier, É.; Bhatnagar, K.; Fuzzen, M.; D'Aoust, P.M.; Hoang, H.-D.; Tian, X.; Towhid, S.T.; Plaza-Diaz, J.; Eid, W.; et al. Near Real-Time Determination of B.1.1.7 in Proportion to Total SARS-CoV-2 Viral Load in Wastewater Using an Allele-Specific Primer Extension PCR Strategy. *Water Res.* **2021**, *205*, 117681. [CrossRef]
104. Corchis-Scott, R.; Geng, Q.; Seth, R.; Ray, R.; Beg, M.; Biswas, N.; Charron, L.; Drouillard, K.D.; D'Souza, R.; Heath, D.D.; et al. Averting an Outbreak of SARS-CoV-2 in a University Residence Hall through Wastewater Surveillance. *Microbiol. Spectr.* **2021**, *9*, e00792-21. [CrossRef]
105. Hubert, C.R.J.; Acosta, N.; Waddell, B.J.M.; Hasing, M.E.; Qiu, Y.; Fuzzen, M.; Harper, N.B.J.; Bautista, M.A.; Gao, T.; Papparis, C.; et al. Tracking Emergence and Spread of SARS-CoV-2 Omicron Variant in Large and Small Communities by Wastewater Monitoring in Alberta, Canada. *Emerg. Infect. Dis.* **2022**, *28*, 1770–1776. [CrossRef] [PubMed]
106. Lawal, O.U.; Zhang, L.; Parreira, V.R.; Brown, R.S.; Chettleburgh, C.; Dannah, N.; Delatolla, R.; Gilbride, K.A.; Graber, T.E.; Islam, G.; et al. Metagenomics of Wastewater Influent from Wastewater Treatment Facilities across Ontario in the Era of Emerging SARS-CoV-2 Variants of Concern. *Microbiol. Resour. Announc.* **2022**, *11*, e00362-22. [CrossRef]
107. Oloye, F.F.; Xie, Y.; Asadi, M.; Cantin, J.; Challis, J.K.; Brinkmann, M.; McPhedran, K.N.; Kristian, K.; Keller, M.; Sadowski, M.; et al. Rapid Transition between SARS-CoV-2 Variants of Concern Delta and Omicron Detected by Monitoring Municipal Wastewater from Three Canadian Cities. *Sci. Total Environ.* **2022**, *841*, 156741. [CrossRef] [PubMed]
108. Peng, J.; Sun, J.; Yang, M.I.; Gibson, R.M.; Arts, E.J.; Olabode, A.S.; Poon, A.F.Y.; Wang, X.; Wheeler, A.R.; Edwards, E.A.; et al. Early Warning Measurement of SARS-CoV-2 Variants of Concern in Wastewaters by Mass Spectrometry. *Environ. Sci. Technol. Lett.* **2022**, *9*, 638–644. [CrossRef]
109. Peterson, S.W.; Lidder, R.; Daigle, J.; Wonitowy, Q.; Dueck, C.; Nagasawa, A.; Mulvey, M.R.; Mangat, C.S. RT-QPCR Detection of SARS-CoV-2 Mutations S 69–70 Del, S N501Y and N D3L Associated with Variants of Concern in Canadian Wastewater Samples. *Sci. Total Environ.* **2022**, *810*, 151283. [CrossRef] [PubMed]
110. Napit, R.; Manandhar, P.; Chaudhary, A.; Shrestha, B.; Poudel, A.; Raut, R.; Pradhan, S.; Raut, S.; Mathema, S.; Rajbhandari, R.; et al. Rapid Genomic Surveillance of SARS-CoV-2 in a Dense Urban Community Using Environmental (Sewage) Samples. *medRxiv* **2021**. [CrossRef]
111. Otero, M.C.B.; Murao, L.A.E.; Limen, M.A.G.; Gaite, P.L.A.; Bacus, M.G.; Acaso, J.T.; Corazo, K.; Knot, I.E.; Sajonia, H.; de los Reyes, F.L.; et al. Wastewater-Based Epidemiology and Whole-Genome Sequencing for Community-Level Surveillance of SARS-CoV-2 in Selected Urban Communities of Davao City, Philippines: A Pilot Study. *medRxiv* **2021**. [CrossRef]
112. Malla, B.; Thakali, O.; Shrestha, S.; Segawa, T.; Kitajima, M.; Haramoto, E. Application of a High-Throughput Quantitative PCR System for Simultaneous Monitoring of SARS-CoV-2 Variants and Other Pathogenic Viruses in Wastewater. *Sci. Total Environ.* **2022**, *853*, 158659. [CrossRef]
113. Erster, O.; Mendelson, E.; Kabat, A.; Levy, V.; Mannasse, B.; Assraf, H.; Azar, R.; Ali, Y.; Bucris, E.; Bar-Ilan, D.; et al. Specific Detection of SARS-CoV-2 Variants B.1.1.7 (Alpha) and B.1.617.2 (Delta) Using a One-Step Quantitative PCR Assay. *Microbiol. Spectr.* **2022**, *10*, e02176-21. [CrossRef]
114. Bar-Or, I.; Indenbaum, V.; Weil, M.; Elul, M.; Levi, N.; Aguvaev, I.; Cohen, Z.; Levy, V.; Azar, R.; Mannasse, B.; et al. National Scale Real-Time Surveillance of SARS-CoV-2 Variants Dynamics by Wastewater Monitoring in Israel. *Viruses* **2022**, *14*, 1229. [CrossRef]
115. Joshi, M.; Kumar, M.; Srivastava, V.; Kumar, D.; Rathore, D.S.; Pandit, R.; Graham, D.W.; Joshi, C.G. Genetic Sequencing Detected the SARS-CoV-2 Delta Variant in Wastewater a Month Prior to the First COVID-19 Case in Ahmedabad (India). *Environ. Pollut.* **2022**, *310*, 119757. [CrossRef]
116. Nag, A.; Arora, S.; Sinha, V.; Meena, E.; Sutaria, D.; Gupta, A.B.; Medicherla, K.M. Monitoring of SARS-CoV-2 Variants by Wastewater-Based Surveillance as a Sustainable and Pragmatic Approach—A Case Study of Jaipur (India). *Water* **2022**, *14*, 297. [CrossRef]
117. Dharmadhikari, T.; Rajput, V.; Yadav, R.; Boargaonkar, R.; Patil, D.; Kale, S.; Kamble, S.P.; Dastager, S.G.; Dharne, M.S. High Throughput Sequencing Based Direct Detection of SARS-CoV-2 Fragments in Wastewater of Pune, West India. *Sci. Total Environ.* **2022**, *807*, 151038. [CrossRef]
118. Barbosa, M.R.F.; Garcia, S.C.; de Castro Bruni, A.; Machado, F.S.; de Oliveira, R.X.; Dropa, M.; da Costa, A.C.; Leal, E.; Brandão, C.J.; da Silva, R.L.O.; et al. One-Year Surveillance of SARS-CoV-2 in Wastewater from Vulnerable Urban Communities in Metropolitan São Paulo, Brazil. *J. Water Health* **2022**, *20*, 471–490. [CrossRef]
119. Masachessi, G.; Castro, G.; Cachi, A.M.; de los Ángeles Marinzalda, M.; Liendo, M.; Pisano, M.B.; Sicilia, P.; Ibarra, G.; Rojas, R.M.; López, L.; et al. Wastewater Based Epidemiology as a Silent Sentinel of the Trend of SARS-CoV-2 Circulation in the Community in Central Argentina. *Water Res.* **2022**, *219*, 118541. [CrossRef] [PubMed]
120. Johnson, R.; Sharma, J.R.; Ramharack, P.; Mangwana, N.; Kinnear, C.; Viraragavan, A.; Glanzmann, B.; Louw, J.; Abdelatif, N.; Reddy, T.; et al. Tracking the Circulating SARS-CoV-2 Variant of Concern in South Africa Using Wastewater-Based Epidemiology. *Sci. Rep.* **2022**, *12*, 1182. [CrossRef] [PubMed]

121. Ahmed, W.; Bivins, A.; Smith, W.J.M.; Metcalfe, S.; Stephens, M.; Jennison, A.V.; Moore, F.A.J.; Bourke, J.; Schlebusch, S.; McMahon, J.; et al. Detection of the Omicron (B.1.1.529) Variant of SARS-CoV-2 in Aircraft Wastewater. *Sci. Total Environ.* **2022**, *820*, 153171. [CrossRef] [PubMed]
122. Bertels, X.; Demeyer, P.; Van den Bogaert, S.; Boogaerts, T.; van Nuijs, A.L.N.; Delputte, P.; Lahousse, L. Factors Influencing SARS-CoV-2 RNA Concentrations in Wastewater up to the Sampling Stage: A Systematic Review. *Sci. Total Environ.* **2022**, *820*, 153290. [CrossRef]
123. Kmush, B.L.; Monk, D.; Green, H.; Sachs, D.A.; Zeng, T.; Larsen, D.A. Comparability of 24-Hour Composite and Grab Samples for Detection of SARS-2-CoV RNA in Wastewater. *FEMS Microbes* **2022**, *3*, xtac017. [CrossRef]
124. Barril, P.A.; Pianciola, L.A.; Mazzeo, M.; Ousset, M.J.; Jaureguiberry, M.V.; Alessandrello, M.; Sánchez, G.; Oteiza, J.M. Evaluation of Viral Concentration Methods for SARS-CoV-2 Recovery from Wastewaters. *Sci. Total Environ.* **2021**, *756*, 144105. [CrossRef]
125. Volz, E.; Hill, V.; McCrone, J.T.; Price, A.; Jorgensen, D.; O'Toole, Á.; Southgate, J.; Johnson, R.; Jackson, B.; Nascimento, F.F.; et al. Evaluating the Effects of SARS-CoV-2 Spike Mutation D614G on Transmissibility and Pathogenicity. *Cell* **2021**, *184*, 64–75.e11. [CrossRef]
126. Tiwari, A.; Ahmed, W.; Oikarinen, S.; Sherchan, S.P.; Heikinheimo, A.; Jiang, G.; Simpson, S.L.; Greaves, J.; Bivins, A. Application of Digital PCR for Public Health-Related Water Quality Monitoring. *Sci. Total Environ.* **2022**, *837*, 155663. [CrossRef]
127. Mazumder, P.; Dash, S.; Honda, R.; Sonne, C.; Kumar, M. Sewage Surveillance for SARS-CoV-2: Molecular Detection, Quantification, and Normalization Factors. *Curr. Opin. Environ. Sci. Health* **2022**, *28*, 100363. [CrossRef]
128. ECDC; WHO. Methods for the Detection and Characterisation of SARS-CoV-2 Variants–Second Update; 2022. Available online: https://www.ecdc.europa.eu/en/publications-data/methods-detection-and-characterisation-sars-cov-2-variants-second-update (accessed on 7 December 2022).
129. Ahmed, W.; Smith, W.J.M.; Metcalfe, S.; Jackson, G.; Choi, P.M.; Morrison, M.; Field, D.; Gyawali, P.; Bivins, A.; Bibby, K.; et al. Comparison of RT-QPCR and RT-DPCR Platforms for the Trace Detection of SARS-CoV-2 RNA in Wastewater. *ACS ES&T Water* **2022**, *2*, 1871–1880. [CrossRef]
130. GISAID. Available online: https://gisaid.org/. (accessed on 7 December 2022).
131. Thakali, O.; Raya, S.; Malla, B.; Tandukar, S.; Tiwari, A.; Sherchan, S.P.; Sherchand, J.B.; Haramoto, E. Pilot Study on Wastewater Surveillance of Dengue Virus RNA: Lessons, Challenges, and Implications for Future Research. *Environ. Chall.* **2022**, *9*, 100614. [CrossRef]
132. Ahmed, W.; Bivins, A.; Stephens, M.; Metcalfe, S.; Smith, W.J.M.; Sirikanchana, K.; Kitajima, M.; Simpson, S.L. Occurrence of Multiple Respiratory Viruses in Wastewater in Queensland, Australia: Potential for Community Disease Surveillance. *Sci. Total Environ.* **2023**, *864*, 161023. [CrossRef] [PubMed]
133. Tiwari, A.; Kurittu, P.; Al-Mustapha, A.I.; Heljanko, V.; Johansson, V.; Thakali, O.; Mishra, S.K.; Lehto, K.-M.; Lipponen, A.; Oikarinen, S.; et al. Wastewater Surveillance of Antibiotic-Resistant Bacterial Pathogens: A Systematic Review. *Front. Microbiol.* **2022**, *13*, 977106. [CrossRef]
134. Adhikari, S.; Halden, R.U. Opportunities and Limits of Wastewater-Based Epidemiology for Tracking Global Health and Attainment of UN Sustainable Development Goals. *Environ. Int.* **2022**, *163*, 107217. [CrossRef]
135. Wolfe, M.K.; Duong, D.; Bakker, K.M.; Ammerman, M.; Mortenson, L.; Hughes, B.; Arts, P.; Lauring, A.S.; Fitzsimmons, W.J.; Bendall, E.; et al. Wastewater-Based Detection of Two Influenza Outbreaks. *Environ. Sci. Technol. Lett.* **2022**, *9*, 687–692. [CrossRef]

water

Review

Effects of Temperature and Water Types on the Decay of Coronavirus: A Review

Ying Guo [1], Yanchen Liu [2], Shuhong Gao [3], Xu Zhou [3], Muttucumaru Sivakumar [1] and Guangming Jiang [1,4,*]

1 School of Civil, Mining, Environmental and Architectural Engineering, University of Wollongong, Wollongong, NSW 2522, Australia
2 State Key Joint Laboratory of Environment Simulation and Pollution Control, School of Environment, Tsinghua University, Beijing 100084, China
3 State Key Laboratory of Urban Water Resource and Environment, School of Civil & Environmental Engineering, Harbin Institute of Technology (Shenzhen), Shenzhen 518055, China
4 Illawarra Health and Medical Research Institute, University of Wollongong, Wollongong, NSW 2522, Australia
* Correspondence: gjiang@uow.edu.au; Tel.: +61-02-4221-3792

Abstract: The analysis of Severe Acute Respiratory Syndrome Coronavirus 2 (SARS-CoV-2) gene copy numbers in wastewater samples can provide quantitative information on Coronavirus Disease-19 (COVID-19) cases within a sewer catchment. However, many wastewater-based epidemiology (WBE) studies have neglected virus decay during the wastewater transportation process in sewers while back-calculating COVID-19 prevalence. Among various sewer condition parameters, wastewater temperature and dilution by fresh/saltwater infiltration may result in a significant change to the virus decay, in terms of both infectivity and Ribonucleic Acid (RNA). This paper reviewed the literature to identify and discuss the effects of temperature and water types (i.e., wastewater, freshwater, and seawater) on coronavirus decay based on the decay rate constants that were collected from published papers. To evaluate the importance of virus decay, a sensitivity analysis was then conducted with decay rates of SARS-CoV-2 RNA based on a WBE back-calculation equation. Finally, the decay rates of coronavirus in wastewater were also compared with those of other viruses to further understand the difference among virus species. The decay of SARS-CoV-2 RNA was found to be less impacted by temperature variation than viable coronaviruses. Nevertheless, WBE back-calculation was still sensitive to the RNA decay rates increased by warm wastewater (i.e., over 26 °C), which could lead to a two-times higher relative variance in estimated COVID-19 prevalence, considering the wastewater temperature variation between 4 and 37 °C in a sewer catchment with a 12-h hydraulic retention time. Comparatively, the sensitivity of the WBE estimation to the enveloped SARS-CoV-2 was greater than nonenveloped enteric viruses, which were less easily degradable in wastewater. In addition, wastewater dilution by stormwater inflow and accompanied cold weather might alleviate the decay of coronavirus infectivity, thus increasing the potential risk of COVID-19 transmission through wastewater. Overall, this paper aims to better understand the impact of in-sewer processes on coronavirus decay and its potential implications for WBE. The outcome could quantitatively inform WBE and improve awareness of the increased risk of COVID-19 infection via wastewater during heavy rainfall events. Given the identified scarcity of data available for coronavirus decay in salt water or with chemical additions, future research on the fate of SARS-CoV-2 subjected to chemical dosing for sewer or wastewater treatment plant operations is recommended.

Keywords: wastewater-based epidemiology; virus decay; SARS-CoV-2; sewer; back-calculation; sensitivity analysis

Citation: Guo, Y.; Liu, Y.; Gao, S.; Zhou, X.; Sivakumar, M.; Jiang, G. Effects of Temperature and Water Types on the Decay of Coronavirus: A Review. *Water* **2023**, *15*, 1051. https://doi.org/10.3390/w15061051

Academic Editor: Robert D. Morris

Received: 24 January 2023
Revised: 26 February 2023
Accepted: 7 March 2023
Published: 9 March 2023

1. Introduction

Severe Acute Respiratory Syndrome Coronavirus 2 (SARS-CoV-2) has spread among human communities all over the world in the past three years and remains a threat to public health and global economy [1]. Providing timely reports on coronavirus disease-19 (COVID-19) cases during outbreaks while using traditional epidemiology approaches is difficult and resource-consuming. Fortunately, wastewater-based epidemiology (WBE) has been demonstrated as an effective method in COVID-19 surveillance at the community level [2,3], since domestic wastewater contains biological information discharged from human bodies, such as loads of bacterial and viral shedding [4]. The detected SARS-CoV-2 concentration in wastewater is proportional to the COVID-19 disease burden in the corresponding catchment [5,6]. By retrospectively linking SARS-CoV-2 concentration in wastewater samples collected at sewer pumping stations or wastewater treatment plants (WWTPs) to virus excretion, WBE can back estimate the prevalence of COVID-19 in populations.

Importantly, WBE back-calculation of COVID-19 cases from SARS-CoV-2 concentration in wastewater needs correction considering the in-sewer virus decay [2]. Failing to include in-sewer decay renders the WBE approach prone to under-predicting the amount of upstream virus shedding. Furthermore, seasonal wastewater temperature variation, seawater intrusion, stormwater inflow, underground water infiltration in sewers, and many other factors [7] can affect the decay kinetics of viruses, thus impacting the infectivity and detectability of SARS-CoV-2 at downstream sampling locations. Wastewater temperature might vary in the range of 4–37 °C depending on the usage in household and weather conditions [2]. Higher temperatures facilitate the virus decay in wastewater compared to lower temperatures [8,9]. During in-sewer transportation, domestic wastewater might be diluted by saline water (i.e., seawater/saline groundwater intrusion, municipal seawater utilization for toilet flushing) or freshwater (i.e., stormwater inflow, or fresh groundwater infiltration), which can have impacts on the decay of viruses in sewers [10]. Correspondingly, the correction factor for WBE back-calculation is the in-sewer decay rate constant, which could be obtained using first-order kinetics [8].

This paper systematically collated the decay rates of SARS-CoV-2 Ribonucleic Acid (RNA) under different water conditions from published literature and evaluated the impact of in-sewer decay on back-calculating COVID-19 cases using sensitivity analysis. In addition, there is still a need to understand the fate of infectious coronaviruses outside the human host after its excretion into wastewater. The decay rates of viral infectivity under various temperatures and water types provide insights into the risk of disease transmission via wastewater exposure, under dynamic weather conditions and sewer environments. Therefore, the decay rates of viable coronaviruses were also collected from published literature through systematic review and compared with the loss of RNA signals in waters.

The SARS-CoV-2 strain, as well as surrogate coronaviruses, were used in the reported decay experiments, being employed as representatives for survivability studies, considering the safety challenges in conducting lab work relating to the highly contagious SARS-CoV-2. The murine hepatitis virus (MHV), transmissible gastroenteritis virus (TGEV), feline infectious peritonitis virus (FIPV), porcine respiratory coronavirus (PRCV), and human coronavirus 229E were accepted as surrogates within the *Coronaviridae* family, given their similarity in size, composition, and morphology [9]. The decay rates of these infectious coronaviruses were included in the collection together with SARS-CoV-2.

Moreover, the persistence of coronaviruses with lipid-containing envelopes might be separate from nonenveloped water-borne viruses that were more commonly investigated in water. To better understand the difference among viral species in environmental fates, the collated decay rates of the coronavirus were further compared to those of norovirus (a widely known nonenveloped enteric virus) and other viruses.

Through systematic review, this study summarized literature data about the decay of coronavirus in municipal wastewater and evaluated the sensitivity of decay rates in WBE back-estimations for COVID-19 prevalence. The results could inform wastewater epidemiology and help with the assessment of public health risks.

2. Methods

2.1. Literature Search and Data Selection

This study compiled the decay rates of coronaviruses from peer-reviewed papers, which were available online before 21 December 2022. Terms "(coronavirus OR SARS-CoV-2) AND (water OR wastewater OR sewer * OR seawater OR stormwater) AND (survival OR decay OR die-off OR inactivat * OR persistence)" were searched in databases available through Web of Science [9]. Initially, the titles of the search results were screened to decide whether their abstract warranted reading. If the title did not contain sufficient information, the abstract would be further evaluated to determine whether the paper can provide numeric decay rates. The papers that passed the screening through the title and/or abstract were subjected to full-text review. The following inclusion criteria were adopted when reviewing the papers: (i) in English, (ii) having decay experiments in dark conditions, (iii) with justifiable methods for enumeration, e.g., using culture methods or reverse transcription quantitative polymerase chain reaction (RT-qPCR), (iv) presentation of extractable decay rates.

For papers providing decay rate constants (k, d^{-1}) directly, the calculation approach of k was confirmed to conform with the first-order kinetics as shown in Equation (1), where C is the coronavirus concentration (gene copy/L, or median tissue culture infectious dose, $TCID_{50}$/L) at time t and C_0 is the initial concentration. If k was determined using the regression slope of $\log(C/C_0)$ versus time (in days), it would be transformed into $\ln(C/C_0)$. For papers providing t_{90} (i.e., time to achieve 90% reduction in concentration), the times were converted to first-order decay rates by applying Chick's law (Equation (2)) [11]. The fitted decay rate constant was included only if the goodness of fit was acceptable ($R^2 > 0.6$). Meta data including temperature, water types, detection methods, and other publication information were also recorded along with the first-order decay rates (Tables S1 and S2).

$$C = C_0 \times e^{-k \times t}, \tag{1}$$

$$k = 2.303/t_{90}, \tag{2}$$

2.2. Data Analysis

The decay rate constants (k, day^{-1}) were grouped according to the types of water, temperature, and detection methods. The k values were imported into R studio (version 1.3.1056, https://www.rstudio.com (accessed on 26 December 2022) for plotting by gglpot2 package and processed using multcomp for multivariate analysis of variance. When necessary, k values were shown in figures after log10-transformation for better visualization of the decay rate constants over multiple orders of magnitude.

The relationship between temperatures and decay rate constants was fitted according to the Arrhenius equation (Equation (3)), where k_0 is the decay rate constant at T_0 °C, k is the decay rate constant at temperature T °C, T_0 is the reference temperature in °C, λ is the temperature correction coefficient indicating the effect of temperature on virus decay rates. The parameters λ and k_0 were estimated by assuming T_0 = 0 °C and fitting Equation (3) to the literature data using SigmaPlot 14 (Systat Software, San Jose, CA, USA).

$$k = k_0 \times \lambda^{(T-T_0)}, \tag{3}$$

The WBE back calculation [2] was adapted as Equation (4), where F is the wastewater flow rate (L/day), C_{RNA} is the concentration of SARS-CoV-2 RNA in wastewater (copies/L), $P_{catchment}$ is the population within the sampled catchment boundary, E is the daily excretion rate (copies/(person·day)). To account for the in-sewer decay, a correction factor R_{sewer} was applied. As the C_{RNA} in Equation (4) should be C_t after decay, R_{sewer} can be derived from the decay kinetics as C_t/C_0 (refers to $e^{-k \times t}$ as in Equation (5)). The number of infected cases $P_{infection}$ thus can be corrected using the decay factor R_{sewer}, where t denotes in-sewer hydraulic retention time during wastewater transportation and k refers to the decay rate constant, as described above.

$$P_{infection} = (F \times C_{RNA})/(P_{catchment} \times R_{sewer} \times E), \quad (4)$$

$$R_{sewer} = C_t/C_0 = e^{-k \times t}, \quad (5)$$

The sensitivity (S) of WBE-calculated infection case number ($P_{infection}$) to the decay rate constant (k) of SARS-CoV-2 RNA is defined in Equation (6). The relative change of P ($\Delta P/P$), due to the change of k by Δk, can be deduced as $\exp(\Delta k \times t) - 1$. The average k values from the collected literature data were taken as inputs into Equation (6), assuming the relative change as 25% of the decay rate constants [10]. The hydraulic residence time in sewer systems was assumed to be 12 h, representing a typical large sewer catchment. Herein, the calculated sensitivity value reflects the percentage of variation in WBE back-calculation caused by the variation of parameter k.

$$S = (\Delta P/P)/(\Delta k/k) = k \times (e^{t \times \Delta k} - 1)/\Delta k, \quad (6)$$

3. Results and Discussion

A total of 79 decay tests for coronavirus in water were identified from 12 unique research articles (Table 1, Tables S1 and S2). One decay test refers to an in-water decay experiment at a certain temperature in one type of water, either for virus infectivity or RNA. These studies employed different temperatures for the decay experiments in water, and the types of water included wastewater, freshwater, and seawater. Wastewater denotes domestic or municipal wastewater, including sterilized, autoclaved, and untreated raw wastewater. Freshwater includes tap water, laboratory water, and river water. The decay rate constant (k) was presented as $\ln(C/C_0)$ per day. Notably, there is just one decay rate in seawater measured by RT-qPCR (0.14 with R^2 as 0.8, at 20 °C) that met the selection criteria, which was reported by Sala-Comorera et al. [12]. Moreover, the numbers of infectivity decay in seawater and SARS-CoV-2 RNA decay in freshwater are limited, with only three for each (Table 1).

Table 1. Summary of identified decay tests of coronavirus in different water types.

Methods	Culture	RT-qPCR	Total
Number of decay tests in wastewater	18	33	51
Number of decay tests in freshwater	21	3	24
Number of decay tests in saline water	3	1	4
Total	42	37	79
References	[12–18]	[14,19–23]	

3.1. *Overview of Coronavirus Decay Rates in Waters*

Table 2 summarizes the collected decay rates from the reviewed papers under different conditions and Table 3 shows the result of the multivariate analysis of variance. The collected dataset contained two category variables (detection methods, type of water) and water temperature as a continuous variable. Multivariate analysis of variance showed that all these three variables led to significant differences in decay rate constants ($p < 0.001$) (Table 3). Most significantly, the collected decay rates were impacted by the different detection methods ($p = 1.68 \times 10^{-8}$). The three SARS-CoV-2 RNA decay rates (0.96–4.32 d^{-1}) reported by Weidhaas et al. [19] were significantly higher than average reports, thus being identified as outliers and excluded from the following modeling analysis. After the exclusion, the decay rate constants of SARS-CoV-2 RNA in wastewater ranged from 0.04 to 0.7 for temperatures 4–37 °C (Table 2), whereas the infectivity of coronaviruses decays at much faster rates (0.066–3.4 d^{-1}), from 4 to 25 °C in wastewater. In freshwater, the decay rates of SARS-CoV-2 RNA (0.039–0.25 d^{-1} at 4–37 °C) were also lower than those of coronavirus infectivity (0.01–1.2 d^{-1} at 4–24 °C), although the difference between two methods was not as much as that in wastewater.

The nucleic acids of viruses may persist longer than viral capsid and remain detectable after losing infectivity; hence, the decay rates of SARS-CoV-2 RNA are supposed to be much lower than those of coronavirus infectivity [14]. The greater stability of genetic fragments makes them suitable candidates for WBE investigations. At the same time, it also indicates that the environmental detection of viral RNA alone does not substantiate the risk of infection. On the other hand, knowledge about the fate of infectious viruses is needed to evaluate the potential disease transmission through urban wastewater systems, especially through sewers, where stormwater infiltration and saltwater intrusion might occur. Considering the scarcity of available data for seawater, the subsequent analysis and discussion mainly focused on wastewater dilution via freshwater inflow.

Table 2. Summary of decay rate constants (d^{-1}).

Target	Water Types	4 °C	10–15 °C	20–26 °C	37 °C
Coronavirus Infectivity	Wastewater	0.066–0.42	0.5–1.4	0.6–3.4	NA
	Freshwater	0.01–0.61	NA	0.2–1.2	NA
	Seawater	1.1	NA	2.0–2.1	NA
SARS-CoV-2 RNA	Wastewater	0.04–0.18	0.08–0.4	0.17–0.70	0.29–0.41
	Freshwater	0.039	NA	0.15	0.25
	Seawater	NA	NA	0.14	NA

Note. NA means not available.

Table 3. Summary of multivariate analysis of variance.

	Df	Sum Sq	Mean Sq	F Value	p (>F)	
Water type	2	7.248	3.624	9.332	0.000262	***
Method	1	15.904	15.904	40.954	1.68×10^{-8}	***
Temperature	1	12.485	12.485	32.151	3.17×10^{-7}	***
Water type: method	1	0.495	0.495	1.275	0.262756	
Water type: temperature	2	1.162	0.581	1.496	0.231343	
Method: temperature	1	4.781	4.781	12.311	0.000804	***
Water type: method: temperature	1	1.098	1.098	2.826	0.097312	
Residuals	68	26.407	0.388			

Note: *** means significant as $p \leq 0.001$.

3.2. *Effect of Wastewater Dilution on Coronavirus Decay*

The decay rates of SARS-CoV-2 RNA in seawater (0.14 d^{-1} at 20 °C) and in freshwater (0.039 d^{-1} at 4 °C, 0.15 d^{-1} at 25 °C, 0.25 d^{-1} at 37 °C) were smaller than those in wastewater (0.04–0.18 d^{-1} at 4 °C, 0.17–0.70 d^{-1} at 20–26 °C, 0.29–0.41 d^{-1} at 37 °C) (Table 2). Thus, the freshwater infiltration and seawater intrusion could alleviate the decay of viral RNA and may alleviate the impact of decay on WBE back-calculations.

On the contrary, the reduction in virus viability was enhanced in seawater, which led to higher decay rates (1.1 d^{-1} at 4 °C and 2.0–2.1 d^{-1} at 20 °C) than in wastewater (0.066–0.42 d^{-1} at 4 °C and 0.6–1.4 d^{-1} at 20 °C). However, such a difference is less confirmative, based on the limited data points. Sun et al. [24] suggested that the virus infectivity was not significantly affected by seawater; however, Lee et al. [25] reported a rapid infectivity loss of SARS-CoV-2 immediately upon being introduced into seawater. It should be noted that Sun et al. [24] examined the stability of the virus in artificial seawater while Lee et al. [25] used real seawater (pH 8, salinity 32‰; Sokcho, Korea). The contradictory results might be caused by the indigenous microbial communities in the marine ecosystem, which might inactivate viruses via proteolytic or nuclease activity [26].

Generally, the salinity and alkalinity of seawater are believed to influence the osmotic pressure and may reduce the survivability of coronaviruses in the ocean, thus helping to eliminate the risk of virus transmission [13,27,28]. Although SARS-CoV-2 is a mammalian virus, the release of viable coronaviruses through municipal sewage into the ocean may affect the marine ecosystem and even cause virion accumulation in seafood such as oysters or fishes through food chains in the same way as norovirus [29]. Interestingly, the COVID-19 outbreak first originated from a seafood market in Wuhan. Later, many infection cases caused by frozen cold chain food products have been reported around the world [30,31]. More consolidated experimental evidence should be involved to confidently exclude the possibility of marine contamination and confirm the safety of the seawater environment even for recreation purposes. Moreover, more experimental investigations in seawater mixed with domestic wastewater are necessary to delineate the effects of salinity on SARS-CoV-2 decay rate constants [32].

Figure 1 illustrates that both coronavirus infectivity and SARS-CoV-2 RNA decay more rapidly in wastewater than in freshwater under different temperatures. The difference at a lower temperature (4 °C) may not be as significant as that at higher temperatures (>20 °C). The increased k of coronaviruses in wastewater compared to freshwater could be attributed to the deactivation from higher extracellular enzymatic activity, eukaryotic predation, or the presence of antiviral chemicals (such as solvents, detergents, etc.), and other organic matters in wastewater [7]. Unlike neutral freshwater, wastewater usually has an acidic pH with fatty acids and other constituents that affect virus decay through the denaturation of capsid proteins and damage of nucleic acids [2]. Wastewater dilution via stormwater could alleviate the inactivation of infectious coronaviruses, thus potentially leading to the release of viable viruses into natural water bodies or city catchments via sewer overflow or untreated discharges [33–35]. Additionally, storm weather is often accompanied by lower temperatures and shorter in-sewer hydraulic retention times, which favor virus survival in waters. Therefore, the risk of spreading infective viruses during heavy rainfall events and urban floods should be carefully evaluated given the ever-increasing virus load in municipal wastewater.

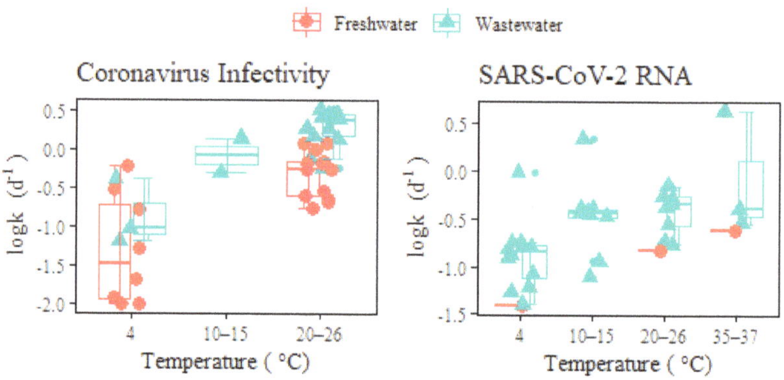

Figure 1. Decay of coronavirus infectivity and SARS-CoV-2 RNA in freshwater and wastewater under different temperatures. The middle lines inside the box represent median k values. The top and bottom borders of the box represent the 75%ile and 25%ile of k values, respectively. The top and bottom whiskers represent the upper and lower limit of k values in the groups. In addition, the smaller dots represent the outliers.

3.3. *Effect of Temperature on Coronavirus Decay*

Figure 2 shows the fitted curves for the numerical relationship between decay rate constants and temperatures in wastewater and freshwater based on the Arrhenius equation (Global goodness of fit: R^2 = 0.84). Separate plots for the four curves are presented in Figure S1. Table 4 shows the estimated parameters of the temperature correction coefficient (λ), decay rate constant (k_{20}) at a reference temperature of 20 °C, and R^2 for each scenario. The poor fitness and wide confidence band (Table 4 and Figure S1) for coronavirus infectivity decay in freshwater might be due to biological differences within coronavirus surrogates and the variability in matrix conditions (tap water versus river water). The even lower fit of SARS-CoV-2 RNA decay in wastewater (R^2 = 0.2) could be a result of the varied detection assays applied for RT-qPCR detection (Table S2) or the wastewater conditions (autoclaved, sterilized, or untreated wastewater sampled from various locations), which made the comparison among fitted parameters λ and k_{20} less convincing. More studies via well-controlled experiments are needed to obtain consistent and reliable conclusions. The detailed reporting of environmental parameters for decay tests should also be encouraged to help understand the influencing factors across different studies.

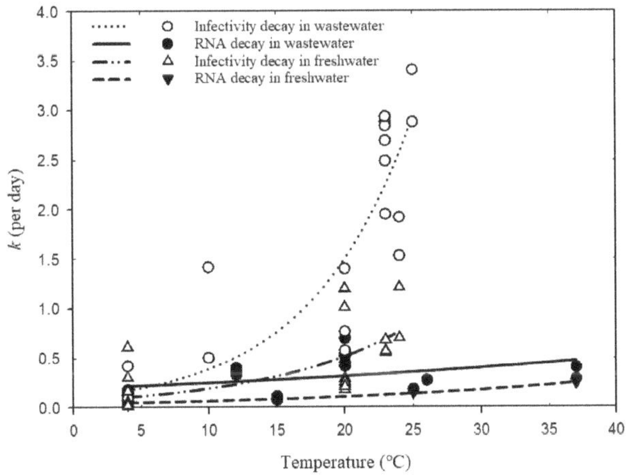

Figure 2. Relationships between decay rate constants of coronavirus and temperatures based on Arrhenius equation (Equation (3)).

Table 4. Parameters in Arrhenius equation fitted by temperatures and decay rates.

Parameters	Water Type	Coronavirus Infectivity	SARS-CoV-2 RNA
k_{20}	Wastewater	1.37 ± 0.10	0.30 ± 0.04
	Freshwater	0.47 ± 0.07	0.11 ± 0.01
λ	Wastewater	1.14 ± 0.05	1.02 ± 0.01
	Freshwater	1.10 ± 0.05	1.05 ± 0.01
R^2	Wastewater	0.72	0.20
	Freshwater	0.42	0.99

According to the estimated k_{20} values in Table 4, the infectivity decay rates were much higher than SARS-CoV-2 RNA, which indicates a greater stability of RNA than the hosting virus body in the water environment [26] and resonates with the previous discussion in Section 3.1. Moreover, both viable coronaviruses and SARS-CoV-2 RNA presented higher stability in freshwater over the wastewater matrix, probably due to the lower abundance of microbial communities and limited biological activity in freshwater, where degradation is alleviated for viruses [36]. This also supports the comparisons made in Section 3.2.

For each group in different water types, the temperature correction coefficient λ was calculated to be greater than one, indicating the influential role of water temperature on decay rates. Higher temperatures would enhance virus decay since the decay rate constants increase with elevated temperatures (Figure 2). In particular, the infectivity decay of coronavirus in wastewater was most significantly enhanced by temperature increase (λ = 1.14 ± 0.05), followed by the infectivity decay in freshwater (λ = 1.10 ± 0.05). Warmer waters normally increase decay rates of microorganisms via activating enzymes and, consequently, the degradation of protein walls or nucleic acids [37]. Exposure of viruses to high temperatures may also inactivate the enzymes and prevent replication. Since wastewater has more biologically and chemically active components than freshwater, the decay of viable viruses would be more vulnerable to temperature variation in wastewater.

For less sensitive SARS-CoV-2 RNA, temperature increments seem to play a similar role in facilitating decay both in freshwater and wastewater, as suggested by the comparable values of λ (Table 4) and the two visually parallel curves in Figure 2. Nucleic acid fragments inside the viral particles are less easily degradable by extracellular enzymes and more persistent than virus infectivity [14,38]. The genetic sequence may degrade gradually into smaller pieces and thus allow the RNA fragments to be prolonged after losing the protection from viral capsid. Hence, their decay rates are less impacted by temperature elevation than those of viable viruses, regardless if in freshwater or wastewater. Although the change in SARS-CoV-2 RNA decay rates in wastewater caused by seasonal temperature variation was much less than those of viable viruses, further assessment regarding its impact on WBE back-estimation should still be useful for understanding the resulting accuracy implications.

3.4. Sensitivity of WBE Back-Calculation to the Decay of SARS-CoV-2

To quantitatively assess the contribution of SARS-CoV-2 RNA decay under varied temperatures to the WBE back-calculation accuracy, the collected k values of SARS-CoV-2 RNA were input to the proposed sensitivity equation (Equation (6)) and the calculated results were presented in Table 5. The average level of sensitivity to 25% change in k is 0.13 over the three water types at 4–37 °C. Wastewater dilution by freshwater or saltwater in sewers might result in less sensitivity; however, the lack of data for SARS-CoV-2 RNA in saltwater made any solid conclusions hard to reach in the current study. More confidently, higher temperatures could increase SARS-CoV-2 RNA decay in sewerage and, correspondingly, the uncertainty of WBE back-estimation. Increased wastewater temperature in summer or tropical regions from 4 to 37 °C could increase the sensitivity considerably from 0.06 to 0.18, leading to a two-fold larger relative variance in the estimation of COVID-19 cases.

Table 5. Sensitivity ratio of predicted COVID-19 cases ($P_{infection}$) to SARS-CoV-2 RNA decay rate constants under different conditions assuming residence time in sewers as 12 h.

Temperature	Wastewater	Freshwater	Sea Water	Average
4 °C	0.06	0.02	NA	0.06
12–15 °C	0.15	NA	NA	0.15
20–26 °C	0.23	0.08	0.07	0.21
37 °C	0.18	0.12	NA	0.16
Average	0.14	0.07	0.07	0.13

Note: NA means not available. The sensitivity is represented as the ratio between the percentage change of estimated COVID-19 cases and 25% change of decay rate constants.

While the case number would possibly be underestimated in hot seasons if not incorporating RNA decay rates in the back-calculation, WBE back-calculation could assess disease prevalence more closely based on wastewater samples with lower temperatures, whether in winter or cold countries, where the importance of considering in-sewer RNA decay is reduced. Since the sensitivity values in Table 5 were calculated with hydraulic retention time fixed as 12 h, the overall sensitivity will be decreased, given a shorter in-sewer travel time, or increased otherwise. Collectively speaking, the in-sewer decay of SARS-CoV-2 RNA could be an influential factor for WBE back-calculation under high environmental temperatures and should be considered for accurate prediction.

3.5. Comparison of Coronavirus Decay Rates in Wastewater to Norovirus and Other Viruses

To put coronavirus decay rates into a broader context, Table 6 compared them with other previously investigated viruses (i.e., noroviruses, Zika virus, Dengue virus, yellow fever virus, Ebolavirus, Human Immunodeficiency Virus, hepatitis A virus, adenovirus). This table was based on a previous compilation [39] with the addition of recent reports. Although SARS-CoV-2 RNA might decay slightly faster at 4–6 °C, its overall decay rate was quite similar to other enveloped single-stranded RNA viruses (Zika, Dengue, and yellow fever virus), and generally ranged from 0 to 0.89 d^{-1} at higher temperatures. As for infectivity, the coronavirus also experienced similar decay rates at room temperatures as other enveloped single-stranded RNA viruses (Ebolavirus and Human Immunodeficiency Virus).

Table 6. Decay rate constants k (d^{-1}) of different viruses in wastewater.

Virus	Method	4–6 °C	10–25 °C	30–37 °C	References
Norovirus	RT-qPCR	0.02–0.06	0.02–0.10	0.03–0.11	[40,41]
Zika	RT-qPCR	0.025–0.046	0.11–0.58	0.27–0.89	[42]
Dengue	RT-qPCR	0.008–0.052	0.50–0.55	0.55–0.61	[43]
Yellow fever	RT-qPCR	0.032–0.047	0.52	0.88	
Murine hepatitis virus	RT-qPCR	0–0.003	0.37	0.45	
Coronavirus	RT-qPCR	0.04–0.18	0.08–0.70	0.29–0.41	[12–23]
	culture	0.066–0.42	0.5–3.4		
Ebolavirus	culture		0.35–1.08		[44]
Human Immunodeficiency Virus	culture		0.80		[45]
Hepatitis A	culture	0.047–0.066	0.10–0.28	0.34	[46]
Adenovirus	culture	0.10	0.10		[47]

It was commonly hypothesized that the lipid-containing envelope that surrounds the coronavirus nucleocapsid made it more susceptible to degradation than nonenveloped enteric viruses (i.e., hepatitis A, adenovirus, and norovirus). The inactivation process for enveloped viruses was observed to be faster than for nonenveloped viruses [18,48], since envelope lipids could be more easily destroyed than other viral parts. As an example, the hepatitis A virus had better survivability than the enveloped coronavirus at 4–25 °C (Table 6). In another instance, the RNA decay rates of a typical nonenveloped enteric waterborne norovirus [10] were significantly lower than coronavirus in wastewater (Figure 3). Such a conclusion contrasts with the opinion of Silverman and Boehm [9], who identified similar persistence between nonenveloped and enveloped viruses in a dark aqueous environment. This discrepancy might be due to the fact that the study combined different water matrices (including wastewaters and natural waters, i.e., fresh, estuarine, and marine waters) together in the comparison. The different water matrices should be considered in view of the significant effect from water types on virus decay (Table 3).

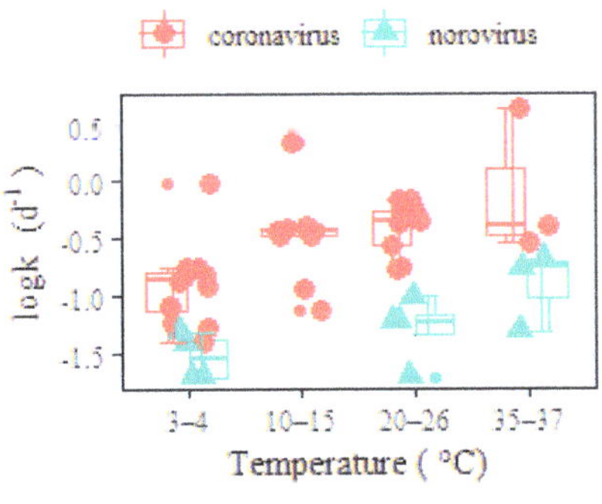

Figure 3. RNA decay rates of coronavirus and norovirus in wastewater at different temperatures. The middle lines inside the box represent median k values. The top and bottom borders of the box represent the 75%ile and 25%ile of k values, respectively. The top and bottom whiskers represent the upper and lower limit of k values in the groups. In addition, the smaller dots represent the outliers.

As a result of the higher SARS-CoV-2 RNA decay rates than norovirus within temperatures ranging between 3 and 37 °C, the sensitivity to COVID-19 back-estimation via WBE (Table 5) was greater than that of norovirus (Table S3) [10], considering both temporal temperature change and wastewater dilution using freshwater. The implications of in-sewer decay should be analyzed in association with specific pathogen species. In addition to the comparison between enveloped and nonenveloped viruses, RNA viruses (i.e., noroviruses, coronaviruses) and DNA viruses (i.e., adenoviruses) should also be differentiated. Furthermore, the difference between viral pathogens and bacterial ones might be even more distinguished [10].

3.6. Implications, Limitations, and Future Perspectives

The decay experiment is time-consuming; hence, the number of decay rates generated from an individual study is limited by resources. Through a systematic literature review and meta-analysis, this research synthesized experimental data from multiple studies and assessed the parameterization of in-sewer coronavirus decay for WBE applications. According to the sensitivity analysis of SARS-CoV-2 RNA decay, in-sewer decay in high temperature conditions should be considered for WBE back-estimation. Thus, recording the wastewater temperature at collection is highly recommended. Moreover, wastewater dilution in sewers induced by stormwater inflow may enhance the survivability of viable coronaviruses and result in an increased risk of infection through wastewater. This should be underlined particularly when the rainfall events are accompanied with cold weather, which is also conducive to the sustaining of virus infectivity. Preventive measures should be taken by the public health sectors to lower such risks.

Unavoidably, this research also has a few limitations regarding the scarcity of collected literature data in seawater, the absence of wastewater dilution experiments by other water types for coronaviruses, and a failure to include the impact of other sewer conditions. More available experimental data would make the statistical analysis consolidated. Further experimental results in wastewater mixed with fresh/saline water at various ratios can help quantify the actual dilution effect on coronavirus decay rates. The mathematical relationship of wastewater dilution at various gradients and the resulting changes in virus decay rates could be utilized for correcting WBE back-estimations. Revealing relevant mechanisms of coronavirus decay is also important for an in-depth understanding of virus fate in wastewater and requires further detailed studies using lab experiments. These outcomes could be extrapolated to other viral species as well.

Furthermore, it is essential to delineate the impact of various other in-sewer processes on coronavirus decay. Most importantly, the decay measured in bulk wastewater (or by in-vial tests) could significantly underestimate the actual decay in sewers [49], largely due to the virus adsorption to sewer biofilms. This has been shown by our sewer reactor studies not only for the enhanced decay of coronavirus infectivity [49], but also for SARS-CoV-2 RNA [50]. Another study [51] also reported the accumulation of SARS-CoV-2 RNA in sewer biofilms to 700 genome copies/cm^2, which demonstrated SARS-CoV-2 retardation in sewers. In addition, the sewer biofilms and sediments (particularly in gravity sewers) may shelter the associated virus particles from degradation and extend their preservation in sewers [52,53]. Later, the attached viruses might then be released into wastewater with biofilm sloughing when subjected to strong stormwater inflow. Given the complexity of a real sewer environment, the sensitivity analysis of WBE back-estimation to SARS-CoV-2 decay in this study is likely limited by the current knowledge gap.

In addition, some of the in-sewer chemical dosing strategies (oxygen, ferric salts, nitrate/nitrite, or free nitrous acid, etc.) may also disinfect microbes and accelerate the degradation of virus RNA at the same time [54]. This might be one of the reasons why there was no infectious SARS-CoV-2 isolated from the influent raw sewage samples and no reported COVID-19 transmission through wastewater exposure to date [55,56]. In-depth knowledge on the effect of the chemicals used in sewers and WWTPs on the survival of coronavirus could help develop effective disinfection strategies which might be applied to other environmental settings rather than wastewater only. Overall, understanding the fate of coronavirus after excretion from human bodies, including its in-sewer decay, is critical to ensure environmental and public health. The actual contribution of in-sewer SARS-CoV-2 RNA decay to overall variance in WBE back-estimation remains unclear due to the research gap in the unrevealed impact of in-sewer processes. To reduce these uncertainties in WBE application, more studies on the unknown in-sewer decay of coronavirus under different conditions are highly encouraged.

4. Conclusions

This paper analyzed the decay rate constants of coronaviruses in three types of water systematically collected from published literature. The impacts of wastewater temperature and dilution in sewers on virus decay and WBE back-estimation were further analyzed and discussed based on the collation of decay rate constants. Moreover, the decay of coronavirus was also compared with those of other viruses in wastewater through a literature survey to understand the differences among species. Accordingly, four major conclusions could be drawn as follows:

- SARS-CoV-2 RNA, as the biomarker for WBE investigations, has much lower decay rates and was less influenced by wastewater temperature variations than viable coronaviruses. However, higher wastewater temperature in summer or tropical regions could still increase the sensitivity of WBE back-calculations considerably from 0.06 to 0.18, resulting in a two-times higher relative variance in back-estimation of COVID-19 cases;
- Wastewater dilution by stormwater inflow might alleviate the decay of coronavirus infectivity. Cold weather along with heavy rainfall events and urban floods could further increase the risk of environmental transmission by improving virus survivability in wastewater;
- As an enveloped RNA virus, coronavirus experienced more rapid decay in wastewater than nonenveloped viruses such as norovirus, which led to the increased sensitivity of WBE back-estimation to in-sewer decay and, consequently, to a greater need to incorporate the decay rate as a correction factor;
- There is a lack of studies on coronavirus decay in saltwater, unlike some other, more extensively studied, waterborne enteric pathogens. Salt groundwater intrusion or the municipal usage of seawater (i.e., toilet flushing) in sewers might enhance the decay of coronavirus infectivity during wastewater transportation to wastewater treatment plants (WWTPs). However, this hypothesis needs further supporting experimental data.

Supplementary Materials: The following supporting information can be downloaded at: https://www.mdpi.com/article/10.3390/w15061051/s1, Table S1: Decay rates of infectious coronaviruses; Table S2: Decay rates of SARS-CoV-2 RNA; Table S3: Sensitivity ratio of predicted cases ($P_{infection}$) to the norovirus decay rate constants under different conditions assuming residence time in sewers as 12 h; Figure S1: Fitted curves using Arrhenius equation for infectious coronaviruses and SARS-CoV-2 RNA in wastewater and freshwater.

Author Contributions: Conceptualization, Y.G. and G.J.; formal analysis, Y.G.; data curation, Y.G.; writing—original draft preparation, Y.G.; writing—review and editing, Y.L., X.Z., S.G. and M.S.; visualization, Y.G.; supervision, G.J.; project administration, G.J.; funding acquisition, G.J. All authors have read and agreed to the published version of the manuscript.

Funding: This research was funded by the ARC Discovery Project, grant number DP190100385. Ying Guo receives a Ph.D. scholarship from the same project.

Data Availability Statement: Data is available in the Supplementary Materials.

Conflicts of Interest: The authors declare no conflict of interest. The funders had no role in the design of the study; in the collection, analyses, or interpretation of data; in the writing of the manuscript; or in the decision to publish the results.

References

1. Paital, B.; Das, K.; Malekdar, F.; Sandoval, M.A.; Niaragh, E.K.; Frontistis, Z.; Behera, T.R.; Balacco, G.; Sangkham, S.; Hati, A.K.; et al. A State-of-the-Art Review on SARS-CoV-2 Virus Removal Using Different Wastewater Treatment Strategies. *Environments* **2022**, *9*, 110. [CrossRef]
2. Li, X.; Zhang, S.; Shi, J.; Luby, S.P.; Jiang, G. Uncertainties in estimating SARS-CoV-2 prevalence by wastewater-based epidemiology. *Chem. Eng. J.* **2021**, *415*, 129039. [CrossRef]
3. Li, X.; Kulandaivelu, J.; Zhang, S.; Shi, J.; Sivakumar, M.; Mueller, J.; Luby, S.; Ahmed, W.; Coin, L.; Jiang, G. Data-driven estimation of COVID-19 community prevalence through wastewater-based epidemiology. *Sci. Total Environ.* **2021**, *789*, 147947. [CrossRef]
4. Mousazadeh, M.; Ashoori, R.; Paital, B.; Kabdaşlı, I.; Frontistis, Z.; Hashemi, M.; Sandoval, M.A.; Sherchan, S.; Das, K.; Emamjomeh, M.M. Wastewater Based Epidemiology Perspective as a Faster Protocol for Detecting Coronavirus RNA in Human Populations: A Review with Specific Reference to SARS-CoV-2 Virus. *Pathogens* **2021**, *10*, 1008. [CrossRef]
5. Ahmed, W.; Angel, N.; Edson, J.; Bibby, K.; Bivins, A.; O'Brien, J.W.; Choi, P.M.; Kitajima, M.; Simpson, S.L.; Li, J.; et al. First confirmed detection of SARS-CoV-2 in untreated wastewater in Australia: A proof of concept for the wastewater surveillance of COVID-19 in the community. *Sci. Total Environ.* **2020**, *728*, 138764. [CrossRef]
6. Li, X.; Zhang, S.; Sherchan, S.; Orive, G.; Lertxundi, U.; Haramoto, E.; Honda, R.; Kumar, M.; Arora, S.; Kitajima, M.; et al. Correlation between SARS-CoV-2 RNA concentration in wastewater and COVID-19 cases in community: A systematic review and meta-analysis. *J. Hazard. Mater.* **2023**, *441*, 129848. [CrossRef]
7. Pinon, A.; Vialette, M. Survival of viruses in water. *Intervirology* **2018**, *61*, 214–222. [CrossRef]
8. Boehm, A.B.; Silverman, A.I.; Schriewer, A.; Goodwin, K. Systematic review and meta-analysis of decay rates of waterborne mammalian viruses and coliphages in surface waters. *Water Res.* **2019**, *164*, 114898. [CrossRef]
9. Silverman, A.I.; Boehm, A.B. Systematic review and meta-analysis of the persistence and disinfection of human coronaviruses and their viral surrogates in water and wastewater. *Environ. Sci. Technol. Lett.* **2020**, *7*, 544–553. [CrossRef]
10. Guo, Y.; Sivakumar, M.; Jiang, G. Decay of four enteric pathogens and implications to wastewater-based epidemiology: Effects of temperature and wastewater dilutions. *Sci. Total Environ.* **2022**, *819*, 152000. [CrossRef]
11. Chick, H. An investigation of the laws of disinfection. *J. Hyg.* **1908**, *8*, 92–158. [CrossRef]
12. Sala-Comorera, L.; Reynolds, L.J.; Martin, N.A.; O'Sullivan, J.J.; Meijer, W.G.; Fletcher, N.F. Decay of infectious SARS-CoV-2 and surrogates in aquatic environments. *Water Res.* **2021**, *201*, 117090. [CrossRef]
13. De Rijcke, M.; Shaikh, H.M.; Mees, J.; Nauwynck, H.; Vandegehuchte, M.B. Environmental stability of porcine respiratory coronavirus in aquatic environments. *PLoS ONE* **2021**, *16*, e0254540. [CrossRef]
14. Bivins, A.; Greaves, J.; Fischer, R.; Yinda, K.C.; Ahmed, W.; Kitajima, M.; Munster, V.J.; Bibby, K. Persistence of SARS-CoV-2 in water and wastewater. *Environ. Sci. Technol. Lett.* **2020**, *7*, 937–942. [CrossRef]
15. Gundy, P.M.; Gerba, C.P.; Pepper, I.L. Survival of coronaviruses in water and wastewater. *Food Environ. Virol.* **2008**, *1*, 10–14. [CrossRef]
16. Casanova, L.; Rutala, W.A.; Weber, D.J.; Sobsey, M.D. Survival of surrogate coronaviruses in water. *Water Res.* **2009**, *43*, 1893–1898. [CrossRef]
17. de Oliveira, L.C.; Torres-Franco, A.F.; Lopes, B.C.; Santos, B.; Costa, E.A.; Costa, M.S.; Reis, M.T.P.; Melo, M.C.; Polizzi, R.B.; Teixeira, M.M.; et al. Viability of SARS-CoV-2 in river water and wastewater at different temperatures and solids content. *Water Res.* **2021**, *195*, 117002. [CrossRef]
18. Ye, Y.; Ellenberg, R.M.; Graham, K.E.; Wigginton, K.R. Survivability, partitioning, and recovery of enveloped viruses in untreated municipal wastewater. *Environ. Sci. Technol.* **2016**, *50*, 5077–5085. [CrossRef]
19. Weidhaas, J.; Aanderud, Z.T.; Roper, D.K.; VanDerslice, J.; Gaddis, E.B.; Ostermiller, J.; Hoffman, K.; Jamal, R.; Heck, P.; Zhang, Y.; et al. Correlation of SARS-CoV-2 RNA in wastewater with COVID-19 disease burden in sewersheds. *Sci. Total Environ.* **2021**, *775*, 145790. [CrossRef]
20. Ahmed, W.; Bertsch, P.M.; Bibby, K.; Haramoto, E.; Hewitt, J.; Huygens, F.; Gyawali, P.; Korajkic, A.; Riddell, S.; Sherchan, S.P.; et al. Decay of SARS-CoV-2 and surrogate murine hepatitis virus RNA in untreated wastewater to inform application in wastewater-based epidemiology. *Environ. Res.* **2020**, *191*, 110092. [CrossRef]
21. Hokajärvi, A.-M.; Rytkönen, A.; Tiwari, A.; Kauppinen, A.; Oikarinen, S.; Lehto, K.-M.; Kankaanpää, A.; Gunnar, T.; Al-Hello, H.; Blomqvist, S.; et al. The detection and stability of the SARS-CoV-2 RNA biomarkers in wastewater influent in Helsinki, Finland. *Sci. Total Environ.* **2021**, *770*, 145274. [CrossRef]
22. Burnet, J.B.; Cauchie, H.M.; Walczak, C.; Goeders, N.; Ogorzaly, L. Persistence of endogenous RNA biomarkers of SARS-CoV-2 and PMMoV in raw wastewater: Impact of temperature and implications for wastewater-based epidemiology. *Sci. Total Environ.* **2023**, *857*, 159401. [CrossRef]
23. Yang, S.; Dong, Q.; Li, S.; Cheng, Z.; Kang, X.; Ren, D.; Xu, C.; Zhou, X.; Liang, P.; Sun, L.; et al. Persistence of SARS-CoV-2 RNA in wastewater after the end of the COVID-19 epidemics. *J. Hazard. Mater.* **2022**, *429*, 128358. [CrossRef]
24. Sun, Z.P.; Yang, S.Y.; Cai, X.; Han, W.D.; Hu, G.W.; Qian, Y.; Wang, Y.Y.; Zhang, R.; Xie, Y.H.; Qu, D. Survival of SARS-CoV-2 in artificial seawater and on the surface of inanimate materials. *J. Med. Virol.* **2022**, *94*, 3982–3987. [CrossRef]
25. Lee, Y.J.; Kim, J.H.; Choi, B.-S.; Choi, J.-H.; Jeong, Y.-I. Characterization of Severe Acute Respiratory Syndrome Coronavirus 2 stability in multiple water matrices. *J. Korean Med. Sci.* **2020**, *35*, e330. [CrossRef]

26. Mahlknecht, J. Presence and persistence of SARS-CoV-2 in aquatic environments: A mini-review. *Curr. Opin. Environ. Sci. Health* **2022**, *29*, 100385. [CrossRef]
27. Mordecai, G.J.; Hewson, I. Coronaviruses in the sea. *Front. Microbiol.* **2020**, *11*, 1795. [CrossRef]
28. Seyer, A.; Şanlıdağ, T. The fate of SARS-CoV-2 in the marine environments: Are marine environments safe from COVID-19? *Erciyes Med. J.* **2021**, *43*, 606–607. [CrossRef]
29. Farkas, K.; Cooper, D.M.; McDonald, J.E.; Malham, S.K.; de Rougemont, A.; Jones, D.L. Seasonal and spatial dynamics of enteric viruses in wastewater and in riverine and estuarine receiving waters. *Sci. Total Environ.* **2018**, *634*, 1174–1183. [CrossRef]
30. Wang, J.; Li, F.; Liu, Z.; Li, N. COVID-19 outbreaks linked to imported frozen food in China: Status and challege. *China CDC Wkly.* **2022**, *4*, 483–487. [CrossRef]
31. Qian, J.; Yu, Q.; Jiang, L.; Yang, H.; Wu, W. Food cold chain management improvement: A conjoint analysis on COVID-19 and food cold chain systems. *Food Control* **2022**, *137*, 108940. [CrossRef]
32. Liang, L.; Goh, S.G.; Gin, K.Y.H. Decay kinetics of microbial source tracking (MST) markers and human adenovirus under the effects of sunlight and salinity. *Sci. Total Environ.* **2017**, *574*, 165–175. [CrossRef]
33. Huang, C.; Hu, Y.; Wang, L.; Wang, Y.; Li, N.; Guo, Y.; Feng, Y.; Xiao, L.; Stams, A.J.M. Environmental transport of emerging human-pathogenic Cryptosporidium species and subtypes through combined sewer overflow and wastewater. *Appl. Environ. Microbiol.* **2017**, *83*, e00682-17. [CrossRef] [PubMed]
34. Steele, J.A.; Blackwood, A.D.; Griffith, J.F.; Noble, R.T.; Schiff, K.C. Quantification of pathogens and markers of fecal contamination during storm events along popular surfing beaches in San Diego, California. *Water Res.* **2018**, *136*, 137–149. [CrossRef]
35. Astrom, J.; Pettersson, T.J.; Reischer, G.H.; Hermansson, M. Short-term microbial release during rain events from on-site sewers and cattle in a surface water source. *J. Water Health* **2013**, *11*, 430–442. [CrossRef]
36. Carratala, A.; Rusinol, M.; Rodriguez-Manzano, J.; Guerrero-Latorre, L.; Sommer, R.; Girones, R. Environmental effectors on the inactivation of human adenoviruses in water. *Food. Environ. Virol.* **2013**, *5*, 203–214. [CrossRef]
37. Aquino de Carvalho, N.; Stachler, E.N.; Cimabue, N.; Bibby, K. Evaluation of phi6 persistence and suitability as an enveloped virus surrogate. *Environ. Sci. Technol.* **2017**, *51*, 8692–8700. [CrossRef]
38. Ye, Y.; Chang, P.H.; Hartert, J.; Wigginton, K.R. Reactivity of enveloped virus genome, proteins, and lipids with free chlorine and UV254. *Environ. Sci. Technol.* **2018**, *52*, 7698–7708. [CrossRef]
39. Guo, Y.; Li, J.; O'Brien, J.; Sivakumar, M.; Jiang, G. Back-estimation of norovirus infections through wastewater-based epidemiology: A systematic review and parameter sensitivity. *Water Res.* **2022**, *219*, 118610. [CrossRef]
40. Kauppinen, A.; Miettinen, I.T. Persistence of Norovirus GII genome in drinking water and wastewater at different temperatures. *Pathogens* **2017**, *6*, 48. [CrossRef]
41. Skraber, S.; Ogorzaly, L.; Helmi, K.; Maul, A.; Hoffmann, L.; Cauchie, H.M.; Gantzer, C. Occurrence and persistence of enteroviruses, noroviruses and F-specific RNA phages in natural wastewater biofilms. *Water Res.* **2009**, *43*, 4780–4789. [CrossRef]
42. Muirhead, A.; Zhu, K.; Brown, J.; Basu, M.; Brinton, M.A.; Costa, F.; Hayat, M.J.; Stauber, C.E. Zika virus RNA persistence in sewage. *Environ. Sci. Technol. Lett.* **2020**, *7*, 659–664. [CrossRef]
43. Chandra, F.; Lee, W.L.; Armas, F.; Leifels, M.; Gu, X.; Chen, H.; Wuertz, S.; Alm, E.J.; Thompson, J. Persistence of Dengue (Serotypes 2 and 3), Zika, yellow fever, and murine hepatitis virus RNA in untreated wastewater. *Environ. Sci. Technol. Lett.* **2021**, *8*, 785–791. [CrossRef]
44. Bibby, K.; Fischer, R.J.; Casson, L.W.; Stachler, E.; Haas, C.N.; Munster, V.J. Persistence of Ebola virus in sterilized wastewater. *Environ. Sci. Technol. Lett.* **2015**, *2*, 245–249. [CrossRef]
45. Slade, J.S.; Pike, E.B.; Eglin, R.P.; Colbourne, J.S.; Kurtz, J.B. The survival of human immunodeficiency virus in water, sewage and sea-water. *Water Sci. Technol.* **1989**, *21*, 55–59. [CrossRef]
46. Deng, M.Y.; Cliver, D.O. Persistence of inoculated hepatitis-a virus in mixed human and animal wastes. *Appl. Environ. Microbiol.* **1995**, *61*, 87–91. [CrossRef]
47. Enriquez, C.E.; Gerba, C.P.; Hurst, C.J. Survival of the enteric adenoviruses 40 and 41 in tap, sea, and wastewater. *Water Res.* **1995**, *29*, 2548–2553. [CrossRef]
48. Farkas, K.; Walker, D.I.; Adriaenssens, E.M.; McDonald, J.E.; Hillary, L.S.; Malham, S.K.; Jones, D.L. Viral indicators for tracking domestic wastewater contamination in the aquatic environment. *Water Res.* **2020**, *181*, 115926. [CrossRef]
49. Shi, J.; Li, X.; Zhang, S.; Sharma, E.; Sivakumar, M.; Sherchan, S.P.; Jiang, G. Enhanced decay of coronaviruses in sewers with domestic wastewater. *Sci. Total Environ.* **2021**, *813*, 151919. [CrossRef]
50. Zhang, S.; Sharma, E.; Tiwaric, A.; Sherchand, S.; Gao, S.; Zhou, X.; Shi, J.; Jiang, G. The Reduction of SARS-CoV-2 RNA concentration in the presence of sewer biofilms. *Int. J. Hyg. Environ. Health*, 2023; *submitted*.
51. Morales Medina, W.R.; D'Elia, S.; Fahrenfeld, N.L. Accumulation of SARS-CoV-2 RNA in sewer biofilms. *ACS EST Water* **2022**, *2*, 1844–1851. [CrossRef]
52. Skraber, S.; Helmi, K.; Willame, R.; Ferreol, M.; Gantzer, C.; Hoffmann, L.; Cauchie, H.M. Occurrence and persistence of bacterial and viral faecal indicators in wastewater biofilms. *Water Sci. Technol.* **2007**, *55*, 377–385. [CrossRef]
53. Brooks, Y.; Aslan, A.; Tamrakar, S.; Murali, B.; Mitchell, J.; Rose, J.B. Analysis of the persistence of enteric markers in sewage polluted water on a solid matrix and in liquid suspension. *Water Res.* **2015**, *76*, 201–212. [CrossRef]

54. Ho, J.; Stange, C.; Suhrborg, R.; Wurzbacher, C.; Drewes, J.E.; Tiehm, A. SARS-CoV-2 wastewater surveillance in Germany: Long-term RT-digital droplet PCR monitoring, suitability of primer/probe combinations and biomarker stability. *Water Res.* **2022**, *210*, 117977. [CrossRef] [PubMed]
55. Amoah, I.D.; Kumari, S.; Bux, F. Coronaviruses in wastewater processes: Source, fate and potential risks. *Environ. Int.* **2020**, *143*, 105962. [CrossRef] [PubMed]
56. Gholipour, S.; Mohammadi, F.; Nikaeen, M.; Shamsizadeh, Z.; Khazeni, A.; Sahbaei, Z.; Mousavi, S.M.; Ghobadian, M.; Mirhendi, H. COVID-19 infection risk from exposure to aerosols of wastewater treatment plants. *Chemosphere* **2021**, *273*, 129701. [CrossRef]

water

MDPI

Communication

MIMS as a Low-Impact Tool to Identify Pathogens in Water

Salvia Sajid [1,2,†], Ishika Aryal [1,†], Suleman Farooq Chaudhri [1], Frants Roager Lauritsen [1], Mikkel Girke Jørgensen [3], Håvard Jenssen [4] and Bala Krishna Prabhala [1,*]

1 Department of Physics, Chemistry, and Pharmacy, University of Southern Denmark, Campusvej 55, 5230 Odense M, Denmark
2 Department of Drug Design and Pharmacology, University of Copenhagen, Universitetsparken 2, 2100 København Ø, Denmark
3 Department of Biochemistry and Molecular Biology, University of Southern Denmark, Campusvej 55, 5230 Odense M, Denmark
4 Institute of Science and Environment, Roskilde University, Universitetsvej 1, 4000 Roskilde, Denmark
* Correspondence: bapra@sdu.dk
† These authors contributed equally to this work.

Abstract: Bacteria produce many kinds of volatile compounds throughout their lifecycle. Identifying these volatile compounds can help to understand bacterial interactions with the host and/or other surrounding pathogens of the same or different species. Some commonly used techniques to detect these volatile compounds are GC and/or LC coupled to mass spectrometric techniques. However, these methods can sometimes become challenging owing to tedious sample preparation steps. Thus, identifying an easier method to detect these volatile compounds was investigated in the present study. Here, Membrane-inlet mass spectrometry (MIMS) provided a facile low-impact alternative to the existing strategies. MIMS was able to differentiate between the pathogenic and nonpathogenic bacterial strains, implying that it can be used as a bioprocess monitoring tool to analyze water samples from either water treatment plants or biotechnological industries.

Keywords: Membrane-inlet mass spectrometry (MIMS); microbial volatome; recombinant protein production; water-treatment plants

check for updates

Citation: Sajid, S.; Aryal, I.; Chaudhri, S.F.; Lauritsen, F.R.; Jørgensen, M.G.; Jenssen, H.; Prabhala, B.K. MIMS as a Low-Impact Tool to Identify Pathogens in Water. *Water* **2023**, *15*, 184. https://doi.org/10.3390/w15010184

Academic Editors: Guangming Jiang, Ryo Honda and Sudipti Arora

Received: 2 November 2022
Revised: 22 December 2022
Accepted: 28 December 2022
Published: 2 January 2023

1. Introduction

Bacteria are prokaryotic organisms, and based on the composition of the cell wall, bacteria can be further classified into two subcategories, i.e., Gram-positive and Gram-negative. The Gram-positive cell type consists of three layers, which are the peptidoglycan, periplasmic space, and a cell membrane, while an additional outer membrane is present in Gram-negative organisms [1].

Throughout their lifecycle, bacteria produce different organic volatile compounds when exposed to different environmental conditions. These volatile compounds are metabolites that can be divided into two categories, primary and secondary metabolites [2]. Primary metabolites are essential for the growth, development, and reproduction of bacteria, while secondary metabolites are produced as part of a defense strategy against other bacteria or fungi, and they generally improve tolerance to environmental stresses. These metabolites can be ketones, aldehydes, alcohols, hydrocarbons, etc. [3–6]. Since different bacteria produce different metabolites, including volatiles, these metabolites may serve as biomarkers for species identification [7]. All volatile metabolites and other volatile organic and inorganic molecules that originate from an organism are included in its volatome [8,9]. Moreover, many these volatile compounds are differentially regulated by certain biochemical processes. These changes in the volatome have been used to study transcriptomes in mushrooms [10]. For instance, significantly altered volatile compounds were observed in mated and unmated strains of mushrooms, which correspond very well to the genomic and proteomic data [11]. These volatile compounds are difficult to analyze as they are released only for a certain period of time to execute a particular biological function. Furthermore, they are released in relatively low amounts complicating the detection of these compounds [12]. Traditionally, GC-MS is commonly used to identify volatile compounds but it possesses several challenges related to sample preparation, which hinder the in-depth coverage of volatiles [13]. On the other hand, membrane-inlet mass spectrometry (MIMS) is a low-impact mass spectrometric technique used for identifying gaseous and volatile organic compounds from solid, liquid, or gaseous samples [14].

MIMS was developed in the 1960s, and the basic principles of MIMS have been described previously [15,16]. Since its development, it has been associated with a wide range of uses for different purposes, such as the analysis of volatomes in soil and oceans, for monitoring in industrial purposes, and for forensic applications [17,18].

Now, the utility of MIMS is gaining even more prominence because it possesses several advantages compared to GC/MS [19], such as the ability to detect samples at low detection limits (lower ppb to ppt) with nearly no sample preparation, whereas in GC-MS, sample preparation is a crucial (and sometimes time-consuming and difficult) step before analysis [14]. In addition, MIMS is also a tool that allows for the rapid analysis of samples in real-time; thus, it has been widely used for the analysis of variety of chemical classes such as semi volatile compounds, organometallic compounds, free radicals, etc. [18]. Additionally, MIMS systems can be purchased at relatively economical prices as compared to LC/GC-MS instruments. Furthermore, the membranes used in MIMS are economical (costing less than 10 USD) and can be recycled for several years, which often is not the case with GC/LC columns (costing 300 USD minimum) [17,20,21]. Finally, the maintenance and running cost of MIMS is significantly lower compared to that of GC/LC-MS instruments.

Until now, MIMS has been used for the analysis of volatile organic metabolites produced by microorganisms [22] and the detection of metabolites secreted by genetically engineered bacteria [23]. MIMS is a well-established method for the mass spectrometric analysis of volatile organic metabolites produced by microorganisms [23], but its use in combination with advanced data treatment technologies has hitherto only been scarcely investigated.

In the present study, it was hypothesized that MIMS, as an analytical technique, would provide a faster and economical way to identify and distinguish bacteria in water. The presence of several pathogenic bacteria has been observed in the drinking water and has been discussed in the literature [24,25]. The aim was thus to investigate the volatomic profile of selected pathogenic bacteria, such as Gram-positive *Staphylococcus epidermidis* and Gram-negative *Escherichia coli* and *Pseudomonas aeruginosa*, as a model system and to explore the use of principal component analysis (PCA) for differentiating between the organisms. Furthermore, we also tested the capacity of MIMS to study a lab strain of *E. coli* expressing a functional recombinant protein and a lab strain of *E. coli* expressing a dysfunctional protein, thus establishing MIMS as an online process monitoring tool for recombinant protein production.

2. Materials and Methods

The overexpression of recombinant protein was performed as described previously [26]. Briefly, *E. coli* BL21(DE3)pLysS cells transformed with plasmids pTTQ18-ydgR (functional protein) and pTTQ18-ydgR mutant (E33Q; dysfunctional protein) were inoculated into 3 mL of lysogeny broth (LB) media containing 100 μg/mL ampicillin and 34 μg/mL chloramphenicol and allowed to grow overnight. The overnight cultures were transferred and diluted 1:50 in 25 mL of LB media with the same concentration of antibiotics. The cells were allowed to grow to an OD_{600} of 0.6 at 37 °C and 160 rpm before induction with 1 mM isopropyl-β-D-thiogalactoside (IPTG, inducing protein expression). After 3 h of incubation at 37 °C and 160 rpm, samples were centrifuged at 10,000 g for 5 min, and supernatant media were collected for testing.

The pathogenic bacterial strains of *P. aeruginosa*, *S. epidermidis* [27], and *E. coli* [28] were grown at 37 °C on LB agar plates overnight. A single colony of each bacterial strain from the plates was transferred to 25 mL of LB broth and was allowed to grow overnight at

37 °C with 160 rpm shaking. The cultures were centrifuged at 4700 g for 15 min and the supernatants were sterile-filtered with a 0.22 μm syringe filter. The filtered samples were stored at −80 °C until analysis.

The samples were recorded using MIMS, as previously reported [29]. The setup for MIMS can be seen in Figure 1. MIMS uses a thermoregulated sample cell, which holds a small aperture exposing the surface of the membrane, which is used as a barrier to separate the liquid phase and the ion source of the mass spectrometer. Gasses and volatile organic compounds evaporate from the liquid through the membrane into the vacuum of the mass spectrometer, where the molecules vaporize and get ionized using electron ionization. The ionized particles are then separated based on their mass-to-charge ratio (m/z) using a quadrupole mass spectrometer, and a 2D-chromatogram with m/z on the x-axis and relative abundance on the y-axis is produced.

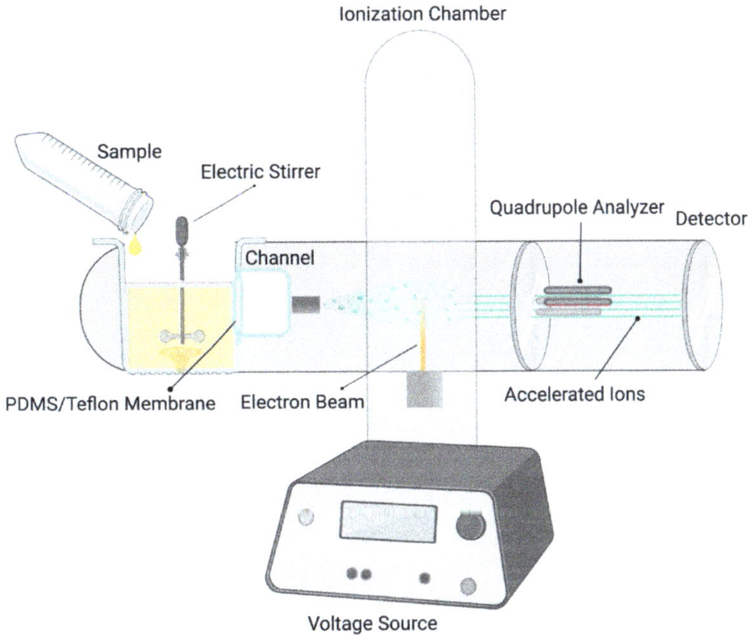

Figure 1. Schematic representation of Membrane-Inlet Mass Spectrometry (MIMS) (Created in BioRender.com (accessed on 22 December 2022)).

Briefly, 20 mL of the LB media samples was poured into a thermoregulated (40 °C) sample cell equipped with an electric stirrer mounted on a Prisma QME200 single quadrupole mass spectrometer (Balzers, Liechtenstein). A 3 mm hole supported by a medical-grade polydimethylsiloxane (PDMS) membrane (125 μm; Sil-Tec Sheeting, USA) is present in the sample cell, allowing molecules (volatile and semi-volatile organic substances) to pass into the ion source. The sample cell was washed twice with water to avoid carryover after each sample. A background spectrum was collected before each sample. The spectra for each sample were recorded in 3 cycles, and the average of these cycles was used to generate the spectrum. The samples were recorded between m/z 50 and 300. The time taken to analyze each sample was approximately 20 min. MIMS data acquisition and data conversion were performed using the Prisma QME200 instrument package. The recorded data were exported to .asc format using the software dispsav and opened into a Microsoft Excel spreadsheet. The heatmaps were prepared using GraphPad Prism 9.0, plotted as an average of three replicates. The Excel files were processed using Python 3.7 (pycharm community edition) for generating principal component plots.

3. Results and Discussion

The MIMS spectra are shown between m/z 50 and 180 as the spectrum depicts signals corresponding to the local environment (e.g., water from media, argon/nitrogen, carbon dioxide etc.) below m/z 50 and there were no significant peaks above m/z 180 (Figure 2A–C).

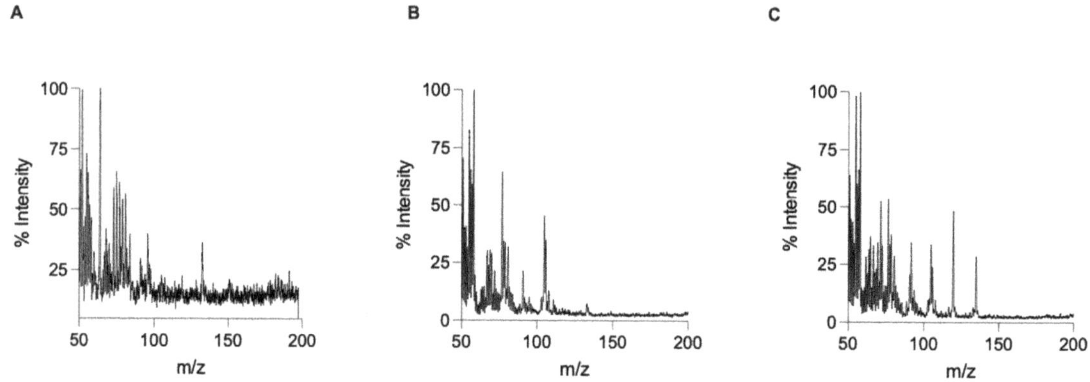

Figure 2. Representative mass spectra of (**A**) control (LB media), (**B**) *S. epidermidis*, and (**C**) *P. aeruginosa* from undiluted overnight cultures.

All samples showed prominent peaks below m/z 150. The bacterial strains showed common fragmentation patterns, such as peaks at m/z 77 corresponding to phenyl fragments; this was expected as many common volatomes detected using GC-MS for bacteria include compounds with a benzene ring [30,31]. Another common fragmentation pattern seen for all strains was a peak at m/z 58 (acetone) and m/z 106 (benzaldehyde). This was also expected because these bacterial strains produce many ketones, carboxylic acids, and aldehydes as their metabolites [32,33]. Using PTR-MS (proton transfer reaction ionization mass spectrometry) [34,35] several volatile markers [indoles (m/z 117), butanol (m/z 57), acetone (m/z 59), and 2-methyl-1-butanol (m/z 89)] from *E. coli* growth culture have previously been detected. MIMS herein utilized electron ionization, and therefore, a direct comparison to the PTR-MS could not be performed. The data obtained from the MIMS have been depicted as heat maps as this gives an instant snapshot of the spectral differences between different samples at a glance. More details of the algorithms, codes, and advantages of presenting mass spectrometric data as heat maps have been discussed in detail elsewhere [36].

Relatively larger amounts of these commonly found volatile compounds are observed in pathogenic species of *E. coli* (Figure 3A) as compared to those in non-pathogenic species. Similarly, higher amounts of indole are present in cells expressing functional protein as compared to those in cells expressing dysfunctional protein (Figure 3B). These differences in the signals obtained from different compounds could be used as biomarkers either for virulence or a metabolic shift in cell systems as a function of protein functionality. Signals corresponding to indoles in bacterial cultures were previously observed using GC-MS [37,38] and MIMS [39]. This shows that MIMS can be a simpler and more economical alternative to GC/LC-MS techniques for biomonitoring. The selectivity of the MIMS can be further improved by utilizing MIMS/MS systems [39,40]. Further investigations will be directed towards identification of the possible volatile signatures, including their quantification, to be utilized as a marker for the detection of protein functionality.

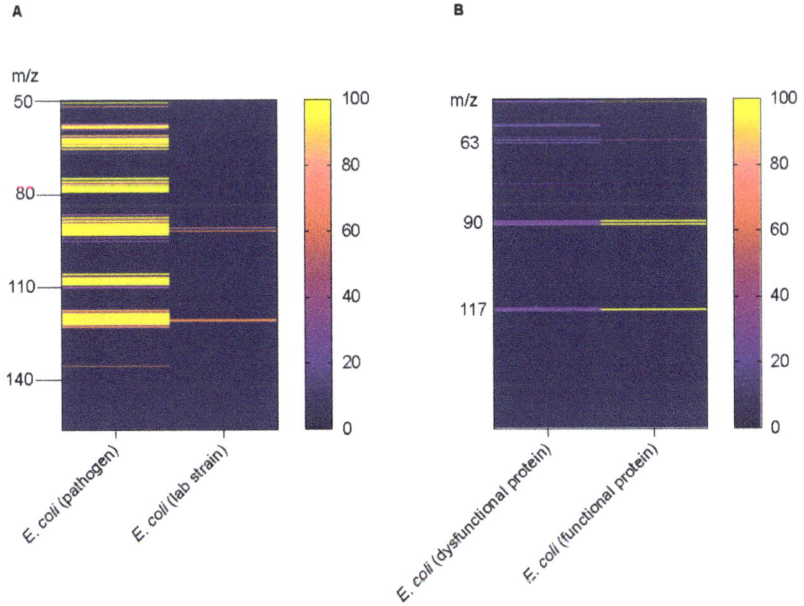

Figure 3. (**A**) Representative heatmap of undiluted overnight cultures from *E. coli* pathogenic and lab strains ($n = 4$). (**B**) Optical density-adjusted cultures harvested 3 h after induction with IPTG ($n = 3$).

It was then investigated if MIMS could differentiate between the bacterial strains based on principal component analysis (PCA) (Figure 4) [29]. PCA was performed using the obtained data. This is an analytical technique used for dimensional reduction and data visualization. The mean and standard deviation of the given features were calculated. The data were then scaled using the calculated mean and standard deviation. The data were then distributed normally such that the mean was zero and had unit standard deviation. The Python inbuilt functionalities of standard scaler and fit transform were used for data standardization. The plotted principal components showed similarity and differences between datasets or samples when grouped together in clusters or when separated apart. The data patterns were visualized by plotting the principal components on the orthogonal axes. PCA allows complex and high-dimensional datasets to be simplified into lower-dimensional data sets. This technique is useful because by lowering the complexity of the dataset, it is easier to visualize patterns, subsequently making data interpretation much easier [41,42]. The interpretation of PCA is that similar samples are closely clustered together, while dissimilar samples are farther from each other [41,42]. The PCA analysis showed that MIMS was able to differentiate between the bacterial strains (Figure 4). This is especially true for *E. coli* where all similar samples were clustered close to each other and were far from the other bacterial strains (Figure 4). The data from PCA also show that pathogenic *S. epidermidis* (Gram-positive) was very well separated from pathogenic *E. coli* (Gram-negative). Furthermore, the lab strain of *E. coli* was very well separated from the respective pathogenic strain further substantiating the capacity of MIMS to differentiate between pathogenic and non-pathogenic species.

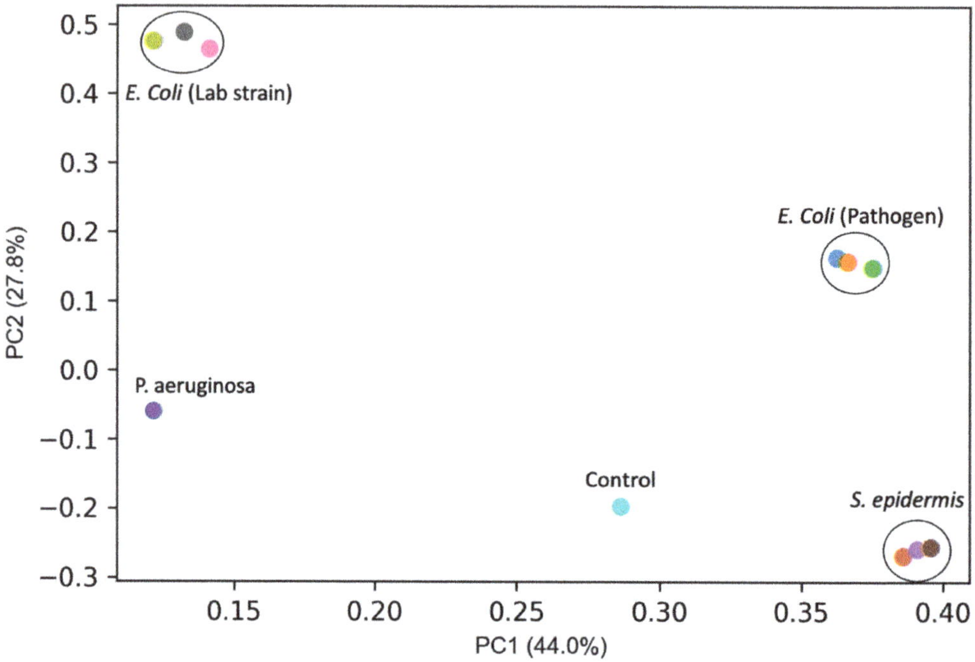

Figure 4. Representative PCA plot illustrating the ability of MIMS to separate the volatile profiles of different bacterial strains and the uninoculated media (control).

To further explore the possibility of utilizing MIMS as a process monitoring tool in biotechnological production facilities, we tested MIMS on a lab strain of *E. coli* with one clone expressing a fully functional protein YdgR encoded by pTTQ18 and a subsequent clone carrying the same vector with a functionally inactive variant of YdgR (mutant). The protein of interest here is a membrane transport protein, which has been routinely utilized in our lab [43,44]. Overexpression of the membrane protein has an impact on cellular systems, significantly affecting the differential regulation of proteins, thus possibly altering the volatile components secreted. This has not been investigated well enough to the best of our knowledge and should be pursued in our future studies.

Based on the PCA plots, it was observed that MIMS was able to distinguish between cells expressing functional protein from the cells expressing a functionally inactive protein (Figure 5). YdgR loses its functionality when mutated at the residue glutamate 30 impairing proton coupling efficiency. This results in the complete loss of YdgR function and has been very well established previously [45].

Differentiating between control strains and strains expressing functionally active and/or functionally inactive proteins can have a marked impact within the biopharmaceutical industry. Specifically, the limitations in oxygen delivery and waste product accumulation, the need for more advanced process control, and the shear sensitivity of cells poses difficulties for the large-scale cultivation of recombinant proteins in cell cultures [46]. These difficulties could be mitigated by following up on the volatile markers of cells, and timely intervention could significantly impact the cell titer obtained from the cell factories. Sometimes cell factories can also get contaminated, and using MIMS as a process monitoring tool can shorten the troubleshooting times within biotechnological industries. The cellular systems need to be thoroughly investigated both qualitatively and quantitatively based on the secreted markers observed to develop algorithms enabling rapid analysis and identification of the problems.

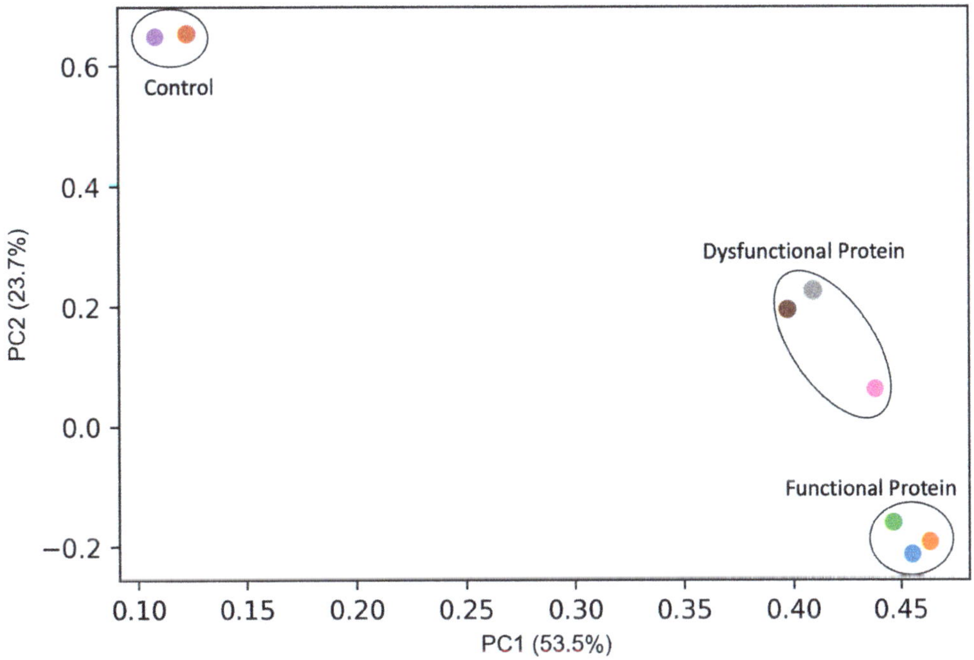

Figure 5. A representative PCA plot illustrating the ability of MIMS to distinguish between the volatile profiles of uninoculated media, i.e., control and growth media with cells expressing the functional YdgR and cells expressing the dysfunctional protein (YdgR mutant).

The focus of the present study was to investigate the possibilities of using MIMS to detect pathogens for water analysis and as an online process monitoring tool. Previous studies have used other microbial indicators to detect the presence of pathogens in water, and based on the findings of this study, it is seen that MIMS can also be used as a microbial indicator [25]. Moreover, the fact that MIMS can separate pathogens from lab strains further substantiates the utility of MIMS for bioprocess monitoring. As of now, we have not focused on the features responsible for the differences in bacteria tested, and a detailed analysis will be presented in future studies. The present paper can be regarded as one of the initial steps toward implementing MIMS in bioprocess monitoring.

Herein, we have utilized only a small sample set with two lab strains and two pathogenic bacteria, but the ability of MIMS to separate the lab strains from the pathogens enhances its applicability in water treatment plants located outside research laboratories and pharmaceutical industries.

4. Conclusions

MIMS has previously been used to study volatomes of genetically engineered bacteria. In this study through the usage of MIMS, we were able to differentiate between the volatile profiles of pathogenic and non-pathogenic *E. coli*. Furthermore, we were able to differentiate between the volatile compounds secreted by two lab strains, where one overexpressed dysfunctional proteins and the other overexpressed a functional protein. Moreover, we also differentiated between the volatomes of Gram-positive and Gram-negative bacteria. Thus, MIMS was identified as a tool that provides a low-impact alternative to existing water monitoring systems and online process monitoring tools to detect the presence of pathogenic bacterial strains.

Author Contributions: Conceptualization, B.K.P., H.J., M.G.J. and F.R.L.; Methodology, S.S. and I.A.; Formal analysis, S.F.C.; Investigation, S.S. and I.A.; Resources, B.K.P., H.J., M.G.J. and F.R.L.; writing—original draft preparation, review and editing, S.S., I.A., S.F.C., H.J., M.G.J., F.R.L. and B.K.P. All authors have read and agreed to the published version of the manuscript.

Funding: This research received no external funding.

Data Availability Statement: The python codes can be made available upon reasonable request to the corresponding author.

Acknowledgments: We wish to acknowledge the Higher Education Commission (HEC), Pakistan, for supporting the salary of Salvia Sajid. Authors also wish to acknowledge Carsten Uhd Nielsen for fruitful discussions.

Conflicts of Interest: The authors declare no conflict of interest.

References

1. Silhavy, T.J.; Kahne, D.; Walker, S. The Bacterial Cell Envelope. *Cold Spring Harbor Persp. Biol.* **2010**, *2*, a000414. [CrossRef] [PubMed]
2. Erb, M.; Kliebenstein, D.J. Plant Secondary Metabolites as Defenses, Regulators, and Primary Metabolites: The Blurred Functional Trichotomy. *Plant Physiol.* **2020**, *184*, 39–52. [CrossRef] [PubMed]
3. Umber, B.J.; Shin, H.-W.; Meinardi, S.; Leu, S.-Y.; Zaldivar, F.; Cooper, D.M.; Blake, D.R. Gas Signatures from *Escherichia coli* and *Escherichia coli*-Inoculated Human Whole Blood. *Clin. Transl. Med.* **2013**, *2*, 13. [CrossRef] [PubMed]
4. Jenkins, C.L.; Bean, H.D. Dependence of the Staphylococcal Volatilome Composition on Microbial Nutrition. *Metabolites* **2020**, *10*, 347. [CrossRef] [PubMed]
5. Vilaplana, L.; Marco, M.-P. Phenazines as Potential Biomarkers of Pseudomonas Aeruginosa Infections: Synthesis Regulation, Pathogenesis and Analytical Methods for Their Detection. *Anal. Bioanal. Chem.* **2020**, *412*, 5897–5912. [CrossRef]
6. Fitzgerald, S.; Holland, L.; Morrin, A. An Investigation of Stability and Species and Strain-Level Specificity in Bacterial Volatilomes. *Front. Microbiol.* **2021**, *12*, 693075. [CrossRef]
7. Nizio, K.D.; Perrault, K.A.; Troobnikoff, A.N.; Ueland, M.; Shoma, S.; Iredell, J.R.; Middleton, P.G.; Forbes, S.L. In Vitro Volatile Organic Compound Profiling Using GC× GC-TOFMS to Differentiate Bacteria Associated with Lung Infections: A Proof-of-Concept Study. *J. Breath Res.* **2016**, *10*, 026008. [CrossRef]
8. Phillips, M.; Cataneo, R.N.; Chaturvedi, A.; Kaplan, P.D.; Libardoni, M.; Mundada, M.; Patel, U.; Zhang, X. Detection of an Extended Human Volatome with Comprehensive Two-Dimensional Gas Chromatography Time-of-Flight Mass Spectrometry. *PLoS ONE* **2013**, *8*, e75274. [CrossRef]
9. Heddergott, C.; Calvo, A.M.; Latgé, J.P. The Volatome of Aspergillus Fumigatus. *Eukaryot. Cell* **2014**, *13*, 1014–1025. [CrossRef]
10. Orban, A.; Weber, A.; Herzog, R.; Hennicke, F.; Rühl, M. Transcriptome of Different Fruiting Stages in the Cultivated Mushroom Cyclocybe Aegerita Suggests a Complex Regulation of Fruiting and Reveals Enzymes Putatively Involved in Fungal Oxylipin Biosynthesis. *BMC Genom.* **2021**, *22*, 324. [CrossRef]
11. Freihorst, D.; Brunsch, M.; Wirth, S.; Krause, K.; Kniemeyer, O.; Linde, J.; Kunert, M.; Boland, W.; Kothe, E. Smelling the Difference: Transcriptome, Proteome and Volatilome Changes after Mating. *Fungal Genet. Biol.* **2018**, *112*, 2–11. [CrossRef] [PubMed]
12. Boots, A.W.; Smolinska, A.; van Berkel, J.J.B.N.; Fijten, R.R.R.; Stobberingh, E.E.; Boumans, M.L.L.; Moonen, E.J.; Wouters, E.F.M.; Dallinga, J.W.; Van Schooten, F.J. Identification of Microorganisms Based on Headspace Analysis of Volatile Organic Compounds by Gas Chromatography–mass Spectrometry. *J. Breath Res.* **2014**, *8*, 027106. [CrossRef] [PubMed]
13. Beale, D.J.; Pinu, F.R.; Kouremenos, K.A.; Poojary, M.M.; Narayana, V.K.; Boughton, B.A.; Kanojia, K.; Dayalan, S.; Jones, O.A.H.; Dias, D.A. Review of Recent Developments in GC-MS Approaches to Metabolomics-Based Research. *Metabolomics* **2018**, *14*, 152. [CrossRef] [PubMed]
14. Nalbantoglu, S. Metabolomics: Basic Principles and Strategies. In *Molecular Medicine*; Nalbantoglu, S., Amri, H., Eds.; IntechOpen: London, UK, 2019. [CrossRef]
15. Kotiaho, T.; Lauritsen, F.R.; Choudhury, T.K.; Cooks, R.G.; Tsao, G.T. Membrane Introduction Mass Spectrometry. *Anal. Chem.* **1991**, *63*, 875A–883A. [CrossRef]
16. Kotiaho, T.; Lauritsen, F.R. Chapter 16 Membrane Inlet Mass Spectrometry. In *Comprehensive Analytical Chemistry*; Elsevier: Amsterdam, The Netherlands, 2002; Volume 37, pp. 531–557.
17. Giannoukos, S.; Brkić, B.; Taylor, S.; France, N. Membrane Inlet Mass Spectrometry for Homeland Security and Forensic Applications. *J. Am. Soc. Mass Spectrom.* **2015**, *26*, 231–239. [CrossRef]
18. Johnson, R.C.; Cooks, R.G.; Allen, T.M.; Cisper, M.E.; Hemberger, P.H. Membrane Introduction Mass Spectrometry: Trends and Applications. *Mass Spectrom. Rev.* **2000**, *19*, 1–37. [CrossRef]
19. Kinani, S.; Richard, B.; Souissi, Y.; Bouchonnet, S. Analysis of Inorganic Chloramines in Water. *Trends Anal. Chem.* **2012**, *33*, 55–67. [CrossRef]

20. Degn, H. Membrane Inlet Mass Spectrometry in Pure and Applied Microbiology. *J. Microbiol. Methods* **1992**, *15*, 185–197. [CrossRef]
21. Burlacot, A.; Burlacot, F.; Li-Beisson, Y.; Peltier, G. Membrane Inlet Mass Spectrometry: A Powerful Tool for Algal Research. *Front. Plant Sci.* **2020**, *11*, 1302. [CrossRef]
22. Douchi, D.; Liang, F.; Cano, M.; Xiong, W.; Wang, B.; Maness, P.-C.; Lindblad, P.; Yu, J. Membrane-Inlet Mass Spectrometry Enables a Quantitative Understanding of Inorganic Carbon Uptake Flux and Carbon Concentrating Mechanisms in Metabolically Engineered Cyanobacteria. *Front. Microbiol.* **2019**, *10*, 1356. [CrossRef]
23. Lauritsen, F.R.; Lloyd, D. Direct Detection of Volatile Metabolites Produced by Microorganisms. In *Mass Spectrometry for the Characterization of Microorganisms*; ACS Symposium Series; American Chemical Society: New York, NY, USA, 1993; Volume 541, pp. 91–106, ISBN 9780841227378.
24. Cabral, J.P.S. Water Microbiology. Bacterial Pathogens and Water. *Int. J. Environ. Res. Public Health* **2010**, *7*, 3657–3703. [CrossRef] [PubMed]
25. Payment, P.; Locas, A. Pathogens in Water: Value and Limits of Correlation with Microbial Indicators. *Ground Water* **2011**, *49*, 4–11. [CrossRef] [PubMed]
26. Jensen, J.M.; Ismat, F.; Szakonyi, G.; Rahman, M.; Mirza, O. Probing the Putative Active Site of YjdL: An Unusual Proton-Coupled Oligopeptide Transporter from E. Coli. *PLoS ONE* **2012**, *7*, e47780. [CrossRef] [PubMed]
27. Saporito, P.; Vang Mouritzen, M.; Løbner-Olesen, A.; Jenssen, H. LL-37 Fragments Have Antimicrobial Activity against *Staphylococcus epidermidis* Biofilms and Wound Healing Potential in HaCaT Cell Line. *J. Pept. Sci.* **2018**, *24*, e3080. [CrossRef]
28. Saporito, P.; Biljana, M.; Løbner Olesen, A.; Jenssen, H. Antibacterial Mechanisms of GN-2 Derived Peptides and Peptoids against *Escherichia coli*. *Biopolymers* **2019**, *110*, e23275. [CrossRef]
29. Hughes, S.S.; Nielsen, M.M.K.; Jonsbo, R.V.; Nielsen, C.U.; Lauritsen, F.R.; Prabhala, B.K. BeerMIMS: Exploring the Use of Membrane-Inlet Mass Spectrometry (MIMS) Coupled to KNIME for the Characterization of Danish Beers. *Eur. J. Mass Spectrom.* **2021**, *27*, 266–271. [CrossRef]
30. Fletcher, J.S.; Vickerman, J.C. Secondary Ion Mass Spectrometry: Characterizing Complex Samples in Two and Three Dimensions. *Anal. Chem.* **2013**, *85*, 610–639. [CrossRef]
31. Smith, F.T.; DeRuiter, J.; Abdel-Hay, K.; Randall Clark, C. GC–MS and FTIR Evaluation of the Six Benzoyl-Substituted-1-Pentylindoles: Isomeric Synthetic Cannabinoids. *Talanta* **2014**, *129*, 171–182. [CrossRef]
32. Morris, H.R.; Taylor, G.W.; Masento, M.S.; Jermyn, K.A.; Kay, R.R. Chemical Structure of the Morphogen Differentiation Inducing Factor from Dictyostelium Discoideum. *Nature* **1987**, *328*, 811–814. [CrossRef]
33. Dickschat, J.S. Capturing Volatile Natural Products by Mass Spectrometry. *Nat. Prod. Rep.* **2014**, *31*, 838–861. [CrossRef]
34. Bunge, M.; Araghipour, N.; Mikoviny, T.; Dunkl, J.; Schnitzhofer, R.; Hansel, A.; Schinner, F.; Wisthaler, A.; Margesin, R.; Märk, T.D. On-Line Monitoring of Microbial Volatile Metabolites by Proton Transfer Reaction-Mass Spectrometry. *Appl. Environ. Microbiol.* **2008**, *74*, 2179–2186. [CrossRef] [PubMed]
35. Rosenthal, K.; Hunsicker, E.; Ratcliffe, E.; Lindley, M.R.; Leonard, J.; Hitchens, J.R.; Turner, M.A. Volatile Atmospheric Pressure Chemical Ionisation Mass Spectrometry Headspace Analysis of *E. coli* and *S. aureus*. *Anal. Methods* **2021**, *13*, 5441–5449. [CrossRef] [PubMed]
36. Key, M. A Tutorial in Displaying Mass Spectrometry-Based Proteomic Data Using Heat Maps. *BMC Bioinform.* **2012**, *13* (Suppl. S16), S10. [CrossRef] [PubMed]
37. Arora, P.K.; Bae, H. Identification of New Metabolites of Bacterial Transformation of Indole by Gas Chromatography-Mass Spectrometry and High Performance Liquid Chromatography. *Int. J. Anal. Chem.* **2014**, *2014*, 239641. [CrossRef]
38. Ratiu, I.-A.; Ligor, T.; Bocos-Bintintan, V.; Al-Suod, H.; Kowalkowski, T.; Rafińska, K.; Buszewski, B. The Effect of Growth Medium on an Escherichia Coli Pathway Mirrored into GC/MS Profiles. *J. Breath Res.* **2017**, *11*, 036012. [CrossRef] [PubMed]
39. Lloyd, D.; Lauritsen, F.R.; Degn, H. The Parasitic Flagellates Trichomonas Vaginalis and Tritrichomonas Foetus Produce Indole and Dimethyl Disulphide: Direct Characterization by Membrane Inlet Tandem Mass Spectrometry. *J. Gen. Microbiol.* **1991**, *137*, 1743–1747. [CrossRef]
40. Etzkorn, J.M.; Davey, N.G.; Thompson, A.J.; Cieba, A.S.; Leblanc, C.W.; Simpson, C.D.; Krogh, E.T.; Gill, C.G. The Use of MIMS-MS-MS in Field Locations as an on-Line Quantitative Environmental Monitoring Technique for Trace Contaminants in Air and Water. *J. Chromatogr. Sci.* **2009**, *47*, 57–66. [CrossRef]
41. Ma, Y.; Zhu, L. A Review on Dimension Reduction. *Int. Stat. Rev.* **2013**, *81*, 134–150. [CrossRef]
42. Simmons, S.; Peng, J.; Bienkowska, J.; Berger, B. Discovering What Dimensionality Reduction Really Tells Us About RNA-Seq Data. *J. Comput. Biol.* **2015**, *22*, 715–728. [CrossRef]
43. Harder, D.; Stolz, J.; Casagrande, F.; Obrdlik, P.; Weitz, D.; Fotiadis, D.; Daniel, H. DtpB (YhiP) and DtpA (TppB, YdgR) Are Prototypical Proton-Dependent Peptide Transporters of *Escherichia coli*. *FEBS J.* **2008**, *275*, 3290–3298. [CrossRef]
44. Prabhala, B.K.; Aduri, N.G.; Iqbal, M.; Rahman, M.; Gajhede, M.; Hansen, P.R.; Mirza, O. Several hPepT1-transported Drugs Are Substrates of the *Escherichia coli* Proton-Coupled Oligopeptide Transporter YdgR. *Res. Microbiol.* **2017**, *168*, 443–449. [CrossRef] [PubMed]

45. Jensen, J.M.; Ernst, H.A.; Wang, X.; Hald, H.; Ditta, A.C.; Ismat, F.; Rahman, M.; Mirza, O. Functional Investigation of Conserved Membrane-Embedded Glutamate Residues in the Proton-Coupled Peptide Transporter YjdL. *Protein Pept. Lett.* **2012**, *19*, 282–287. [CrossRef] [PubMed]
46. Chu, L.; Robinson, D.K. Industrial Choices for Protein Production by Large-Scale Cell Culture. *Curr. Opin. Biotechnol.* **2001**, *12*, 180–187. [CrossRef] [PubMed]

water

MDPI

Article

Comparison of Detecting and Quantitating SARS-CoV-2 in Wastewater Using Moderate-Speed Centrifuged Solids versus an Ultrafiltration Method

Maria Hasing [1], Jiaao Yu [1], Yuanyuan Qiu [1], Rasha Maal-Bared [2], Sudha Bhavanam [1], Bonita Lee [3], Steve Hrudey [1] and Xiaoli Pang [1,4,*]

1 Department of Laboratory Medicine and Pathology, University of Alberta, Edmonton, AB T6G 2B7, Canada; hasing@ualberta.ca (M.H.); jiaao3@ualberta.ca (J.Y.); yuanyuan@ualberta.ca (Y.Q.); bhavanam@ualberta.ca (S.B.); shrudey@ualberta.ca (S.H.)
2 EPCOR Water Services Inc., Edmonton, AB T5H 0E8, Canada; rmaal-bared@epcor.com
3 Department of Pediatrics, University of Alberta, Edmonton, AB T6G 1C9, Canada; bonita.lee@albertahealthservices.ca
4 Public Health Laboratory, Alberta Precision Laboratories, Edmonton, AB T6G 2J2, Canada
* Correspondence: xiao-li.pang@albertaprecisionlabs.ca; Tel.: +1-780-407-3483

Abstract: Mounting evidence suggests that solids are a reliable matrix for SARS-CoV-2 detection in wastewater, yet studies comparing solids-based methods and common concentration methods using the liquid fraction remain limited. In this study, we developed and optimized a method for SARS-CoV-2 detection in wastewater using moderate-speed centrifuged solids and evaluated it against an ultrafiltration reference method. SARS-CoV-2 was quantified in samples from 12 wastewater treatment plants from Alberta, Canada, using RT-qPCR targeting the N2 and E genes. PCR inhibition was examined by spiking salmon DNA. The effects of using different amounts of solids, adjusting the sample pH to 9.6–10, and modifying the elution volume at the final step of RNA extraction were evaluated. SARS-CoV-2 detection rate in solids from 20 mL of wastewater showed no statistically significant difference compared to the ultrafiltration method (97/139 versus 90/139, $p = 0.26$, McNemar's mid-p test). The optimized wastewater solids-based method had a significantly lower rate of samples with PCR inhibition versus ultrafiltration (3% versus 9.5%, $p = 0.014$, Chi-square test). Our optimized moderate-speed centrifuged solids-based method had similar sensitivity when compared to the ultrafiltration reference method but had the added advantages of lower costs, fewer processing steps, and a shorter turnaround time.

Keywords: COVID-19; SARS-CoV-2; real-time PCR; solids; virus; ultrafiltration; wastewater

Citation: Hasing, M.; Yu, J.; Qiu, Y.; Maal-Bared, R.; Bhavanam, S.; Lee, B.; Hrudey, S.; Pang, X. Comparison of Detecting and Quantitating SARS-CoV-2 in Wastewater Using Moderate-Speed Centrifuged Solids versus an Ultrafiltration Method. *Water* **2021**, *13*, 2166. https://doi.org/10.3390/w13162166

Academic Editors: Guangming Jiang, Ryo Honda and Sudipti Arora

Received: 14 July 2021
Accepted: 3 August 2021
Published: 6 August 2021

1. Introduction

The current COVID-19 pandemic has prompted a global surge in monitoring SARS-CoV-2 in wastewater as an additional and supplementary surveillance tool to inform public health authorities on disease burden. An important advantage of wastewater-based surveillance (WBS) over indicators such as clinical cases and hospitalization rates is that it provides a comprehensive snapshot of SARS-CoV-2 presence that includes symptomatic, asymptomatic, and pre-symptomatic carriers with the analysis of a single sample. SARS-CoV-2 WBS also has the potential to allow early detection of changes in disease burden in advance to clinical testing data and hospitalization rates [1,2]. Moreover, SARS-CoV-2 WBS is not affected by policies and delivery of clinical diagnostics provision, which can be limited by resource allocation and availability. For these characteristics, SARS-CoV-2 WBS represents a valuable tool particularly in locations where clinical testing has reached maximum capacity and test reports are lagging or not representative of actual disease burden. WBS has gained increasing importance since the start of the COVID-19 pandemic and as of the time of writing this manuscript, at least 55 countries have set up over 2300 sites for monitoring SARS-CoV-2 in wastewater [3].

While WBS historically has proven to be a valuable tool for monitoring the re-emergence of polio [4,5] and multiple other human viruses [6,7], wastewater is a complex and challenging matrix especially for molecular detection and quantification of pathogens. Wastewater composition includes a high content of organic matter, a wide range of PCR inhibitors [8,9], and it can vary significantly within hours [10,11]. In addition, most methods for virus detection from wastewater were developed and optimized for non-enveloped viruses which are more stable in aqueous environments than enveloped viruses like coronaviruses [12–14].

Most extraction methods use the liquid fraction of wastewater samples and, since viruses are typically found at low levels in wastewater, they frequently require virus concentration steps. There are multiple methods for virus concentration including but not limited to electropositive filtration, electronegative filtration, polyethylene glycol (PEG) precipitation, and ultrafiltration [15–18]. Even if these time-consuming, laborious, and costly virus concentration steps are used, virus recovery rates are frequently low. We have identified median recovery rates of 3% to 50% for different enteric viruses using ultrafiltration [14]. An interlaboratory reproducibility study by Pecson et al. [19], involving 32 laboratories and 36 different methods for quantification of SARS-CoV-2 in wastewater, reported recoveries for betacoronavirus OC43 with interlaboratory median values of 10%, 5.8%, 3.3%, and 3.6%, respectively, for methods using either no concentration steps, electronegative filtration, PEG precipitation, or ultrafiltration.

For virus detection purposes, it is not uncommon for some methods to remove wastewater solids prior to virus nucleic acids extraction to avoid filter clogging issues [16,20]. However, the partitioning and stability of viruses in solid or liquid fractions of wastewater samples can vary significantly by virus type. Growing evidence suggests that a considerable proportion of coronaviruses in wastewater are adsorbed on the solids fraction. Methods that include solids as well as liquid fractions reportedly perform better than those that exclude solids [16] and up to 26% of enveloped viruses versus 6% of nonenveloped viruses have been found to adsorb onto wastewater solids [21]. Based on analysis of primary sludge collected at a period of high COVID-19 activity, Peccia et al. [22] reported concentrations 1000 times higher than previously reported by others analyzing influent wastewater. At the time of manuscript preparation, several other groups have described the use of solids for the detection of SARS-CoV-2 in wastewater [17,18,23–26]. However, only a few studies provide first-hand data on the performance of solids-based methods compared to more conventional methods [17,25]. In this study, we present method optimization data toward improving the sensitivity of SARS-CoV-2 detection in moderate-speed centrifuged solids from post-grit influent wastewater. We also compared our solids-based method with a previously validated ultrafiltration reference method to assess detection rate, costs, simplicity, and turnaround times.

2. Materials and Methods

2.1. Wastewater Samples

Post-grit influent wastewater samples (400 mL, 24-h composite samples) were collected from 12 different wastewater treatment plants (WWTP) across 10 cities of Alberta, Canada: Edmonton, Calgary, Canmore, Red Deer, Banff, Fort Saskatchewan, Lethbridge, Grande Prairie, Medicine Hat and High River. Samples from each WWTP were collected between 16 October 2020, and 20 December 2020, at a frequency of three samples per week. Samples were stored at −20 °C in 500 mL bottles upon collection until weekly shipment and were immediately processed once received by the laboratory. Storage at −20 °C might have reduced detectable virus loads [27] but given that samples were run in parallel using both solids-based and ultrafiltration methods, it would have no impact on our comparative analyses. Prior to testing, samples were thoroughly mixed and quickly poured to assure that aliquots represented a homogeneous suspension for both, solids-based and ultrafiltration methods.

2.2. Ultrafiltration Method

Virus potentially present in wastewater samples was extracted and concentrated as previously described [14] with modifications as follows. Briefly, 100 mL of wastewater sample were spiked with 100 µL of a suspension of human coronavirus (HCoV) strain 229E (10^5 IU/mL titrated by $TCID_{50}$) for monitoring virus recovery. The spiked samples were adjusted to pH 9.6–10 with 5N NaOH and mixed vigorously for 30 s. Samples were then centrifuged at 4500× *g* for 10 min to pellet solids; the liquid fraction was transferred into a new container and adjusted to pH 7–7.5 with 1.2N HCl. The viral particles present in the liquid fraction were concentrated by ultrafiltration using a Centricon Plus-70™ filter with a pore size or Nominal Molecular Weight Limit (NMWL) of 30 KDa (Merck Millipore, Carrigtwohill, Ireland) according to the manufacturer's instructions, except the pre-rinse step which was eliminated. Filters were loaded with 70 mL of sample and centrifuged at 3000× *g* for 10 min at room temperature. The filtrate was discarded, and the same procedure was repeated for the rest of the sample for a total volume of 100 mL. The filtrate collection cup was removed, and the concentration cup was placed on top of the sample filter cup. The Centricon™ filter was then inverted carefully and centrifuged at 800× *g* for 2 min. The sample was collected from the concentration cup and adjusted to a final volume of 1 mL by adding phosphate-buffered saline (PBS). The concentrated samples were stored at −70 °C until RNA extraction.

2.3. Moderate-Speed Centrifuged Solids Method

A total of 100 mL of wastewater was spiked with 100 µL of a suspension of HCoV 229E (10^5 IU/mL titrated by $TCID_{50}$) and centrifuged at 4500× *g* for 10 min at 4 °C. The supernatant was discarded, and the weight of the remaining wet solids fraction was recorded to calculate the number of viral genome copies present in 100 mL of wastewater. Different wet weights of solids (50 mg, 100 mg, 200 mg, 300 mg, or 400 mg) were aliquoted for RNA extraction; when required, an additional 100 mL aliquot of wastewater was spiked and centrifuged to supply sufficient solids for testing. For tests involving the effect of pH adjustment, the same solids obtained under *Ultrafiltration method* were used for analysis. For tests using solids from 20 mL of wastewater, a 20 mL aliquot of sample was centrifuged at 4500× *g* for 10 min at 4 °C. The supernatant was discarded, the weight of the resulting wet solids fraction was recorded, and the entire solids fraction was used for RNA extraction.

2.4. RNA Extraction

Viral RNA was extracted from ultrafiltration wastewater concentrates and solids fractions using the MagMAX™-96 Viral RNA isolation kit (Thermo Fisher Scientific Baltics UAB, Vilnius, Lithuania) and the King Fisher™ Flex Purification System (Thermo Fisher Scientific, Vantaa, Finland). The extraction process was performed according to the MagMAX manufacturer's instructions with the following modifications. Input sample was either A) 400 µL of wastewater concentrate prepared as described under *Ultrafiltration method* mixed with 502 µL of Lysis/Binding Solution or B) wet solids (50 mg, 100 mg, 200 mg, 300 mg, 400 mg, or all solids obtained from 20 mL of wastewater, as described under *Moderate-speed centrifuged solids method*) were mixed with 200 µL of PBS and 502 µL of Lysis/Binding Solution (250 µL of Lysis/Binding Solution Concentrate, 2 µL of carrier RNA and 250 µL of 100% isopropanol); the resulting solids-PBS-Lysis/Binding Solution mixture was vortexed for 20 s and centrifuged at 14,000× *g* for 2 min and the entire supernatant (approximately 670 µL) was used for RNA extraction. The extraction process was set up in the King Fisher™ Flex system loading input sample (either A or B described above), 20 µL of Bead Mix and 5 µL of salmon testes DNA (0.5 µg/mL) in the sample plate, two washes of 300 µL each of Wash Solution 1 and two washes of 450 µL each of Wash Solution 2. In the final step, nucleic acids from solids were eluted either in 50 μL or 100 μL of elution buffer; all extracts from the ultrafiltration method were eluted with 100 μL of elution buffer.

2.5. RT-qPCR

SARS-CoV-2 detection and quantification was performed by RT-qPCR with primers and probes targeting genes E and N2 previously described by others [28,29] and summarized in Table S1. One-step RT-qPCR was carried out for both targets with a final reaction volume of 10 μL, containing 2.5 μL of 4× TaqMan™ Fast Virus 1-Step RT-PCR Master Mix (Thermo Fisher Scientific Baltics UAB, Vilnius, Lithuania), 0.4 μL of 20 μM each forward and reverse primers, 0.2 μL of 10 μM TaqMan probe (Applied Biosystems, Foster City, CA, USA), 1.5 μL PCR grade H_2O, and 5 μL of RNA extract. Thermocycler conditions consisted of 5 min at 50 °C for reverse transcription reaction, followed by 20 s at 95 °C for retrotranscriptase inactivation, and then 45 PCR amplification cycles of 3 s at 95 °C and 30 s at 60 °C. Amplification data were collected and analyzed with the 7500 Software version 2.0.5 (Applied Biosystems, Foster City, CA, USA). SARS-CoV-2 RNA levels were quantified from a standard curve included in each PCR run, prepared with a 10-fold dilution series ranging from 166 copies to 1.66×10^6 copies per PCR reaction of a SARS-CoV-2 RNA fragment enclosing the RT-qPCR gene E target, and were expressed as viral genome copies per 100 mL of wastewater. The same E fragment was used to create a standard curve for gene N2 using the same primers and probe sequences except that the probe had FAM as reporter dye and BHQ1 as a quencher. Samples were tested in duplicate for each gene. A sample was considered positive if at least two out of the total four RT-qPCR reactions tested positive. If salmon DNA tested positive or had a delayed Ct, SARS-CoV-2 quantification was attempted again on a 1:10 or 1:5 dilution of RNA extract. The limit of detection (95% LOD) of this RT-qPCR assay is two copies per reaction for the E gene [28].

2.6. Quality Control Processes

Virus recovery rates were evaluated by RT-qPCR quantification of spiked HCoV 229E in samples vs. a baseline control consisting of a mixture of 100 μL of 10^5 IU of 229E/mL and 900 μL of water. The reaction and thermocycler conditions were set up similarly as described above for SARS-CoV-2 but with primers and probes (Applied Biosystems, Foster City, CA, USA) by Vijgen et al. [30] (Table S1), which target the membrane protein M of HCoV 229E. A standard curve of 10-fold serial dilutions starting from 166 genome copies per PCR reaction of the E gene was included in each PCR run for virus quantification.

The detection of salmon testes DNA as a control for PCR inhibition was carried out as previously described [31], using a final reaction volume of 10 μL containing 5 μL of 2× Taqman™ Fast Universal PCR Master Mix, 0.5 μL of primer/probe mix (18 μM of each primer and 5 μM probe listed in Table S1), 2 μL of PCR grade water, and 2.5 μL of RNA extract. Thermocycler conditions were initial denaturation for 20 s at 95 °C, followed by 45 PCR amplification cycles of 3 s at 95 °C and 30 s at 60 °C. The presence of an inhibitory effect was defined as a delay by at least three cycles as compared to a distilled water control spiked with the same concentration of salmon DNA.

A negative control in the form of distilled water was included in each RNA extraction and RT-qPCR runs.

2.7. PMMoV Quantification

Pepper Mild Mottle Virus (PMMoV) levels in wastewater were determined to characterize it as a potential internal fecal biomarker for the normalization of wastewater dilution. The RT-qPCR reactions were set up with the primers and probes by Zhang et al. [32] (Table S1) and using 2.5 μL of 4× Taqman™ Fast Virus 1-Step RT-PCR Master Mix (Thermo Fisher Scientific Baltics UAB, Vilnius, Lithuania), 0.5 μL of 10 μM each forward and reverse primers, 0.2 μL of 10 μM TaqMan probe (Applied Biosystems, Foster City, CA, USA), 1.3 μL PCR grade H_2O, and 5 μL of RNA extract. Thermocycler conditions were the same as described above for SARS-CoV-2. For PMMoV quantification, a standard curve from a

SARS-CoV-2 RNA fragment enclosing the E gene target was generated, as a substitute for a PMMoV target, using the primers and probe specified on Table S1, except that the probe had FAM as reporter dye and BHQ1 as a quencher.

2.8. Statistical Analysis

All statistical analyses were performed using R software (version 4.0.3). The association between solids weight and SARS-CoV-2 qPCR Ct values was evaluated using linear regression. The statistical significance in differences in detection rate between wastewater matrices and pH treatments were calculated as previously described [33], using McNemar's mid-*p* test when discordant cases were <25 and the McNemar's test without correction in other instances. The difference in proportions of samples with delayed or negative salmon DNA Ct values was calculated using the Chi-square test with Yates' continuity correction and the alternative hypothesis that PCR inhibition rate is greater in solids versus ultrafiltration, except for solids from 20 mL, where the alternative hypothesis was that the proportion of samples with PCR inhibition is less in solids from 20 mL versus ultrafiltration. Comparisons of SARS-CoV-2 quantification between solids-based and ultrafiltration methods were evaluated by Spearman's correlation analysis. Differences in yield of wet solid in terms of mg per 100 mL of wastewater across WWTP and differences in log genome copies of PMMoV per 100 mL of wastewater were analyzed using ANOVA and Tukey HSD for post hoc tests. In all tests, a *p*-value below 0.05 was considered statistically significant.

3. Results

3.1. Optimization of the Weight of Moderate-Speed Centrifuged Wastewater Solids Used as Input and Effect of pH Adjustment

To determine the weight of wet solids that can allow SARS-CoV-2 detection with minimum or no PCR inhibition, we first carried out an exploratory evaluation with 100 mg, 200 mg, 300 mg, and 400 mg of wet solids for three samples (Figure S1). Ct values for SARS-CoV-2 did not change as solids weight increased ($p = 0.57$ for gene E and $p = 0.94$ for gene N2, linear regression slope), however, Ct values for spiked salmon DNA indicated a high level of PCR inhibition, therefore, we continued further testing using no more than 100 mg of wet solids.

Our ultrafiltration reference method includes a step of pH adjustment to 9.6–10 prior virus concentration to release some of the viruses adsorbed on wastewater solids into the surrounding liquid fraction. While this pH adjustment increases the number of viral genomic targets when the liquid fraction is used for analysis, we hypothesized that it may negatively impact SARS-CoV-2 detection when analyzing the remaining solids fraction. Thus, we compared the performance of using 50 mg and 100 mg of solids as well as the effect of not adjusting versus adjusting the pH of the wastewater. In terms of detection rate, none of the four tested weight-pH adjustment combinations were statistically superior to the ultrafiltration method (Table 1). All wastewater samples included in this study had original pH values with a median of 7.6 (interquartile range 7.3–7.8). Adjusting the pH resulted in a lower detection rate compared to not adjusting the pH for 100 mg of solids ($p = 0.04$, McNemar's *p*-mid test) but not for 50 mg solids ($p = 0.13$, McNemar's test). The proportions of samples with delayed Ct values for salmon DNA, with each of the four solids treatments ranging from 37% to 56%, were significantly higher compared to that observed with the ultrafiltration method which had 10% ($p < 0.001$, Chi-square test) (Table 2).

Table 1. Effect of different amounts of wastewater solids and pH adjustment on SARS-CoV-2 detection compared to an ultrafiltration-based method.

Method a	Method b	Concordant Positive	Concordant Negative	Discordant a + b −	Discordant a − b +	*n*	*p*-Value [1]
50 mg solids, pH not adjusted, 50µL elution	**Ultrafiltration**	14	37	10	12	73	0.68
50 mg solids, pH 9.6–10, 50 µL elution	**Ultrafiltration**	9	39	7	17	72	0.04
100 mg solids, pH not adjusted, 50 µL elution	**Ultrafiltration**	15	28	14	11	68	0.55
100 mg solids, pH 9.6–10, 50 µL elution	**Ultrafiltration**	13	37	7	13	70	0.19
100 mg solids, pH not adjusted 100 µL elution	**Ultrafiltration**	57	13	17	9	96	0.12
Solids from 20 mL, pH not adjusted 100 µL elution	**100 mg solid,** pH not adjusted 100 µL elution	45	6	4	10	65	0.12
Solids from 20 mL, pH not adjusted 100 µL elution	**Ultrafiltration**	74	26	23	16	139	0.26

[1] The statistical significance of differences between methods a and b were calculated using McNemar's mid-*p* test if discordant cases were < 25 or the McNemar's test without correction if discordant cases were ≥ 25.

Table 2. Comparison of PCR inhibition rates.

Method	Fraction of Samples with Negative or Delayed (≥34) Salmon DNA Ct Values	%	Chi-Square Statistic	*p*-Value
50 mg solids, pH not adjusted, 50 µL elution	32/73	43.8	43.77	<0.001
50 mg solids, pH 9.5–10, 50 µL elution	28/72	38.9	33.303	<0.001
100 mg solids, pH not adjusted, 50 µL elution	38/68	55.9	69.691	<0.001
100 mg solids, pH 9.5–10, 50 µL elution	26/70	37.1	29.468	<0.001
100 mg solids, pH not adjusted, 100 µL elution	10/96	10.4	0.0040	0.47
Solids from 20 mL, pH not adjusted, 100 µL elution	4/139	3.0	4.8815	0.014
Ultrafiltration	23/243	9.5	-	-

3.2. *Effect of Increasing Nucleic Acid Elution Volume*

With the intention to further dilute the PCR inhibitors present in the nucleic acid extracts from wastewater solids, we explored the effect of doubling the volume of buffer used at the final elution step of nucleic acid extraction, from 50 µL to 100 µL. The SARS-CoV-2 detection rate from nucleic acid extracts prepared from 100 mg of solids and eluted in 100 µL of elution buffer was similar to that of extracts prepared by the ultrafiltration method ($p = 0.12$, McNemar's test, Table 1). The percentage of samples with qPCR inhibition was also similar for both 100 mg of solids and the ultrafiltration method (10.4% and 9.5%, respectively, Table 2).

3.3. Input Sample Units of Measure: Weight versus Volume

To further simplify the solids-based method, we explored the alternative of using a fixed volume of wastewater for analysis instead of a fixed weight of solids. The median of wet solids yield (n = 176) was 492 (interquartile range, 69–1376) mg per 100 mL of post-grit wastewater with differences across WWTPs ($p < 0.001$, ANOVA) (Figure S2). Given that 100 mg represents about a fifth of the average weight present in 100 mL of wastewater, we performed comparison tests between 100 mg of solids and solids from 20 mL of wastewater. The yield of wet solids from 20 mL of wastewater (n = 143) had a median of 97 (interquartile range, 66–140) mg. Comparison in terms of SARS-CoV-2 detection rate showed no statistical differences between 100 mg of solids versus solids from 20 mL ($p = 0.12$, McNemar's p-mid test) (Table 1). Among the discordant 20 mL-positive-100 mg-negative and 20 mL-negative-100 mg-positive samples, there was no significant difference in the proportion of samples yielding less than 100 mg of solids from 20 mL (69% versus 65%, respectively, Chi-square, $p = 0.5$).

Compared to the ultrafiltration reference method, analysis of solids from 20 mL showed no differences in terms of SARS-CoV-2 detection rates ($p = 0.26$, McNemar's test) (Table 1). The rate of samples with PCR inhibition was significantly lower in solids from 20 mL (3%) versus the ultrafiltration method (9.5%) ($p = 0.014$, Chi-square) (Table 2).

3.4. 229 E Virus Recovery Data

Our ultrafiltration reference method for SARS-CoV-2 detection includes spiking of HCoV 229E to control for virus recovery. The rates of 229E recovery in wastewater solids were considerably lower compared to the ultrafiltration method, with a median of 0.07% (interquartile range, 0.03–0.16%) for 50 mg solids and a median of 0.09% (interquartile range, 0.05–0.17%) for 100 mg solids, whereas the ultrafiltration method had a median of 4.65% (interquartile range, 2.31–7.50%). Because SARS-CoV-2 is secreted in stools, a significant proportion of the virus is expected to be embedded in wastewater solids. This contrasts with the spiked HCoV 229E, which was added to the sample as a liquid suspension and could freely disperse into the liquid phase. Given that is likely that the partition of HCoV 229E was highly biased toward the liquid phase, HCoV 229E recovery rates were considered not to be representative of SARS-CoV-2 recovery and therefore, not measured in wastewater solids in subsequent tests. None of the data presented in this manuscript (solids or ultrafiltration methods) were corrected for HCoV 229E recovery rates.

3.5. Evaluation of Wastewater Moderate-Speed Centrifuged Solids for Quantification of SARS-CoV-2 Levels

Next, we examined how different concentrations of SARS-CoV-2 quantified using solids correlate with those quantified using ultrafiltration. SARS-CoV-2 log 10 copies per 100 mL of wastewater quantified using 100 mg solids, as well as solids from 20 mL of wastewater, correlated with those quantified using the ultrafiltration method with significant ($p < 0.001$) Spearman's correlation coefficients ranging from 0.4 to 0.6. The coefficients were indicative of a moderate positive relationship for both genes, E and N2, meaning that as levels detected using solids increase, levels detected by ultrafiltration also tend to increase, although not in all instances (Figure 1).

As evidenced in all comparison plots, the distribution of discordant results ranged from 2 to above 3 logs, suggesting that these were not associated with low virus levels. Visual inspection of timeline plots of SARS-CoV-2 levels by WWTP did not show any trends or difference in performance between 100 mg of solids, solids from 20 mL of wastewater, and the liquid fraction (Figure 2).

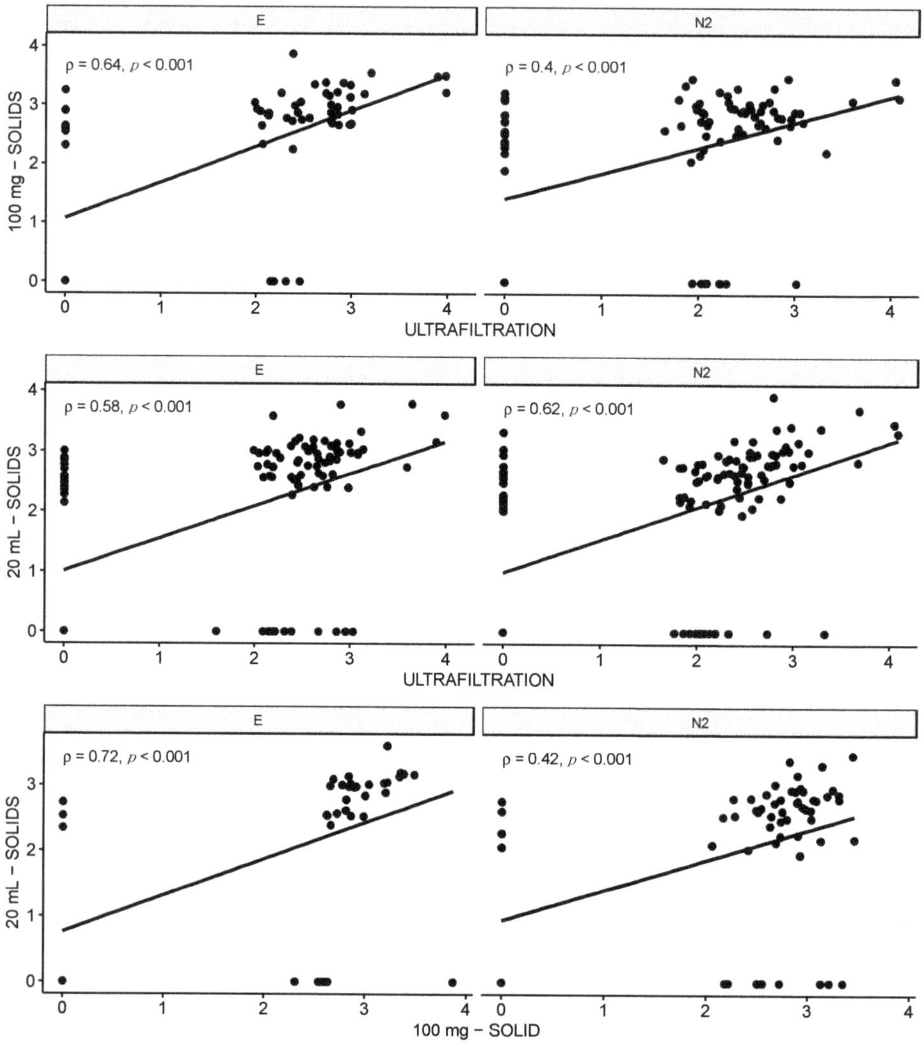

Figure 1. Correlation of SARS-CoV-2 levels in wastewater quantified using moderate-speed centrifuged solids versus an ultrafiltration reference method. The x- and y-axis scales represent the logarithm (log 10) of SARS-CoV-2 genome copies per 100 mL of wastewater. Spearman's correlation coefficients (ρ) and the corresponding p-values are shown for all pairwise comparisons (gene E and N2 data) between 100 mg of solids, solids from 20 mL, and the ultrafiltration reference method.

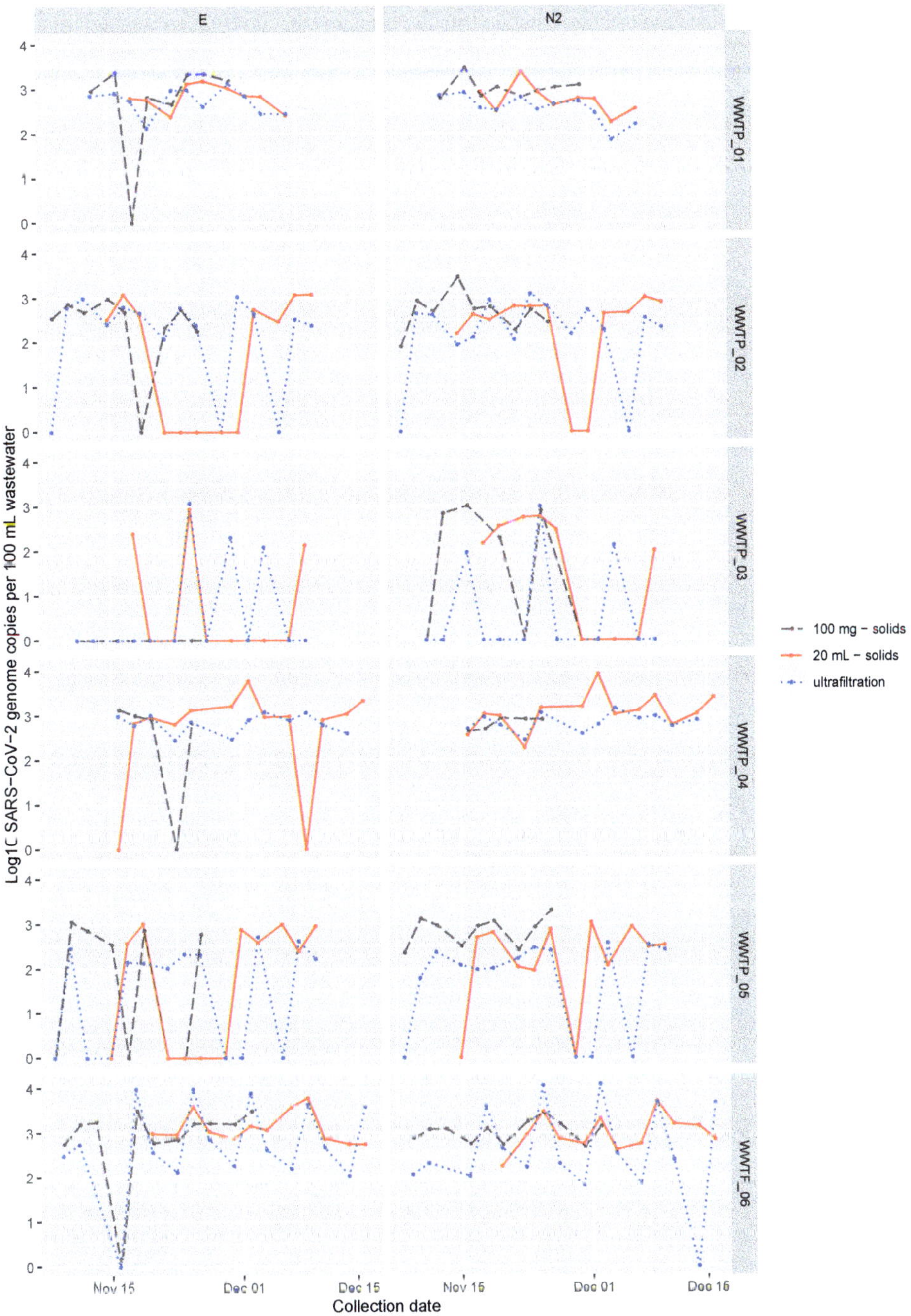

Figure 2. *Cont.*

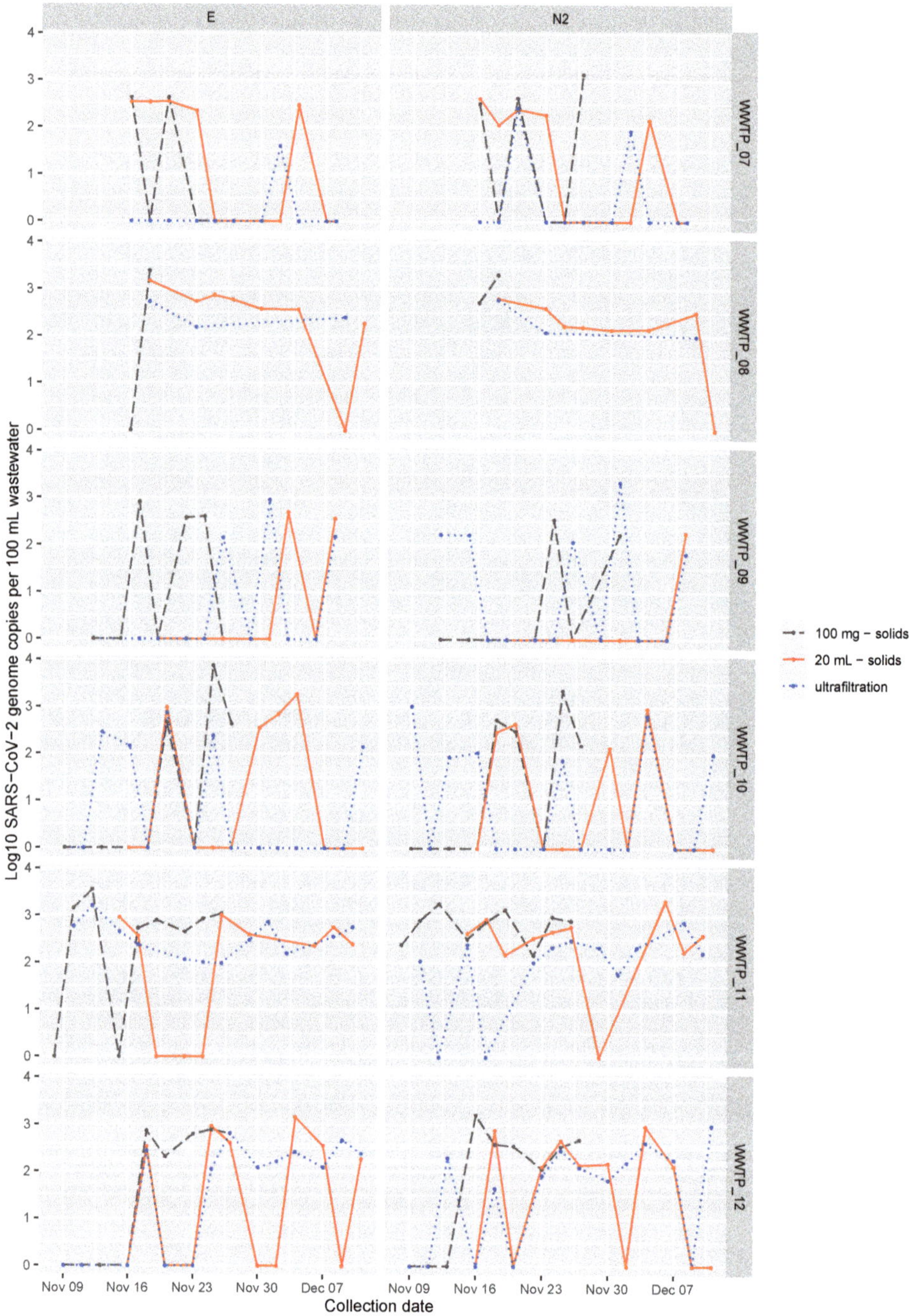

Figure 2. Timeline trends of SARS-CoV-2 levels in post-grit influent wastewater as quantified using moderate-speed centrifuged solids and an ultrafiltration reference method. Only a subset of samples was tested using 100 mg solids because weighting wet solids is labor-intensive, time-consuming, and not practical for routine analysis.

3.6. Comparison of PMMoV Levels

PMMoV in wastewater solids was also quantified to assess its potential use as a biomarker for data normalization. PMMoV levels differed significantly when quantified in wastewater solids compared to the ultrafiltration method ($p < 0.001$, ANOVA), with median values of 8.98×10^6 (interquartile range, 6.38×10^6–1.20×10^7) copies per 100 mL when using the ultrafiltration method, 1.71×10^6 (interquartile range, 1.52×10^6–2.58×10^6) copies per 100 mL when using 100 mg solids and 8.34×10^5 (interquartile range, 5.14×10^5–1.10×10^6) copies per 100 mL when using solids from 20 mL (Figure 3). PMMoV levels quantified using ultrafiltration were significantly higher compared to those quantified using solids ($p < 0.001$ in all pairwise comparisons with solids, Tukey HSD). Quantification of PMMoV using 100 mg of solids also resulted in significantly higher virus load versus using solids from 20 mL of wastewater ($p < 0.001$, Tukey HSD).

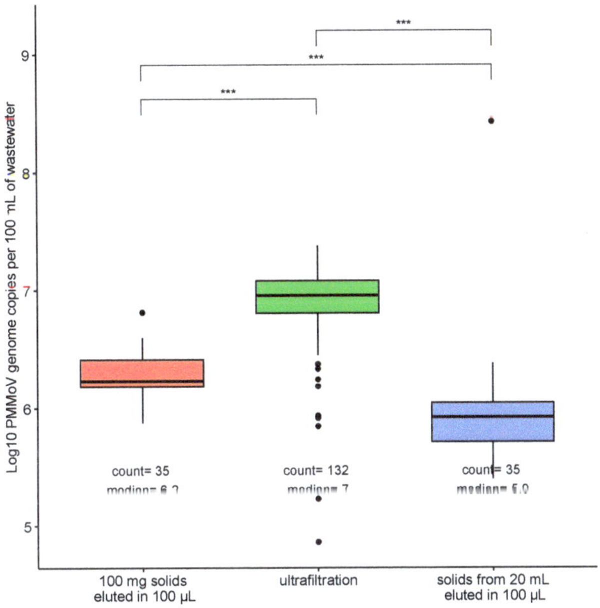

Figure 3. PMMoV levels in wastewater quantified by different methods. Differences in PMMoV Log 10 copies per 100 mL of wastewater were identified (ANOVA, $p < 0.001$). Results from Tukey HSD post hoc test are shown with horizontal bars and significance code: $p < 0.001$ '***'.

3.7. Cost and Turnaround Time Comparison between Moderate-Speed Centrifuged Solids-Based Method and Ultrafiltration

A comparison of reagent costs and turnaround times associated with our solids-based method using 20 mL of wastewater versus the ultrafiltration reference method is presented in Table 3. The average cost per sample is 90 CAD for the ultrafiltration method and 45 CAD for the solids-based method. The cost of the solids-based method represents about 50% of the costs of the ultrafiltration method and the difference is largely attributed to the cost of the Centricon™ filter alone.

Table 3. Comparison of consumables costs and turnaround times.

Step	Ultrafiltration	Moderate-Speed Centrifuged Solids [1]
Costs per sample (CAD):		
Virus extraction and plastic ware (pipette tips, microfuge tubes, etc.)	5	4
Virus concentration (Centricon™ filter)	44	0
RNA extraction reagents	6	6
RT-PCR and PCR reagents [2]	35	35
Total cost of consumables	90	45
Minimum turnaround time [3]:		
Separation of solids/liquid fractions	15 min	15 min
Virus extraction and concentration: pH adjustments and filtration through Centricon™ column	1 h 50 min	Not applicable
RNA extraction	1 h	1 h
RT-PCR	1 h 15 min	1 h 15 min
Total minimum turnaround time	4 h 20 min	2 h 30 min

[1] Using 20 mL of wastewater as input sample. [2] Includes five targets: SARS-CoV-2 genes E and N2, HuCoV 229E, PMMoV, salmon DNA. [3] Estimated time for running one sample and controls; times might increase depending on the number of samples.

Besides an initial centrifugation step to separate wastewater solids, the solids-based method does not require additional steps prior to RNA extraction. In contrast, the ultrafiltration method entails a set of extra steps: adjusting the pH of the sample to 9.6–10 prior to solids separation, then adjusting sample pH back to 7–7.5 after solids separation, and lastly, a virus concentration step using the Centricon™ filter. The extra steps also result in an increased turnaround time. The solids method is faster, taking 2 h 30 min to complete, whereas the ultrafiltration method requires no less than 4 h 20 min (Table 3).

4. Discussion

While solid-based assays for detection of SARS-CoV-2 in wastewater have been previously developed and generally claimed to outperform influent-based methods, only a few studies [17,23,26] provide first-hand data in terms of detection rate and PCR inhibition. In addition, a majority of solids-based methods are coupled with PEG precipitation (Table 4) and thus require ultracentrifugation, which might not be readily available in most laboratories. In this study, we successfully optimized a method for SARS-CoV-2 detection and quantification in post-grit influent wastewater solids that does not require special consumables or equipment. Our data indicate that, compared to a reference ultrafiltration method, the performance of our solids-based method was equivalent in terms of detection rate and with significantly lower rates of PCR inhibition. Lower PCR inhibition rates should translate to more accurate quantification of SARS-CoV-2 levels and higher virus detection rates. Accordingly, with negative controls in all RT-qPCR showing no false-positive results, we identified more positive samples using solids from 20 mL compared to ultrafiltration, even though the difference was not statistically significant. A limitation of our study is that the probability of type II errors cannot be calculated since the assay sensitivity, a key factor in sample size calculation for McNemar's test, for wastewater samples is not known and cannot be extrapolated from sensitivity determined by assay validation in clinical diagnostics [34–36]. On the other hand, the sample size in our study is considerably larger (median = 72, ranges from 65 to 139) than previous studies (median = 42, range 32 to 89) [17,23,25,26] and we included more wastewater treatment plants for sample variations.

Table 4. Previous studies describing solids-based methods for SARS-CoV-2 detection in wastewater.

Research Group	Solids Type	Method for Separation of Solids [1]	PEG Addition	PCR Inhibition Test
Peccia et al. [1]	Primary sludge	None, sludge used directly for RNA extraction	No	Samples (n = 5) were spiked with a target RNA and tested undiluted and diluted (5× and 25×). No significant differences in Ct values were identified between diluting versus no diluting samples
Balboa et al. [26]	Primary sludge, biological sludge, thickened sludge, digested sludge	None, sludge used directly for PEG precipitation	Yes	Used RT-qPCR Allplex™ internal control. Inhibition was identified in 2 out of 50 samples
Kitamura et al. [17]	Raw influent solids	1840× g for 30 min	No	Used PMMoV as indicator of RT-PCR inhibition. PMMoV levels were similar between WWTPs but were lower in solids compared to those in the supernatant/liquid fraction
D'Aoust et al. [2,18]	Influent post grit solids, primary clarified sludge	Gravity settled for 1h at 4 °C	Yes	Inhibition identified by comparing RT-qPCR and RT-ddPCR results and after testing diluted samples (2× and 5×)
Graham et al. [23] Wolfe et al. [37]	Primary settled solids	24,000× g for 30 min at 4 °C	Yes	Identified at high frequency after testing undiluted and diluted (10× and 50×) samples.
Kocamemi et al. [24]	Primary sludge, waste activated sludge	7471× g for 30 min at 4 °C	Yes	Did not report inhibition tests
Tomasino et al. [25]	Raw influent solids	4700× g for 30 min at 4 °C	No	Did not report inhibition tests

[1] Prior PEG, if PEG was added.

Several factors might be facilitating SARS-CoV-2 detection in wastewater solids. SARS-CoV-2 is secreted in feces with reported concentrations ranging between 5 × 10^3 to $10^{7.6}$ copies/mL [38] thus, wastewater solids represent a significant source or start point matrix where the virus would distribute to the aqueous fraction. In addition, coronaviruses appear to have an intrinsic preference to bind the wastewater solids. Based on experiments using spiked viruses, Ye et al. reported that about 26% of murine hepatitis virus (MHV, a member of the Coronavirus genus), and 22% of cystovirus ϕ6 (another enveloped virus), adsorbed to the solids fraction of wastewater compared to only 6% of non-enveloped bacteriophage MS2 (Levivirus) [21]. It has been hypothesized that the hydrophobic nature of the coronavirus envelope may be responsible for a reduced solubility of the virus in water, promoting virus adsorption to solids [39]. In support of this hypothesis, a study comparing seven different methods for SARS-CoV-2 detection in wastewater, reported that the only two methods that included the solids fraction outperformed those that used the liquid fraction [16].

The trade-off between sensitivity and PCR inhibition was a major hurdle in optimizing the adequate weight of the input sample for the solids-based method as procuring a higher number of viral RNA genome copies in the sample also leads to an increased content of PCR inhibitors. Sample dilution is a widely known strategy to decrease the effects of PCR inhibitors [8]. In our case, a concomitant modification of both, the amount of input sample as well as the RNA extract elution volume, allowed us to reduce the effect of PCR inhibitors present in wastewater to a level below that observed with the ultrafiltration reference method without compromising sensitivity. The LOD of the RT-qPCR assay used for virus quantification is two copies per reaction, which would be equivalent to 100 and 200 copies of SARS-CoV-2 per 100 mL of wastewater for the ultrafiltration reference method and our solids-20mL method, respectively.

The present study involved 12 different WWTPs and a total of 176 samples were used for the validation of our method for SARS-CoV-2 detection. This aspect provides robustness to our solids-based approach given that wastewater composition and solids content may vary drastically between WWTPs. Still, different WWTPs may present larger variations in solids content thus, the volume of input sample used for the solids-based method is a variable that may need to be adjusted on a case-by-case basis. Importantly, the performance of our method reported in this study is applicable to post-grit wastewater, and we strongly recommend further validation when implementing it for different wastewater matrices. We have identified that our method had lower performance when testing raw wastewater samples collected directly from manholes from long-term care facilities (data not shown), resulting in lower detection rates compared to our ultrafiltration reference method.

Our data support previous observations that the distribution of coronaviruses added for "spike-and-recovery" measurements to adjust for losses occurring throughout sample processing can be significantly lower in the solids-associated versus the aqueous fraction [20]. In our experiments, the human coronavirus strain 229E was recovered at a considerably lower rate when quantified in solids compared to the ultrafiltration method which includes the liquid fraction. The reasons for such a difference were unknown and beyond the focus of this study. It is possible that the relatively short period of time between spiking and removal of the aqueous fraction was not sufficient to allow 229E to penetrate, adsorb, and reach equilibrium in the solids phase to the same extent as SARS-CoV-2. In contrast to spiked HCoV 229E, a considerable fraction of SARS-CoV-2 is expected to remain associated with solids because the virus is secreted in stools from where it can progressively disperse into the liquid fraction of wastewater.

Interestingly, PMMoV was found at significantly lower levels of about one log difference when quantified using our solids-based method compared to the ultrafiltration reference method, providing further evidence that a majority of conventional methods for virus detection in wastewater are more efficient for non-enveloped viruses. The 100 mg solids matrix also had significantly more PMMoV genome copies compared to solids from 20 mL. This finding is compatible with our results for SARS-CoV-2 showing that 100 mg solids had a higher, although not statistically significant, detection rate than solids from 20 mL.

Lastly, we identified highly significant correlations between the quantitative levels of SARS-CoV-2 determined using moderate-speed centrifuged solids versus the ultrafiltration reference method. Our data, however, did not encompass a time frame long enough to allow comparison against indicators of COVID-case data (e.g., active cases, percentage of positive tests, hospitalization rates). SAR-CoV-2 levels in primary-clarified sludge have been reportedly used successfully by others to model disease dynamics at the community level [1,2], thus our post-grit influent solids-based method may also find application in tracking and predicting disease progression in populations.

5. Conclusions

Our study identified moderate-speed centrifuged solids of post-grit influent as a reliable sample matrix for the detection of SARS-CoV-2 in wastewater. In terms of detection rate, our optimized solids-based method performed comparably to the ultrafiltration reference method and has fewer processing steps, lower costs, and a shorter turnaround time. Our work contributes to the development of improved detection methods for enveloped viruses in wastewater and provides an optimal alternative in settings with limited resources and equipment. Future studies should be focused on studying the feasibility of using the data from this solids-based method to model and predict disease dynamics.

Supplementary Materials: The following are available online at https://www.mdpi.com/article/10.3390/w13162166/s1, Figure S1: Ct values of SARS-CoV-2 RT-qPCR detection by wet weight of wastewater solids, Figure S2: Yield of wet solids per 100 mL of post-grit influent by wastewater treatment plant, Table S1: Primers used in this study, Table S2: qPCR and solids (wet) weight data.

Author Contributions: Conceptualization, X.P.; methodology, J.Y., M.H., Y.Q., S.B., X.P.; validation, J.Y., M.H., S.B.; formal analysis, M.H.; investigation, J.Y., M.H., S.B.; resources, X.P.; data curation, M.H.; writing—original draft preparation, M.H.; writing—review and editing, M.H., R.M.-B., J.Y., S.H., B.L., X.P.; visualization, M.H., X.P.; supervision, X.P.; project administration, X.P.; funding acquisition, X.P., B.L. All authors have read and agreed to the published version of the manuscript.

Funding: This research was funded by research grants from the Canadian Institutes of Health Research, Alberta Innovates and Alberta Health [RES0051047].

Institutional Review Board Statement: Not applicable.

Informed Consent Statement: Not applicable.

Data Availability Statement: Data from qPCR tests and solids (wet) weight are available under Table S2.

Acknowledgments: We thank Emma Zwaigenbaum and Melissa Wilson for their technical support. We would also like to thank the Public Health Laboratory, Alberta Precision Laboratory for their assistance with samples transportation. We thank Nicholas Ashbolt (Faculty of Science and Engineering, Southern Cross University, Lismore, New South Wales, Australia) for his support in conceiving the project. We would also like to thank Norma Ruecker and staff from the City of Calgary, Jeff Charrois and Stephen Craik from EPCOR Water Services Inc., and all other participating wastewater treatment plants for the collection of sewage samples in this study.

Conflicts of Interest: The authors declare no conflict of interest.

References

1. Peccia, J.; Zulli, A.; Brackney, D.E.; Grubaugh, N.D.; Kaplan, E.H.; Casanovas-Massana, A.; Ko, A.I.; Malik, A.A.; Wang, D.; Wang, M.; et al. Measurement of SARS-CoV-2 RNA in Wastewater Tracks Community Infection Dynamics. *Nat. Biotechnol.* **2020**, *38*, 1164–1167. [CrossRef] [PubMed]
2. D'Aoust, P.M.; Graber, T.E.; Mercier, E.; Montpetit, D.; Alexandrov, I.; Neault, N.; Baig, A.T.; Mayne, J.; Zhang, X.; Alain, T.; et al. Catching a Resurgence: Increase in SARS-CoV-2 Viral RNA Identified in Wastewater 48 h before COVID-19 Clinical Tests and 96 h before Hospitalizations. *Sci. Total Environ.* **2021**, *770*, 145319. [CrossRef] [PubMed]
3. Naughton, C.; Roman, F.; Alvarado, A.; Tariqi, A.; Deeming, M.; Bibby, K.; Bivins, A.; Rose, J.; Medema, G.; Ahmed, W.; et al. COVID19 Poops Dashboard. Available online: https://www.covid19wbec.org/covidpoops19 (accessed on 27 July 2021).
4. Manor, Y.; Shulman, L.M.; Kaliner, E.; Hindiyeh, M.; Ram, D.; Sofer, D.; Moran-Gilad, J.; Lev, B.; Grotto, I.; Gamzu, R.; et al. Intensified Environmental Surveillance Supporting the Response to Wild Poliovirus Type 1 Silent Circulation in Israel, 2013. *Eurosurveillance* **2014**, *19*, 20708. [CrossRef]
5. Poyry, T.; Stenvik, M.; Hovi, T. Viruses in Sewage Waters during and after a Poliomyelitis Outbreak and Subsequent Nationwide Oral Poliovirus Vaccination Campaign in Finland. *Appl. Environ. Microbiol.* **1988**, *54*, 371–374. [CrossRef]
6. Fumian, T.M.; Fioretti, J.M.; Lun, J.H.; dos Santos, I.A.L.; White, P.A.; Miagostovich, M.P. Detection of Norovirus Epidemic Genotypes in Raw Sewage Using next Generation Sequencing. *Environ. Int.* **2019**, *123*, 282–291. [CrossRef] [PubMed]
7. Aw, T.G.; Gin, K.Y.H. Environmental Surveillance and Molecular Characterization of Human Enteric Viruses in Tropical Urban Wastewaters. *J. Appl. Microbiol.* **2010**, *109*, 716–730. [CrossRef]
8. Schrader, C.; Schielke, A.; Ellerbroek, L.; Johne, R. PCR Inhibitors—Occurrence, Properties and Removal. *J. Appl. Microbiol.* **2012**, *113*, 1014–1026. [CrossRef]
9. Sims, N.; Kasprzyk-Hordern, B. Future Perspectives of Wastewater-Based Epidemiology: Monitoring Infectious Disease Spread and Resistance to the Community Level. *Environ. Int.* **2020**, *139*, 105689. [CrossRef]
10. Ahmed, W.; Bivins, A.; Bertsch, P.M.; Bibby, K.; Gyawali, P.; Sherchan, S.P.; Simpson, S.L.; Thomas, K.V.; Verhagen, R.; Kitajima, M.; et al. Intraday Variability of Indicator and Pathogenic Viruses in 1-h and 24-h Composite Wastewater Samples: Implications for Wastewater-Based Epidemiology. *Environ. Res.* **2021**, *193*, 110531. [CrossRef] [PubMed]
11. Friedler, E.; Butler, D.; Alfiya, Y. Wastewater composition. In *Source Separation and Decentralization for Wastewater Management*; Larse, T., Udert, K., Lienert, J., Eds.; IWA Publishing: London, UK, 2013; ISBN 9781780401072.
12. La Rosa, G.; Bonadonna, L.; Lucentini, L.; Kenmoe, S.; Suffredini, E. Coronavirus in Water Environments: Occurrence, Persistence and Concentration Methods—A Scoping Review. *Water Res.* **2020**, *179*, 115899. [CrossRef] [PubMed]
13. Haramoto, E.; Katayama, H.; Ohgaki, S. Detection of Noroviruses in Tap Water in Japan by Means of a New Method for Concentrating Enteric Viruses in Large Volumes of Freshwater. *Appl. Environ. Microbiol.* **2004**, *70*, 2154–2160. [CrossRef]
14. Qiu, Y.; Lee, B.E.; Ruecker, N.J.; Neumann, N.; Ashbolt, N.; Pang, X. A One-Step Centrifugal Ultrafiltration Method to Concentrate Enteric Viruses from Wastewater. *J. Virol. Methods* **2016**, *237*, 150–153. [CrossRef] [PubMed]
15. Alygizakis, N.; Markou, A.N.; Rousis, N.I.; Galani, A.; Avgeris, M.; Adamopoulos, P.G.; Scorilas, A.; Lianidou, E.S.; Paraskevis, D.; Tsiodras, S.; et al. Analytical Methodologies for the Detection of SARS-CoV-2 in Wastewater: Protocols and Future Perspectives. *TrAC Trends Anal. Chem.* **2021**, *134*, 116125. [CrossRef]
16. Ahmed, W.; Bertsch, P.M.; Bivins, A.; Bibby, K.; Farkas, K.; Gathercole, A.; Haramoto, E.; Gyawali, P.; Korajkic, A.; McMinn, B.R.; et al. Comparison of Virus Concentration Methods for the RT-QPCR-Based Recovery of Murine Hepatitis Virus, a Surrogate for SARS-CoV-2 from Untreated Wastewater. *Sci. Total Environ.* **2020**, *739*, 139960. [CrossRef] [PubMed]
17. Kitamura, K.; Sadamasu, K.; Muramatsu, M.; Yoshida, H. Efficient Detection of SARS-CoV-2 RNA in the Solid Fraction of Wastewater. *Sci. Total Environ.* **2021**, *763*, 144587. [CrossRef]

18. D'Aoust, P.M.; Mercier, E.; Montpetit, D.; Jia, J.-J.; Alexandrov, I.; Neault, N.; Baig, A.T.; Mayne, J.; Zhang, X.; Alain, T.; et al. Quantitative Analysis of SARS-CoV-2 RNA from Wastewater Solids in Communities with Low COVID-19 Incidence and Prevalence. *Water Res.* **2021**, *188*, 116560. [CrossRef] [PubMed]
19. Pecson, B.M.; Darby, E.; Haas, C.N.; Amha, Y.M.; Bartolo, M.; Danielson, R.; Dearborn, Y.; di Giovanni, G.; Ferguson, C.; Fevig, S.; et al. Reproducibility and Sensitivity of 36 Methods to Quantify the SARS-CoV-2 Genetic Signal in Raw Wastewater: Findings from an Interlaboratory Methods Evaluation in the U.S. *Environ. Sci. Water Res. Technol.* **2021**, *7*, 504–520. [CrossRef]
20. Chik, A.H.S.; Glier, M.B.; Servos, M.; Mangat, C.S.; Pang, X.-L.; Qiu, Y.; D'Aoust, P.M.; Burnet, J.-B.; Delatolla, R.; Dorner, S.; et al. Comparison of Approaches to Quantify SARS-CoV-2 in Wastewater Using RT-QPCR: Results and Implications from a Collaborative Inter-Laboratory Study in Canada. *J. Environ. Sci.* **2021**, *107*, 218–229. [CrossRef]
21. Ye, Y.; Ellenberg, R.M.; Graham, K.E.; Wigginton, K.R. Survivability, Partitioning, and Recovery of Enveloped Viruses in Untreated Municipal Wastewater. *Environ. Sci. Technol.* **2016**, *50*, 5077–5085. [CrossRef]
22. Peccia, J.; Zulli, A.; Brackney, D.E.; Grubaugh, N.D.; Kaplan, E.H.; Casanovas-Massana, A.; Ko, A.I.; Malik, A.A.; Wang, D.; Wang, M.; et al. SARS-CoV-2 RNA Concentrations in Primary Municipal Sewage Sludge as a Leading Indicator of COVID-19 Outbreak Dynamics. *medRxiv* **2020**. [CrossRef]
23. Graham, K.E.; Loeb, S.K.; Wolfe, M.K.; Catoe, D.; Sinnott-Armstrong, N.; Kim, S.; Yamahara, K.M.; Sassoubre, L.M.; Mendoza Grijalva, L.M.; Roldan-Hernandez, L.; et al. SARS-CoV-2 RNA in Wastewater Settled Solids Is Associated with COVID-19 Cases in a Large Urban Sewershed. *Environ. Sci. Technol.* **2021**, *55*, 488–498. [CrossRef]
24. Kocamemi, B.A.; Kurt, H.; Sait, A.; Sarac, F.; Saatci, A.M.; Pakdemirli, B. SARS-CoV-2 Detection in Istanbul Wastewater Treatment Plant Sludges. *medRxiv* **2020**. [CrossRef]
25. Tomasino, M.P.; Semedo, M.; Vieira e Moreira, P.; Ferraz, E.; Rocha, A.; Carvalho, M.F.; Magalhães, C.; Mucha, A.P. SARS-CoV-2 RNA Detected in Urban Wastewater from Porto, Portugal: Method Optimization and Continuous 25-Week Monitoring. *Sci. Total Environ.* **2021**, *792*, 148467. [CrossRef]
26. Balboa, S.; Mauricio-Iglesias, M.; Rodriguez, S.; Martínez-Lamas, L.; Vasallo, F.J.; Regueiro, B.; Lema, J.M. The Fate of SARS-COV-2 in WWTPS Points out the Sludge Line as a Suitable Spot for Detection of COVID-19. *Sci. Total Environ.* **2021**, *772*, 145268. [CrossRef] [PubMed]
27. Markt, R.; Mayr, M.; Peer, E.; Wagner, A.O.; Lackner, N.; Insam, H. Detection and Stability of SARS-CoV-2 Fragments in Wastewater: Impact of Storage Temperature. *medRxiv* **2021**. [CrossRef]
28. Pabbaraju, K.; Wong, A.A.; Douesnard, M.; Ma, R.; Gill, K.; Dieu, P.; Fonseca, K.; Zelyas, N.; Tipples, G.A. Development and Validation of Reverse Transcriptase-PCR Assays for the Testing of SARS-CoV-2. *Off. J. Assoc. Med. Microbiol. Infect. Dis. Can.* **2021**, *6*, 16–22. [CrossRef]
29. Lu, X.; Wang, L.; Sakthivel, S.K.; Whitaker, B.; Murray, J.; Kamili, S.; Lynch, B.; Malapati, L.; Burke, S.A.; Harcourt, J.; et al. US CDC Real-Time Reverse Transcription PCR Panel for Detection of Severe Acute Respiratory Syndrome Coronavirus 2. *Emerg. Infect. Dis.* **2020**, *26*, 8. [CrossRef] [PubMed]
30. Vijgen, L.; Keyaerts, E.; Moës, E.; Maes, P.; Duson, G.; van Ranst, M. Development of One-Step, Real-Time, Quantitative Reverse Transcriptase PCR Assays for Absolute Quantitation of Human Coronaviruses OC43 and 229E. *J. Clin. Microbiol.* **2005**, *43*, 5452–5456. [CrossRef]
31. Haugland, R.A.; Siefring, S.C.; Wymer, L.J.; Brenner, K.P.; Dufour, A.P. Comparison of Enterococcus Measurements in Freshwater at Two Recreational Beaches by Quantitative Polymerase Chain Reaction and Membrane Filter Culture Analysis. *Water Res.* **2005**, *39*, 559–568. [CrossRef]
32. Zhang, T.; Breitbart, M.; Lee, W.H.; Run, J.Q.; Wei, C.L.; Soh, S.W.; Hibberd, M.L.; Liu, E.T.; Rohwer, F.; Ruan, Y. RNA Viral Community in Human Feces: Prevalence of Plant Pathogenic Viruses. *PLoS Biol.* **2006**, *4*, e3. [CrossRef]
33. Pembury Smith, M.Q.R.; Ruxton, G.D. Effective Use of the McNemar Test. *Behav. Ecol. Sociobiol.* **2020**, *74*, 133. [CrossRef]
34. Lachenbruch, P.A. On the Sample Size for Studies Based upon Mcnemar's Test. *Stat. Med.* **1992**, *11*, 1521–1525. [CrossRef]
35. Vuolo, M.; Uggen, C.; Lageson, S. Statistical Power in Experimental Audit Studies: Cautions and Calculations for Matched Tests With Nominal Outcomes. *Sociol. Methods Res.* **2015**, *45*, 260–303. [CrossRef]
36. Lu, Y.; Bean, J.A. On the Sample Size for One-Sided Equivalence of Sensitivities Based upon McNemar's Test. *Stat. Med.* **1995**, *14*, 1831–1839. [CrossRef] [PubMed]
37. Wolfe, M.K.; Archana, A.; Catoe, D.; Coffman, M.M.; Dorevich, S.; Graham, K.E.; Kim, S.; Grijalva, L.M.; Roldan-Hernandez, L.; Silverman, A.I.; et al. Scaling of SARS-CoV-2 RNA in Settled Solids from Multiple Wastewater Treatment Plants to Compare Incidence Rates of Laboratory-Confirmed COVID-19 in Their Sewersheds. *Environ. Sci. Technol. Lett.* **2021**, *8*, 398–404. [CrossRef]
38. Foladori, P.; Cutrupi, F.; Segata, N.; Manara, S.; Pinto, F.; Malpei, F.; Bruni, L.; la Rosa, G. SARS-CoV-2 from Faeces to Wastewater Treatment: What Do We Know? A Review. *Sci. Total Environ.* **2020**, *743*, 140444. [CrossRef] [PubMed]
39. Gundy, P.M.; Gerba, C.P.; Pepper, I.L. Survival of Coronaviruses in Water and Wastewater. *Food Environ. Virol.* **2008**, *1*, 10. [CrossRef]

water

MDPI

Article

Evaluation of Pre-Analytical and Analytical Methods for Detecting SARS-CoV-2 in Municipal Wastewater Samples in Northern Italy

Laura Pellegrinelli [1,*], Sara Castiglioni [2], Clementina E. Cocuzza [3], Barbara Bertasi [4], Valeria Primache [1], Silvia Schiarea [2], Giulia Salmoiraghi [2], Andrea Franzetti [5], Rosario Musumeci [3], Michela Tilola [4], Elisa Galuppini [4], Giorgio Bertanza [6], Marialuisa Callegari [7], Fabrizio Stefani [8], Andrea Turolla [9], Emanuela Ammoni [10], Danilo Cereda [10], Elena Pariani [1], Sandro Binda [1] and the WBE Study Group †

1 Department of Biomedical Sciences of Health, University of Milan, 20133 Milan, Italy; valeria.primache@unimi.it (V.P.); elena.pariani@unimi.it (E.P.); sandro.binda@unimi.it (S.B.)
2 Department of Environmental Sciences, Istituto di Ricerche Farmacologiche Mario Negri IRCCS, 20156 Milan, Italy; sara.castiglioni@marionegri.it (S.C.); silvia.schiarea@marionegri.it (S.S.); giulia.salmoiraghi@marionegri.it (G.S.)
3 Department of Medicine and Surgery, University of Milano-Bicocca, 20900 Monza, Italy; clementina.cocuzza@unimib.it (C.E.C.); rosario.musumeci@unimib.it (R.M.)
4 Istituto Zooprofilattico Sperimentale della Lombardia e dell'Emilia-Romagna "B. Ubertini", 25124 Brescia, Italy; barbara.bertasi@izsler.it (B.B.); michela.tilola@izsler.it (M.T.); elisa.galuppini@izsler.it (E.G.)
5 Department of Earth and Environmental Sciences, University of Milano-Bicocca, 20126 Milano, Italy; andrea.franzetti@unimib.it
6 Dipartimento di Ingegneria Civile, Architettura, Territorio, Ambiente e di Matematica, Università di Brescia, 25123 Brescia, Italy; giorgio.bertanza@unibs.it
7 Università Cattolica Sacro Cuore Dipartimento di Scienze e Tecnologie Alimentari per una Filiera Agro-Alimentare Sostenibile (DiSTAS), 29122 Piacenza, Italy; marialuisa.callegari@unicatt.it
8 Water Research Institute-National Research Council (IRSA-CNR), 20861 Brugherio, Italy; stefani@irsa.cnr.it
9 Department of Civil and Environmental Engineering (DICA)—Environmental Section, Politecnico di Milano, 20133 Milan, Italy; andrea.turolla@polimi.it
10 DG Welfare, Regione Lombardia, 20124 Milan, Italy; emanuela_ammoni@regione.lombardia.it (E.A.); danilo_cereda@regione.lombardia.it (D.C.)
* Correspondence: laura.pellegrinelli@unimi.it
† Members of the 'WBE Study Group' are listed in acknowledgments.

Citation: Pellegrinelli, L.; Castiglioni, S.; Cocuzza, C.E.; Bertasi, B.; Primache, V.; Schiarea, S.; Salmoiraghi, G.; Franzetti, A.; Musemeci, R.; Tilola, M.; et al. Evaluation of Pre-Analytical and Analytical Methods for Detecting SARS-CoV-2 in Municipal Wastewater Samples in Northern Italy. *Water* **2022**, *14*, 833. https://doi.org/10.3390/w14050833

Academic Editors: Guangming Jiang, Ryo Honda and Sudipti Arora

Received: 9 February 2022
Accepted: 4 March 2022
Published: 7 March 2022

Abstract: (1) Background: The surveillance of SARS-CoV-2 RNA in urban wastewaters allows one to monitor the presence of the virus in a population, including asymptomatic and symptomatic individuals, capturing the real circulation of this pathogen. The aim of this study was to evaluate the performance of different pre-analytical and analytical methods for identifying the presence of SARS-CoV-2 in untreated municipal wastewaters samples by conducting an inter-laboratory proficiency test. (2) Methods: three methods of concentration, namely, (A) Dextran and PEG-6000 two-phase separation, (B) PEG-8000 precipitation without a chloroform purification step and (C) PEG-8000 precipitation with a chloroform purification step were combined with three different protocols of RNA extraction by using commercial kits and were tested by using two primers/probe sets in three different master mixes. (3) Results: PEG-8000 precipitation without chloroform treatment showed the best performance in the SARS-CoV-2 recovery; no major differences were observed among the protocol of RNA extraction and the one-step real-time RT-PCR master mix kits. The highest analytic sensitivity was observed by using primers/probe sets targeting the N1/N3 fragments of SARS-CoV-2. (4) Conclusions: PEG-8000 precipitation in combination with real-time RT-PCR targeting the N gene (two fragments) was the best performing workflow for the detection of SARS-CoV-2 RNA in municipal wastewaters.

Keywords: environmental surveillance; urban wastewater; PEG-8000 precipitation; SARS-CoV-2

1. Introduction

Amid the pandemic of SARS-CoV-2 [1], communities have faced the rapid spread of the virus and its related disease—called COVID-19—affecting the testing capacity of public health systems and microbiological laboratories [2,3]. Strong evidence has shown the utility of viral RNA monitoring in municipal wastewater samples (sewage) for SARS-CoV-2 infection surveillance at a population-wide level—according to the wastewater-based epidemiology (WBE) approach [4–6]. Since SARS-CoV-2 is shed by feces in the early stage of infection and can cause asymptomatic infection in a large proportion of individuals, it is an ideal target for WBE. This strategy may allow to: (i) estimate the real prevalence of SARS-CoV-2 infection at a population level, (ii) monitor SARS-CoV-2 spread after the implementation of containment measures and restrictions and (iii) provide an early warning of virus re-introduction [4–6]. Moreover, surveillance of SARS-CoV-2 in sewage may provide timely indications on SARS-CoV-2 infection dynamics, overcoming the lag in monitoring exclusively COVID-19 symptoms and tests, since the onset of symptoms might be 2 weeks apart from viral infection [7–9]. Moreover, this approach of surveillance can overcome test availability and indications that can result under pressure during the surge of new outbreaks.

Several proof-of-principle studies on SARS-CoV-2 monitoring in municipal wastewater samples were designed and conducted through a number of different pre-analytical and analytical protocols, encompassing sewage concentration, RNA extraction and SARS-CoV-2 molecular detection, making it difficult to compare inter-laboratory results [5,9–11]. However, methods optimization and quality control are crucial for generating reliable public health information among countries and over time [4–6], as demonstrated in the surveillance of poliovirus in environmental wastewater samples in the framework of the global polio eradication initiative [12].

In Italy, a WBE network in the Lombardy Region (a region in Northern Italy accounting for nearly 10 million inhabitants) was recently established [13] in order to provide local support in SARS-CoV-2 infection surveillance in one of the Italian epidemic hot-spots. Different research institutions (co-authoring this work) in the Lombardy Region have collaborated to develop a common protocol of analysis by optimizing and standardizing the methods for the pre-analytical and analytical workflow in order to make results of inter-laboratory analysis comparable and applicable on a wider scale.

Initially, in order to evaluate the sensitivity and turn-around time of the different pre-analytical and analytical methods for detection of SARS-CoV-2 in municipal wastewater samples, an inter-laboratory proficiency test (PT) was carried out by the laboratories that participated in the WBE network, allowing researchers to identify the best-performing laboratory protocol to be included in the WBE network pipelines. The optimized protocol will be adopted for future regional and national surveillance studies in order to improve the quality and reproducibility of the results.

2. Materials and Methods

2.1. Generation of Wastewater Samples Stock

Two composite 24 h raw, untreated urban wastewater samples were collected at the inlet of two wastewater treatment plants in the Lombardy Region. The wastewater treatment plants are in a high-density urban setting in Milan, serving a population of nearly 1 million inhabitants each and receiving mainly municipal waste. Sampling was done in volume- or time-proportional mode, depending on the automatic sampler available. After the collection, samples were immediately processed for viral concentration or were stored at −80 °C until analysis.

The first sample was collected in March 2019, in Milan municipality, almost one year in advance of the COVID-19 pandemic onset and was considered as the blank negative control (NC); this sample was tested for the presence of SARS-CoV-2 RNA by carrying out real-time RT-PCR assays targeting the ORF-1ab and the N gene in triplicate in four different laboratories. This sample was analyzed following a preservation step at −80 °C.

The second sample, collected in December 2020, in Monza-Brianza municipality, was analyzed following a preservation step at −80 °C, and was split into two separate untreated wastewater aliquots: one was spiked with SARS-CoV-2 culture supernatant (SARS-CoV-2 viral load; 4.7×10^7 copies/mL; cycle threshold [Ct] 20) and was considered a positive control (PC); one was directly processed as an "unknown sample" in terms of the presence of SARS-CoV-2, but was expected to be weak positive. These sewage samples were then split into identical aliquots to be tested in parallel by the WBE Lombardy Network collaborating laboratories.

2.2. Pre-Analytical Process: Concentration of Sewage Samples

Untreated urban wastewater samples were processed using three different protocols for sample concentration:

(1) Dextran and polyethylene glycol-6000 (PEG) two-phase separation according to the 2003 WHO Guidelines for Environmental Surveillance of Poliovirus protocol [14], omitting the chloroform treatment to preserve the integrity of the SARS-CoV-2 envelope, as described, firstly, by La Rosa, G. et al. [15]. Briefly, 250 mL of wastewater sample was centrifuged for 30 min at 4500× *g* to pellet the wastewater solids, retaining the pellet for further processing. The clarified wastewater was mixed with dextran and PEG-6000 (19.8 mL of 22% dextran, 143.5 mL 29% PEG 6000, 17.5 mL 5 N NaCl); after a constant agitation for 30 min using a horizontal shaker, the mixture was left to stand overnight at 4 °C in a separation funnel. The bottom layer and the interphase were then collected drop-wise; this concentrate was added to the wastewater solids [14].

(2) PEG-8000 precipitation of 90-mL sewage, modified from Wu, F. et al. [16] and described, firstly, by Castiglioni, S. et al. [17], as follows:

A total of 80 mL of wastewater sample was centrifuged for 30 min at 4500× *g* and 4 °C without break to pellet the wastewater solids. Two aliquots of 40 mL of the clarified wastewater was mixed with 4 g PEG-8000 and 0.9 g sodium chloride (Carlo Erba, Milan, Italy) and were left in a shaker for 15 min at room temperature to dissolve the PEG-8000. Samples were centrifuged for 2 h at 12,000× *g* and 4 °C without break. After centrifugation, the supernatant was discarded and the tubes were returned to the centrifuge at 4 °C for a second centrifugation step at 12,000× *g* for 5 min. The pellet in each tube was suspended in 750 µL of Tryzol (Life Technologies, Monza and Brianza, Italy) and stored at −20 °C until RNA extraction.

(3) PEG-8000 precipitation of 250 mL of sewage, modified from Wu, F. et al. [16], as follows:

A 250 mL wastewater sample was centrifuged for 30 min at 1200× *g* and 4 °C with break to pellet the wastewater solids. Four aliquots of 50 mL of the clarified wastewater were mixed with 4 g PEG-8000 and 0.9 g sodium chloride (Carlo Erba, Milan, Italy) and were left in a shaker for 60 min at room temperature to dissolve the PEG-8000. Samples were centrifuged for 30 min at 10,000× *g* and 4 °C with break. After centrifugation, the supernatant was discarded and the tubes were returned to the centrifuge at 4 °C for a second centrifuge step at 10,000× *g* for 5 min. The pellet in each tube was suspended in 5 mL of PBS (Life Technologies, Monza and Brianza, Italy), treated with chloroform (1:4 *v*/*v*) and centrifuged for 10 min at 1000× *g* and 4 °C. The supernatant was stored at −20 °C until RNA extraction.

A UV treatment of samples (30 min) or a heat treatment (56 °C, 30 min) of the wastewater sample was included before all concentration processes to increase the safety for the laboratory personnel during sample manipulation.

2.3. Pre-Analytical Process: RNA Extraction from Concentrated Sewage Samples

RNA was extracted by means of two commercial kits according to manufacturer's instructions and by combining three different protocols, as follows:

(A) QIAamp MinElute Virus Spin Kit (QIAGEN, Hilden, Germany) with an input of 400 μL of sample and an elution volume of 60 μL, as previously described [17]. (B) NucliSens EasyMag (bioMerieux, Marcy-l'Étoile, France), with an input of 400 uL and 500 μL of sample and an elution volume of 100 μL, (C) NucliSens EasyMag (bioMerieux, Marcy-l'Étoile, France), with an input of 4 mL of sample and an elution volume of 100 μL.

2.4. Analytical Process: Real-Time RT-PCR Assays

The primer/probe sets used in this study targeted two different regions of the nucleocapsid (N) gene, namely, N1 and N3, as listed by the CDC (USA) (2020), and ORF-1b-nsp14, according to the methods described by La Rosa, G. et al. [15]. Three different one-step RT-PCR assays for SARS-CoV-2 were performed using: (1) AgPath-ID One-Step RT-PCR™ kit (Thermofisher Scientific, Waltham, MA, USA), (2) TaqMan™ Fast Virus 1-Step Master Mix (Thermofisher Scientific, Waltham, MA, USA), and (3) QScript XLT 1-Step RT-PCR ToughMix® (QuantaBio, Beverly, MA, USA). Primers and probes were obtained from Eurofins genomics (Eurofins Genomics Germany GmbH, Ebersberg, Germany).

To determine any potential contamination and/or inhibition, specific positive (EURM-019) and negative (DNAse/RNAse-free distilled water) controls were included in each real-time RT-PCR run. A sample was considered positive for SARS-CoV-2 when N1 or N3 or both viral targets showed a cycle threshold (Ct) $\leq$ 39. Real-time RT-PCR runs were performed by using the QuantStudio 5 Real-time RT-PCR system (thermofisher Scientific, Waltham, MA, USA), the ABI PRISM 7500 Sequence Detection System (Thermofisher Scientific, Henogen, Saudi Arabia) and the CFX96 BIo-Rad Detection System (Bio-Rad, Milan, Italy). All samples were tested in triplicate and in three different runs. Since all semi-quantitative assays were performed in triplicate, the reported Ct corresponded to the mean value of the three triplicates.

To minimize contamination risk, RNA extraction, molecular assays set-up and real-time RT-PCR runs were performed in separate rooms, according to good laboratory practice for molecular assays.

2.5. Pre-Analytical and Analytical Workflows

In the preliminary PT workflow, 8 different combinations of methods were run, as shown in Figure 1. Briefly, the three different methods of concentration, namely, (A) Dextran and PEG-6000 two-phase separation, (B) PEG-8000 precipitation without a chloroform purification step and (C) PEG-8000 precipitation with a chloroform purification step were combined with the three different protocols of RNA extraction by using commercial kits.

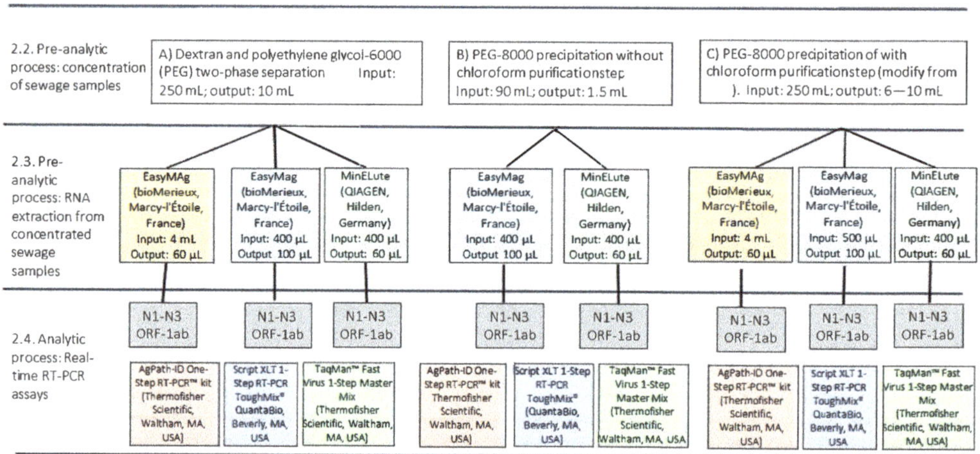

Figure 1. Flowchart of the pre-analytical and analytical workflow in this proficiency test [14,16,17].

2.6. Evaluation of SARS-CoV-2 Recovery Efficiency

The SARS-CoV-2 recovery efficiency of each replicate for each concentration method was calculated based upon the copies of RNA quantified by RT-PCR as follows:

$$\text{Recovery Efficienty (\%)} = \text{SARS-CoV-2 copies}/\mu\text{L recovered}/\text{SARS-CoV-2 copies}/\mu\text{L seeded} \quad (1)$$

$$\text{Recovery Efficiency}(\%) = \frac{\text{SARS-CoV-2 copies recovered}}{\text{SARS-CoV-2 copies seeded}} \quad (2)$$

For each concentration method, the mean and standard deviation were calculated.

2.7. Evaluation of SARS-CoV-2 RT-PCR Assays Efficiency

Once the best pre-analytical protocol in the tested workflows was assessed, it was implemented in all research laboratories involved in the WBE network in Lombardy. In order to evaluate the analytical processes, to explore SARS-CoV-2 RT-PCR assays performance and to calibrate RT-PCR methods, the standard curves were constructed using the SARS-CoV-2 Research Grade Test Reference Material (RGTM 10169) from the National Institute of Standards and Technology (NIST). It consists of a synthetic RNA fragment from the SARS-CoV-2 genome (Fragment 1—Total length: 3985 nt, SARS-CoV-2 sequence: 25,949–29,698,) with a concentration of approximately 5×10^6 copies/µL.

Evaluation of the analytical processes was conducted by comparing AgPath-ID One-Step RT-PCR™ kit (Thermofisher Scientific, Waltham, MA, USA), TaqMan™ Fast Virus 1-Step Master Mix (Thermofisher Scientific, Waltham, MA, USA) and QScript XLT 1-Step RT-PCR ToughMix® (QuantaBio, Beverly, MA, USA) efficiencies based on SARS CoV 2 standard curves generated for both the N1 and N3 target sequences, using the following amplification efficiency formula (Wong and Medrano, 2005):

$$\text{Efficiency} = [10(-1/\text{slope})] - 1 \quad (3)$$

2.8. Data Analysis

The QuantStudio 5 Real-time RT-PCR system (Thermofisher Scientific, Waltham, MA, USA), the ABI PRISM 7500 Sequence Detection System (Thermo Fisher Scientific, Henogen, Saudi Arabia) and the CFX96 Bio-Rad Detection System (Bio-Rad, Milan, Italy). were used to analyze all RT-PCR tests; data were collected and managed using Microsoft Excel (Microsoft Corp., Redmond, WA, USA). Samples with reduced fluorescence, as evident in the RT-PCR curves, were considered inhibited; samples with non-exponential multiplication were considered false positives.

The delta ct value (ΔCt) was calculated by comparing the mean value of the Ct of N1 recovered vs. seeded.

All samples with a Ct $\leq$ 39 cycles were considered positive.

The one-way analysis of variance (ANOVA) was used to determine whether there was a difference in SARS-CoV-2 recovery among the concentration methods tested.

3. Results

3.1. Performance of the Concentration Methods

Method A of concentration allowed researchers to recover nearly 10 mL of samples' concentrates from 250 mL of untreated sewage, with a turnaround time (TAT) of 14–16 h; method B allowed researchers to recover 1.5 mL of concentrated sample from 80 mL of untreated sewage with a TAT of 3–4 h; method C allowed researchers to recover from 6 to 10 mL of concentrated sample from 250 mL of untreated sewage with a TAT of 3–4 h. Overall, N1, N3 and ORF-1b-nsp14 were identified in sewage samples spiked with 2.5×10^9 copies/mL of SARS-CoV-2 by using all the evaluated workflows, with 100% of positive replicates (Table 1). Method B for wastewater concentration (the PEG-8000 precipitation without chloroform purification step) showed the best ΔCt values, which resulted in −1.9 by using the QIAamp MinElute Virus Spin Kit (QIAGEN, Hilden, Germany) and −0.6 by using the NucliSens EasyMag (bioMerieux, Marcy-l'Étoile, France) ($p > 0.05$) (Table 1). In all considered wastewater concentration methods, the Ct values of ORF-1b-nsp14 were shown to be statistically lower ($p < 0.005$) than those of N1 and N3, with a mean ΔCt between ORF-1b-nsp14 and N1–N3 of 4.71 (SD: ±1.56).

Considering the sewage "unknown sample" for the presence of SARS-CoV-2, the detection of N1, N3 and ORF-1b-nsp14 with 100% of positive replicates was identified only by using method B of concentration, with mean Ct values of N1 ranging from 33 (SD: ±0.4) by using the QIAamp MinElute Virus Spin Kit, QIAGEN to 34 (SD: ±0.2) by using NucliSens EasyMag, bioMerieux. The mean Ct values of ORF-1b-nsp14 of 37 (SD: ±0.3) were obtained by using the QIAamp MinElute Virus Spin Kit (QIAGEN, Hilden, Germany) and of 37 (SD: ±0.6) by using NucliSens EasyMag (bioMerieux, Marcy-l'Étoile, France). When the sewage sample was concentrated by using method A and C, ORF-1b-nsp14 always tested negative; the N1 positive replicates ranged from 17% (1/6) to 83% (5/6), with the Ct values ranging from 37.3 (SD: undeterminable) and 38.7 (SD: ±0.3) (Table 1); N3 positive replicates resulted in 67%, with the Ct values ranging from 36.8 (SD: ±0.5) and 37.9 (SD: ±0.1) (Table 1).

3.2. *SARS-CoV-2 Recovery Efficiency*

For the 500 mL untreated wastewater sample seeded with SARS-CoV-2, method B (i.e., PEG-8000 precipitation without chloroform purification step) provided the highest ($p < 0.001$) SARS-CoV-2 recovery of 76% by using the QIAamp MinElute Virus Spin Kit (QIAGEN, Hilden, Germany) and 31.4% by using NucliSens EasyMag (bioMerieux, Marcy-l'Étoile, France) (Table 1). The other concentration methods showed a SARS-CoV-2 recovery efficiency < 18% (Table 1) and were, thus, excluded from the workflow.

3.3. *Real-Time RT-PCR Efficiency*

The calculated efficiencies were significantly lower for the TaqMan™ Fast Virus 1-Step Master Mix (N1 = 85.2%, N3 = 90.8%) compared to the AgPath-ID One-Step RT-PCR™ kit (N1 = 98.4%, N3 = 98.2%) and to QScript XLT 1-Step RT-PCR ToughMix® (N1 = 98.8%, N3 = 99.5%) (Table 2, Figure 2).

Table 1. Results from the different workflows implemented in this inter-laboratory proficiency test (PT).

									Spiked Sample by SARS-CoV-2 Viral Load: 4.7 × 10^7 Copies/mL					Unknown Sample		
Volume of Sewage to be Concentrated	Methods of Concentration	Volume of Concentrating Sewage	RNA Extraction Kit	Extraction Input Elution Volume	RT-qPCR Kit	RT-PCR Instrument	Tournaround Time	Target	Positive Replicates	Mean Ct Value	SD Ct Value	ΔCt (Recoverded vs. Seeded)	Mean Recovery Efficiency (%)	Positive Replicates	Mean Ct Value	SD Ct Value
250 mL	(A) Dextran and PEG-6000 two-phase separation	nearly 10 mL	QIAamp MinElute Virus Spin Kit (QIAGEN)	400 µL 60 µL	AGPATH-ID ONE-STEP RT-PCR (Thermo Fisher)	Applied Biosystems 7500 Real-Time PCR System (Thermo Fisher)	16-h	N1	100%	29.2	0.62	7.2	12%	17%	37.27	/
								N3	100%	30.4	0.85			und	und	/
								ORF	100%	36.5	0.79			und	und	/
			NucliSens EasyMag (bioMerieux)	400 µL 100 µL	AGPATH-ID ONE-STEP RT-PCR (Thermo Fisher)		16-h	N1	100%	26.2	0.38	4.2		67%	37.9	0.45
								N3	100%	25.6	0.41			und	und	/
								ORF	100%	28.9	0.23			und	und	0.02
			NucliSens EasyMag (bioMerieux)	4 mL 60 µL	AGPATH-ID ONE-STEP RT-PCR (Thermo Fisher)			N1	100%	27.6	0.25	5.6		33%	38.2	0.02
								N3	100%	27	0.22			67%	37.6	0.2
								ORF	100%	30.4	0.17			und	und	/
80 mL	(B) PEG-8000 precipitation without chloroform purification step	1.5 mL	QIAamp MinElute Virus Spin Kit (QIAGEN)	400 µL 60 µL	AGPATH-ID ONE-STEP RT-PCR (Thermo Fisher)	Applied Biosystems 7500 Real-Time PCR System (Thermo Fisher)	8-h	N1	100%	20.1	0.13	−1.9	76%	100%	33.1	0.4
								N3	100%	20.5	0.20			100%	33.4	0.34
								ORF	100%	24.6	0.30			100%	37.5	0.75
			NucliSens EasyMag (bioMerieux)	500 µL 100 µL	AGPATH-ID ONE-STEP RT-PCR (Thermo Fisher)		8-h	N1	100%	21.4	0.38	0.6		100%	34.1	0.22
								N3	100%	19.7	0.31			100%	32.6	0.19
								ORF	100%	24.6	0.43			100%	37.2	0.34
250 mL	(C) PEG-8000 precipitation with chloroform purification step	6–10 mL	QIAamp MinElute Virus Spin Kit (QIAGEN)	400 µL 60 µL	QScript XLT 1-Step RT-PCR ToughMix (QuantaBio)	CFX96 BioRad real-time PCR System (Biorad)	8-h	N1	100%	25.6	1.27	3.6	18%	83%	37.7	1.69
								N3	100%	26.9	1.05			67%	36.8	0.46
								ORF	100%	32.8	1.08			und	und	/
			NucliSens EasyMag (bioMerieux)	500 µL 100 µL	QScript XLT 1-Step RT-PCR ToughMix (QuantaBio)		8-h	N1	100%	29.2	0.33	7.2		50%	38.7	0.3
								N3	100%	29.5	0.43			67%	37.4	0.01
								ORF	100%	32.5	0.57			und	und	/
			NucliSens EasyMag (bioMerieux)	4 mL 60 µL	QScript XLT 1-Step RT-PCR ToughMix (QuantaBio)		8-h	N1	100%	29.3	0.5	7.3	1%	17%	38.03	/
								N3	100%	26.1	0.38			67%	37.9	0.46
								ORF	100%	33.1	1.41			und	und	/

Table 2. Comparison of AgPath-ID One-Step RT-PCR, QScript XLT 1-Step RT-PCR ToughMix® and TaqMan™ Fast Virus 1-Step master mix efficiencies.

	Target	Standard Curve	R^2	Efficiency
AgPath-ID One-Step RT-PCR™ kit	N1	y = −3.3863x + 37.009	0.9984	98.4%
	N3	y = −3.3677x + 38.426	0.9982	98.2%
QScript XLT 1-Step RT-PCR ToughMix®	N1	y = −3.279x + 39.076	0.9988	98.8%
	N3	y = −3.3073x + 40.08	0.9995	99.5%
TaqMan™ Fast Virus 1-Step Master Mix	N1	y = −3.7356x + 41.786	0.9985	85.2%
	N3	y = −3.5652x + 38.426	0.9971	90.8%

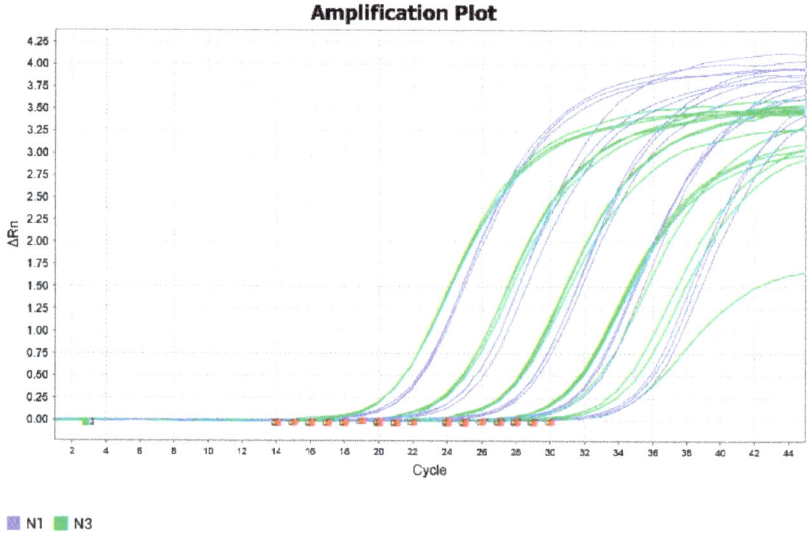

Figure 2. Real-Time RT-PCR amplification plot of the assay targeting N1 and N3 of the 10-fold dilutions of the SARS-CoV-2 Research Grade Test Reference Material (RGTM 10169) by using AgPath-ID One-Step RT-PCR™ kit.

4. Discussion

The development of a surveillance system through the implementation of the WBE approach may serve to monitor viral transmission in the community and to act as an early-warning system, allowing timely interventions to face new pathogens that may threaten human health [18]. The WBE approach has been used for decades to detect poliovirus and to track other viruses—able to persist long enough in untreated wastewater to allow reliable detection—in consideration that the sewage system can blend viral shedding variation among single individuals and over the course of their infection, into an average amount that represents the entire community under investigation [19–23]. Recently, several studies have reported the detection of SARS-CoV-2 RNA in wastewater samples worldwide [16,24–28] and have also shown a good correlation between the number of active COVID-19 cases and the SARS-CoV-2 RNA concentration in wastewater samples from different cities in Europe, Asia, the USA and Australia [16,24,25].

The sensitive detection of SARS-CoV-2 RNA in wastewater and, thus, the identification of SARS-CoV-2 infections within a community, depends on both the wastewater concentration pre-analytical phase and the molecular methods employed for the analysis, which are often different and lack standardization. Considering the pre-analytical process, SARS-CoV-2 concentration methods are particularly important because the concentration of this virus in wastewater samples is expected to be low at the onset or at the offset of the COVID-19 epidemic curve [16,17,24,25]; thus, the concentration methods must be sensitive enough to detect a very low concentration of SARS-CoV-2 in an environmental matrix to provide an effective early warning system and to track in a real-time manner the introduction of SARS-CoV-2 in a community.

Nowadays, a number of virus concentration methods have been developed for the detection of enteric viruses in water and wastewater matrices [4,18]. In this study, nine different workflows, including pre-analytical and analytical processes, were evaluated to then be implemented in the WBE of the Lombardy Region; these included three different methods of concentration, three different protocols of RNA extraction and three different one step real-time RT-PCR reagents. In this study, the method showing the best performance in the recovery of SARS-CoV-2, from both mock and unseeded samples, was that carried out by using PEG-8000 precipitation without chloroform treatment. In particular, this method allowed for a better recovery efficiency of SARS-CoV-2 when compared to Dextran and polyethylene glycol-6000 (PEG) two-phase separation, in contrast to other Italian preliminary results [15]. PEG-8000 precipitation without chloroform has also shown a good performance in concentrating SARS-CoV-2 from wastewater matrices in other published studies [11,29,30]. In the study from Ahmed, W. et al., the mean $\pm$ SD of the recovery of murine hepatitis virus (as a proxy of SARS-CoV-2) was shown to be 44.0% $\pm$ 27.7, similar to that observed in our study, where the recovery of spiked SARS-CoV-2 ranged from 31.4% to 76% by using PEG-8000 precipitation [30]. The only equipment needed to carry out PEG-8000 precipitation is a centrifuge that reaches up to 12,000$\times$ g, thus, resulting in a relatively simple and inexpensive protocol; moreover, it allows one to process larger volumes (e.g., 1 L) of wastewater and to concentrate SARS-CoV-2 from both the solid and the liquid phases, as well as being non-time consuming (3–4 h). On the other hand, this method requires handling of hazardous chemicals (such as Tryzol) that, however, could be replaced by elution in phosphate-buffered saline (PBS).

In respect to the evaluation of the three different protocols of RNA extraction by commercially available kits, no major differences were identified, as observed elsewhere [4,31]. Regarding SARS-CoV-2 real-time RT-PCR assays efficiency, in combination with all evaluated concentration methods, we observed that the Ct values for ORF-1b-nsp14 were statistically higher than those for N1 and N3, with a mean ΔCt between ORF-1b-nsp14 and N1-N3 of nearly 5, meaning a loss of sensitivity of nearly 2 Log when using ORF-1b-nsp14 instead of N viral targets, as also described in other studies that investigated the analytical sensitivity and efficiency of different SARS-CoV-2 real-time RT-PCR primer-probe sets [31,32].

Recently, the Water Research Foundation (WRF) released a question survey via social media to collect information on the development of methods for the detection of genes that indicate the presence of SARS-CoV-2 in wastewater samples [33]; feedback was obtained by 35 countries, with results showing that the concentration methods most frequently used were the PEG-8000 precipitation, followed by nucleic acid extraction and assay for primarily nucleocapsid gene targets (N1, N2, and/or N3) [33]. These results from WRF are in line with the output of our study.

A limitation of this study is that there was a limited number of replicates of the tested methods due to the time required for processing and restrictions on people and laboratory spaces during the pandemic, which is when this experimental work was carried out.

5. Conclusions

In conclusion, a new pre-analytical and analytical workflow to detect SARS-CoV-2 from wastewater samples was implemented in the framework of the WBE laboratories' network in the Lombardy Region.

The main innovation of this surveillance approach relies on the fact that it can overcome the testing availability, rates and indications and that it can capture the viral spread from symptomatic and asymptomatic individuals, offering a comprehensive and cost-effective solution for SARS-CoV-2 surveillance and providing a strong and independent signal of how much the virus is circulating in a given community. All these aspects make the WBE an innovative real-time cost-effective tool for community-based surveillance that can also be used for other emerging pathogens of concern for human health, to track outbreaks and guide public health interventions of prevention and control.

Author Contributions: Conceptualization, L.P., S.C., C.E.C., B.B., E.P. and S.B.; data curation, R.M., E.G., G.B., M.C. and A.T.; formal analysis, L.P., S.C., A.F. and A.T.; investigation, V.P., S.S., G.S., R.M., M.T. and F.S.; methodology, V.P., A.F., M.T., F.S. and E.P.; project administration, S.B.; supervision, L.P., E.P. and S.B.; validation, A.F., R.M. and E.P.; visualization, V.P., S.S., G.S., E.G., M.C. and F.S.; writing—original draft, L.P., S.C. and S.B.; writing—review and editing, C.E.C., B.B., E.A., D.C. and E.P. All authors have read and agreed to the published version of the manuscript.

Funding: This research received no external funding.

Institutional Review Board Statement: Not relevant for this study.

Informed Consent Statement: Not applicable.

Data Availability Statement: Not applicable.

Acknowledgments: The authors are grateful to WWTP personnel for sample collection and technical support with information for data analysis. In particular, we acknowledge the valid assistance of Uniacque S.p.A, Società SAL srl, A2A Ciclo Idrico S.p.A., Metropolitana Milanese S.p.A., Padania Acque S.p.A. Members of the 'WBE Study Group' involved in this study were: Cristina Galli, Laura Bubba and Arlinda Seiti, Department of Biomedical Sciences of Health, University of Milan, Milan, Italy; Ettore Zuccato, Department of Environmental Sciences, Istituto di Ricerche Farmacologiche Mario Negri IRCCS; Marianna Martinelli, Claudio Molteni and Chiara Giubbi, Department of Medicine and Surgery, University of Milano-Bicocca; Francesca Malpei, Manuela Antonelli and Arianna Azzelino, Department of Civil and Environmental Engineering (DICA)—Environmental Section, Politecnico di Milano.

Conflicts of Interest: The authors declare no conflict of interest.

References

1. Zhu, N.; Zhang, D.; Wang, W.; Li, X.; Yang, B.; Song, J.; Zhao, X.; Huang, B.; Shi, W.; Lu, R.; et al. A Novel Coronavirus from Patients with Pneumonia in China, 2019. *N. Engl. J. Med.* **2020**, *382*, 727–733. [CrossRef] [PubMed]
2. Vandenberg, O.; Martiny, D.; Rochas, O.; van Belkum, A.; Kozlakidis, Z. Considerations for diagnostic COVID-19 tests. *Nat. Rev. Microbiol.* **2021**, *19*, 171–183. [CrossRef] [PubMed]
3. Durant, T.J.S.; Peaper, D.R.; Ferguson, D.; Schulz, W.L. Impact of COVID-19 Pandemic on Laboratory Utilization. *J. Appl. Lab. Med.* **2020**, *5*, 1194–1205. [CrossRef] [PubMed]
4. Ahmed, W.; Bivins, A.; Bertsch, P.M.; Bibby, K.; Choi, P.M.; Farkas, K.; Gyawali, P.; Hamilton, K.A.; Haramoto, E.; Kitajima, M.; et al. Surveillance of SARS-CoV-2 RNA in wastewater: Methods optimisation and quality control are crucial for generating reliable public health information. *Curr. Opin. Environ. Sci. Health* **2020**, *17*, 82–93. [CrossRef]
5. Michael-Kordatou, I.; Karaolia, P.; Fatta-Kassinos, D. Sewage analysis as a tool for the COVID-19 pandemic response and management: The urgent need for optimised protocols for SARS-CoV-2 detection and quantification. *J. Environ. Chem. Eng.* **2020**, *8*, 104306. [CrossRef]
6. Medema, G.; Been, F.; Heijnen, L.; Petterson, S. Implementation of environmental surveillance for SARS-CoV-2 virus to support public health decisions: Opportunities and challenges. *Curr. Opin. Environ. Sci. Health* **2020**, *17*, 49–71. [CrossRef]
7. Peccia, J.; Zulli, A.; Brackney, D.E.; Grubaugh, N.D.; Kaplan, E.H.; Casanovas-Massana, A.; Ko, A.I.; Malik, A.A.; Wang, D.; Wang, M.; et al. Measurement of SARS-CoV-2 RNA in wastewater tracks community infection dynamics. *Nat. Biotechnol.* **2020**, *38*, 1164–1167. [CrossRef]
8. Larsen, D.A.; Wigginton, K.R. Tracking COVID-19 with wastewater. *Nat. Biotechnol.* **2020**, *38*, 1151–1153. [CrossRef]
9. Agrawal, S.; Orschler, L.; Lackner, S. Long-term monitoring of SARS-CoV-2 RNA in wastewater of the Frankfurt metropolitan area in Southern Germany. *Sci. Rep.* **2021**, *11*, 5372. [CrossRef]
10. Alygizakis, N.; Markou, A.N.; Rousis, N.I.; Galani, A.; Avgeris, M.; Adamopoulos, P.G.; Scorilas, A.; Lianidou, E.S.; Paraskevis, D.; Tsiodras, S.; et al. Analytical methodologies for the detection of SARS-CoV-2 in wastewater: Protocols and future perspectives. *Trends Anal. Chem.* **2021**, *134*, 116125. [CrossRef]
11. Philo, S.E.; Keim, E.K.; Swanstrom, R.; Ong, A.Q.; Burnor, E.A.; Kossik, A.L.; Harrison, J.C.; Demeke, B.A.; Zhou, N.A.; Beck, N.K.; et al. A comparison of SARS-CoV-2 wastewater concentration methods for environmental surveillance. *Sci. Total Environ.* **2021**, *760*, 144215. [CrossRef] [PubMed]
12. Word Health Organization (WHO). Contributions of the Polio Network to COVID-19 Response. Available online: https://apps.who.int/iris/bitstream/handle/10665/336261/9789240011533-eng.pdf?sequence=1&isAllowed=y (accessed on 5 March 2022).

13. Regione Lombardia. Regione Lombardia, Delibera N.4127—Accordo di Collaborazione con il Ministero Della Salute-Centro Nazionale per la Prevenzione e il Controllo delle Malattie e Convenzioni con l'Istituto Superiore di Sanità e l'ATS Milano Città Metropolitana per la Realizzazione del Progetto CCM 2020—Area Progettuale "Epidemiologia delle Acque Reflue: Implementazione del Sistema di Sorveglianza per L'identificazione Precoce di Agenti Patogeni, con Particolare Riferimento al SARS-CoV2". Available online: https://www.dati.lombardia.it/api/views/q639-trxu/rows.xml?accessType=DOWNLOAD (accessed on 5 March 2022).
14. Word Health Organization (WHO). WHO Guidelines for Environmental Surveillance of Poliovirus Protocol—WHO/V&B/03.03. Available online: https://polioeradication.org/wp-content/uploads/2016/07/WHO_V-B_03.03_eng.pdf (accessed on 5 March 2022).
15. la Rosa, G.; Iaconelli, M.; Mancini, P.; Ferraro, G.B.; Veneri, C.; Bonadonna, L.; Lucentini, L.; Suffredini, E. First detection of SARS-CoV-2 in untreated wastewaters in Italy. *Sci. Total Environ.* **2020**, *736*, 139652. [CrossRef] [PubMed]
16. Wu, F.; Xiao, A.; Zhang, J.; Moniz, K.; Endo, N.; Armas, F.; Bushman, M.; Chai, P.R.; Duvallet, C.; Erickson, T.B.; et al. Wastewater surveillance of SARS-CoV-2 across 40 U.S. states from February to June 2020. *Water Res.* **2021**, *202*, 117400. [CrossRef] [PubMed]
17. Castiglioni, S.; Schiarea, S.; Pellegrinelli, L.; Primache, V.; Galli, C.; Bubba, L.; Mancinelli, F.; Marinelli, M.; Cereda, D.; Ammoni, E.; et al. SARS-CoV-2 RNA in urban wastewater samples to monitor the COVID-19 pandemic in Lombardy, Italy (March-June 2020). *Sci. Total Environ.* **2022**, *806 Pt 4*, 150816. [CrossRef]
18. Haramoto, E.; Kitajima, M.; Hata, A.; Torrey, J.R.; Masago, Y.; Sano, D.; Katayama, H. A review on recent progress in the detection methods and prevalence of human enteric viruses in water. *Water Res.* **2018**, *135*, 168–186. [CrossRef]
19. Bisseux, M.; Colombet, J.; Mirand, A.; Roque-Afonso, A.-M.; Abravanel, F.; Izopet, J.; Archimbaud, C.; Peigue-Lafeuille, H.; Debroas, D.; Bailly, J.-L.; et al. Monitoring human enteric viruses in wastewater and relevance to infections encountered in the clinical setting: A one-year experiment in central France, 2014 to 2015. *Eurosurveillance* **2018**, *23*, 17-00237. [CrossRef]
20. Pellegrinelli, L.; Bubba, L.; Primache, V.; Pariani, E.; Battistone, A.; Delogu, R.; Fiore, S.; Binda, S. Surveillance of poliomyelitis in Northern Italy: Results of acute flaccid paralysis surveillance and environmental surveillance, 2012–2015. *Hum. Vaccines Immunother.* **2016**, *13*, 332–338. [CrossRef]
21. Hellmér, M.; Paxéus, N.; Magnius, L.O.; Enache, L.; Arnholm, B.; Johansson, A.M.; Bergström, T.; Norder, H. Detection of Pathogenic Viruses in Sewage Provided Early Warnings of Hepatitis A Virus and Norovirus Outbreaks. *Appl. Environ. Microbiol.* **2014**, *80*, 6771–6781. [CrossRef]
22. Battistone, A.; Buttinelli, G.; Fiore, S.; Amato, C.; Bonomo, P.; Patti, A.M.; Vulcano, A.; Barbi, M.; Binda, S.; Pellegrinelli, L.; et al. Sporadic Isolation of Sabin-Like Polioviruses and High-Level Detection of Non-Polio Enteroviruses during Sewage Surveillance in Seven Italian Cities, after Several Years of Inactivated Poliovirus Vaccination. *Appl. Environ. Microbiol.* **2014**, *80*, 4491–4501. [CrossRef]
23. Pellegrinelli, L.; Galli, C.; Binda, S.; Primache, V.; Tagliacarne, C.; Pizza, F.; Mazzini, R.; Pariani, E.; Romanò, L. Molecular Characterization and Phylogenetic Analysis of Enteroviruses and Hepatitis A Viruses in Sewage Samples, Northern Italy, 2016. *Food Environ. Virol.* **2019**, *11*, 393–399. [CrossRef]
24. Ahmed, W.; Angel, N.; Edson, J.; Bibby, K.; Bivins, A.; O'Brien, J.W.; Choi, P.M.; Kitajima, M.; Simpson, S.L.; Li, J.; et al. First confirmed detection of SARS-CoV-2 in untreated wastewater in Australia: A proof of concept for the wastewater surveillance of COVID-19 in the community. *Sci. Total Environ.* **2020**, *728*, 138764. [CrossRef] [PubMed]
25. Randazzo, W.; Truchado, P.; Cuevas-Ferrando, E.; Simón, P.; Allende, A.; Sánchez, G. SARS-CoV-2 RNA in wastewater anticipated COVID-19 occurrence in a low prevalence area. *Water Res.* **2020**, *181*, 115942. [CrossRef] [PubMed]
26. Haramoto, E.; Malla, B.; Thakali, O.; Kitajima, M. First Environmental Surveillance for the Presence of SARS-CoV-2 RNA in Wastewater and River Water in Japan. *Sci. Total Environ.* **2020**, *737*, 140405. [CrossRef] [PubMed]
27. Sherchan, S.P.; Shahin, S.; Ward, L.M.; Tandukar, S.; Aw, T.G.; Schmitz, B.; Ahmed, W.; Kitajima, M. First detection of SARS-CoV-2 RNA in wastewater in North America: A study in Louisiana, USA. *Sci. Total Environ.* **2020**, *743*, 140621. [CrossRef] [PubMed]
28. Westhaus, S.; Weber, F.-A.; Schiwy, S.; Linnemann, V.; Brinkmann, M.; Widera, M.; Greve, C.; Janke, A.; Hollert, H.; Wintgens, T.; et al. Detection of SARS-CoV-2 in raw and treated wastewater in Germany—Suitability for COVID-19 surveillance and potential transmission risks. *Sci. Total Environ.* **2021**, *751*, 141750. [CrossRef]
29. Torii, S.; Oishi, W.; Zhu, Y.; Thakali, O.; Malla, B.; Yu, Z.; Zhao, B.; Arakawa, C.; Kitajima, M.; Hata, A.; et al. Comparison of five polyethylene glycol precipitation procedures for the RT-qPCR based recovery of murine hepatitis virus, bacteriophage phi6, and pepper mild mottle virus as a surrogate for SARS-CoV-2 from wastewater. *Sci. Total Environ.* **2021**, *807*, 150722. [CrossRef]
30. Ahmed, W.; Bertsch, P.M.; Bivins, A.; Bibby, K.; Farkas, K.; Gathercole, A.; Haramoto, E.; Gyawali, P.; Korajkic, A.; McMinn, B.R.; et al. Comparison of virus concentration methods for the RT-qPCR-based recovery of murine hepatitis virus, a surrogate for SARS-CoV-2 from untreated wastewater. *Sci. Total Environ.* **2020**, *739*, 139960. [CrossRef]
31. Lázaro Perona, F.; Rodriguez Antolín, C.; Alguacil Guillén, M.; Gutiérrez-Arroyo, A.; Mingorance, J.; García-Rodriguez, J.; SARS-CoV-2 Working Group. Evaluation of two automated low-cost RNA extraction protocols for SARS-CoV-2 detection. *PLoS ONE* **2021**, *16*, e0246302. [CrossRef]

32. Vogels, C.B.F.; Brito, A.F.; Wyllie, A.L.; Fauver, J.R.; Ott, I.M.; Kalinich, C.C.; Petrone, M.E.; Casanovas-Massana, A.; Muenker, M.C.; Moore, A.J.; et al. Analytical sensitivity and efficiency comparisons of SARS-CoV-2 RT-qPCR primer-probe sets. *Nat. Microbiol.* **2020**, *5*, 1299–1305. [CrossRef]
33. Zhou, N.A.; Tharpe, C.; Meschke, J.S.; Ferguson, C.M. Survey of rapid development of environmental surveillance methods for SARS-CoV-2 detection in wastewater. *Sci. Total Environ.* **2021**, *769*, 144852. [CrossRef]

water

MDPI

Communication

The Reduction of SARS-CoV-2 RNA Concentration in the Presence of Sewer Biofilms

Shuxin Zhang [1], Elipsha Sharma [1], Ananda Tiwari [2], Yan Chen [1], Samendra P. Sherchan [3,4], Shuhong Gao [5], Xu Zhou [5], Jiahua Shi [6] and Guangming Jiang [1,6,*]

1 School of Civil, Mining and Environmental Engineering, University of Wollongong, Wollongong 2520, Australia; sz877@uowmail.edu.au (S.Z.); es788@uowmail.edu.au (E.S.); yc742@uowmail.edu.au (Y.C.)
2 Finnish Institute for Health and Welfare, 00271 Helsinki, Finland; ananda.tiwari@thl.fi
3 Department of Environmental Health Sciences, Tulane University, New Orleans, LA 70118, USA; samendra.sherchan@morgan.edu
4 Bioenvironmental Science Program, Morgan State University, Baltimore, MD 21251, USA
5 State Key Laboratory of Urban Water Resource and Environment, School of Civil & Environmental Engineering, Harbin Institute of Technology, Shenzhen 518055, China; gaoshuhong@hit.edu.cn (S.G.); zhouxu@hit.edu.cn (X.Z.)
6 Illawarra Health and Medical Research Institute (IHMRI), University of Wollongong, Wollongong 2520, Australia; jiahua@uow.edu.au
* Correspondence: gjiang@uow.edu.au; Tel.: +61-02-4221-3792

Abstract: Wastewater surveillance has been widely used to track the prevalence of SARS-CoV-2 in communities. Although some studies have investigated the decay of SARS-CoV-2 RNA in wastewater, understanding about its fate during wastewater transport in real sewers is still limited. This study aims to assess the impact of sewer biofilms on the dynamics of SARS-CoV-2 RNA concentration in naturally contaminated real wastewater (raw influent wastewater without extra SARS-CoV-2 virus/gene seeding) using a simulated laboratory-scale sewer system. The results indicated that, with the sewer biofilms, a 90% concentration reduction of the SARS-CoV-2 RNA was observed within 2 h both in wastewater of gravity (GS, gravity-driven sewers) and rising main (RM, pressurized sewers) sewer reactors. In contrast, the 90% reduction time was 8–26 h in control reactors without biofilms. The concentration reduction of SARS-CoV-2 RNA in wastewater was significantly more in the presence of sewer biofilms. In addition, an accumulation of *c.a.* 260 and 110 genome copies/cm^2 of the SARS-CoV-2 E gene was observed in the sewer biofilm samples from RM and GS reactors within 12 h, respectively. These results confirmed that the in-sewer concentration reduction of SARS-CoV-2 RNA in wastewater was likely caused by the partition to sewer biofilms. The need to investigate the in-sewer dynamic of SARS-CoV-2 RNA, such as the variation of RNA concentration in influent wastewater caused by biofilm attachment and detachment, was highlighted by the significantly enhanced reduction rate of SARS-CoV-2 RNA in wastewater of sewer biofilm reactors and the accumulation of SARS-CoV-2 RNA in sewer biofilms. Further research should be conducted to investigate the in-sewer transportation of SARS-CoV-2 and their RNA and evaluate the role of sewer biofilms in leading to underestimates of COVID-19 prevalence in communities.

Keywords: SARS-CoV-2; sewer; wastewater; biofilm; decay; adsorption

check for updates

Citation: Zhang, S.; Sharma, E.; Tiwari, A.; Chen, Y.; Sherchan, S.P.; Gao, S.; Zhou, X.; Shi, J.; Jiang, G. The Reduction of SARS-CoV-2 RNA Concentration in the Presence of Sewer Biofilms. *Water* **2023**, *15*, 2132. https://doi.org/10.3390/w15112132

Academic Editor: Jesus Gonzalez-Lopez

Received: 17 April 2023
Revised: 30 May 2023
Accepted: 1 June 2023
Published: 4 June 2023

1. Introduction

Since the first COVID-19 outbreak in 2019, the RNA of severe acute respiratory syndrome coronavirus 2 (SARS-CoV-2) was detected in the wastewater of many countries. Wastewater-based epidemiology (WBE) for COVID-19 monitoring attracted unprecedented attention and has been employed worldwide [1–4]. Based on the SARS-CoV-2 wastewater surveillance data from around the world, several modelling and back-estimation studies were carried out to estimate COVID-19 case numbers in communities [5–8]. Data-driven black-box models were developed for COVID-19 prevalence calculation, due to unknowns like the transportation and decay of SARS-CoV-2 RNA in sewers [8–12].

The correction of SARS-CoV-2 RNA concentration in influent wastewater includes several aspects such as the evaluation of virus/RNA loss caused by the virus decay in wastewater and the potential virus attachment onto sewer biofilms during in-sewer transportation. Several previous studies reported that, at room temperature (20~25 °C), it takes above three days for one log decay of SARS-CoV-2 RNA concentration in untreated wastewater [13,14]. At 4 °C, the one log decay of SARS-CoV-2 RNA concentration is more than 27 days [13,15]. In comparison to the wastewater residence time (i.e., <10 h for 95% of wastewater treatment plants (WWTPs)) [16,17] and temperature (i.e., 75% of the world's global wastewater were between 6.9–34.4 °C) [18] in sewers, the reported SARS-CoV-2 RNA persistence implies no need of significant correction for its decay in typical sewers for accurate WBE applications unless the sewer has long hydraulic retention time (HRT) or abnormal high temperature [19]. However, other factors that might induce RNA loss in the wastewater phase during the in-sewer transportation are unknown.

Based on the flow regimes, sewer systems could be divided into two main types, including the rising mains (RM) and the gravity (GS) sewers. Rising main pipes lift wastewater to higher elevations and run under anaerobic conditions because the pipes are full of wastewater. In contrast, gravity pipes transport wastewater to lower elevations by gravity and are partially filled with wastewater, thus containing both aerobic and anaerobic conditions [20]. Studies have confirmed that different sewer conditions affect the stability of various health and diet biomarkers, illicit drugs and pharmaceutical compounds [5,21–25]. Our recent study shows that the decay of infectivity of coronaviruses including feline infectious peritonitis virus (FIPV) and infectious human coronavirus 229E (HCoV-229E) was enhanced in sewer conditions as compared to wastewater [26]. In addition, another recent study reported that, during the 4-week study period of high COVID-19 prevalence in a community, an accumulation of $\sim$700 genome copies/cm^2 was observed in simulated sewer biofilms [10]. Therefore, in comparison to the low decay rate of SARS-CoV-2 RNA in wastewater, the effect of sewer biofilms (and sediments when present in gravity sewers) might be a higher contributing factor for the reduction of SARS-CoV-2 RNA concentration in influent wastewater.

In this study, the in-sewer dynamics of SARS-CoV-2 RNA concentrations, including N and E genes, were investigated by evaluating the variation in both wastewater and biofilm samples in a laboratory-scale sewer reactor system comprising both rising main and gravity sewers. Intrinsically positive wastewater with SARS-CoV-2 RNA was used in batch tests. Sewer and control reactors, with and without biofilm, respectively, were employed to determine the role of partition of different sewer biofilms and the role of RNA decay itself in causing the reduction of RNA concentration during the in-sewer transportation. Concentrations of SARS-CoV-2 N and E genes were detected by the RT-qPCR method at different time points during a period of two pumping cycles (i.e., 12 h), which covers most in-sewer residence time of wastewater. Biofilm samples from the sewer reactors were also extracted to determine the attached SARS-CoV-2 during their normal operation. The results are expected to delineate the role of sewer biofilms in the reduction of SARS-CoV-2 RNA in wastewater.

2. Materials and Methods

2.1. SARS-CoV-2 RNA Positive Wastewater

Two influent wastewater samples from a local wastewater treatment plant (WWTP) in Wollongong, Australia, on 11 and 25 January 2022 were used in this study for batch tests. Both samples were confirmed as SARS-CoV-2 RNA positive by using China Center for Disease Control (CCDC) developed assay (CCDC-N RT-qPCR assay) and E-Sarbeco RT-qPCR assay (Table S1), as described previously [27]. These two untreated influent wastewater samples were used without SARS-CoV-2 RNA/virus seeding. The characteristics of the collected wastewater was provided in Table S2. The wastewater samples were stored at

−80 °C (11 January 2022) or 4 °C (25 January 2022) until the experiments were conducted within a two-week period. According to other studies about the storage of SARS-CoV-2 RNA, the T_{90} of SARS-CoV-2 RNA decay in wastewater was around 28 to 52 days at 4 °C, and no decay was observed at −75 °C [15].

2.2. Laboratory-Scale Sewer System

A laboratory-scale sewer reactor system was adopted in this study (Figure 1) to simulate the real sewer system. This system was composed by connecting two types of reactors: two rising main (RM) reactors and two gravity sewer (GS) reactors, both having a diameter of 80 mm and a water height of 150 mm. All reactors were made of Perspex™ because of the good chemical stability to act as the substratum for biofilm development. These sewer reactors have been reported to represent typical sewer conditions by many previous studies [23,28,29]. The characteristics of the reactors were also described in our previous study [26]. The microbial community and abundance characterization were also reported in our previous publications [30].

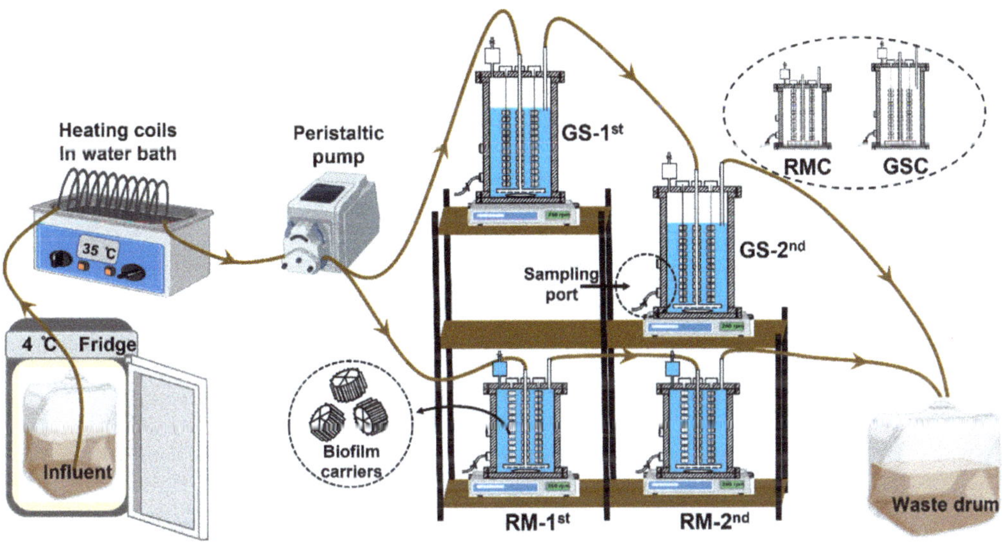

Figure 1. The laboratory-scale sewer system is composed of two rising main (RM) and two gravity sewer (GS) bioreactors.

Briefly, the sewer reactor system was fed with raw wastewater (collected from a local WWTP in Wollongong, Australia) for biofilm cultivation in 2020. 90 L of domestic wastewater was collected fortnightly and was stored at 4 °C. The wastewater was pumped by using a peristaltic pump (Masterflex L/S) every 6 h (a typical wastewater hydraulic retention time (HRT) and was pumped through heating coils in a water bath to warm the wastewater to room temperature (22 °C). Each feed pumping event lasted 2 min to provide one reactor working volume wastewater (0.75 L) into each reactor. The overall biofilm area of each reactor, including the carrier surface and reactor wall, is around 0.05 m^2. Hence, the biofilm area to wastewater volume ratio (A/V) was around 70.9 m^2/m^3. A magnetic stirrer (MLS8, VELP Scientific, Italy) was used to continuously provide mixing (250 rpm) to generate a moderate shear force, calculated as 1.7 Pa, on the inner surface of the reactor wall and to avoid the settling of solids at the bottom. To confirm that the reactors reached their semi-steady states, batch tests were conducted to determine biofilm activity including the sulfate reduction rate and chemical oxygen demand (COD) reduction rate in the RM and GS reactors [5,26,29,31,32]. The characteristics were listed in Table S3. During the

6 h pumping circle, the pH of wastewater was stable in both RM and GS sewer reactors and was between 6.8–7.2. The sulfate concentration decreased from 27 mg/L to 14 mg/L within two hours in RM. While in the GS reactor, no obvious decrease was observed. The dissolved oxygen (DO) was around 0.65 mg/L in the GS reactor. The RM reactor was filled with wastewater, so the DO in the RM reactor should be zero. The soluble chemical oxygen demand (COD) of the RM reactor reduced from 116 to 84 mg/L and was varied between 60–110 mg/L in the GS reactor. These data are within the normal range test by our previous study [5].

2.3. Sewer Reactor Tests and Sampling Schemes

Four reactors, including the first RM and GS sewer reactors (RM-1st and GS-1st) in the lab-scale sewer system, and two control reactors (RMC and GSC), were used in the batch tests. The control reactors are plain reactors that have the same structure of each type of biofilm reactors. The control reactors were not fed with wastewater to grow biofilm before the batch tests. For each batch test, wastewater was pumped continuously into the sewer system for 5 min. At the same time, a volume of 0.75 L wastewater was added to the RMC and GSC control reactors for parallel blank tests. The temperature of the wastewater was adjusted to around 22 °C. The experimental schema was shown in Figure S1. Briefly, a volume of 50 mL wastewater was sampled from RM-1st, GS-1st, RMC and GSC reactors at time points of 0 h, 0.5 h, 1 h, 3 h and 6 h. After the 6 h sampling, the sewer reactors were fed for 2 min as per the normal operation. Then, a volume of 50 mL wastewater was collected from the RMC and GSC reactors at 12 h. In addition, sewer biofilms (one biocarrier at each time) were sampled from the first RM and GS sewer reactors (RM-1st and GS-1st) at 0 h and 12 h, and the second RM and GS sewer reactors (RM-2nd and GS-2nd) at 12 h. The total surface area of one biocarrier was around 3.17 cm^2. One biocarrier with attached biofilms was extracted from the sewer reactors for analysis as described below. The above sewer reactor experiment was conducted in duplicate. Test 2 is a technical repeat of Test 1 and was started from the same timepoint (8:30 am) as Test 1 on the next day of Test 1 (Figure S1). All samples were immediately stored at −80 °C after sampling until further analysis.

2.4. Sample Processing and RNA Extraction

In previous studies, an adsorption-extraction method (no need for pH adjustment or addition of $MgCl_2$) that utilized an electronegative membrane was found to be effective in recovering a human coronavirus surrogate, murine hepatitis virus (MHV), with a recovery rate of 60.5 ± 22.2% [33]. Therefore, this method was adopted in this study to recover SARS-CoV-2 RNA from both wastewater and bio-carrier samples [33]. A volume of 50 mL wastewater was filtered through electronegative membranes with a 0.45 μm pore size and 90 mm diameter (HAWP09000; Merck Millipore Ltd., Sydney, Australia). The filter paper was then folded and transferred to a 5 mL bead tube of the RNeasy PowerWater Kit (Qiagen, Hilden, Germany). For the biofilm samples (biocarriers), each of the plastic biocarrier was soaked and brushed in 50 mL of 4 °C phosphate-buffered saline (PBS, pH 7.4) and vortexed for 5 min to detach the biofilm. (Plastic biocarrier was used because of its high stability, and the mature biofilm on the surface of plastic material could be easily sampled.) The 50 mL wash solution was also filtered with electronegative membranes. The filter paper and the biocarrier were then transferred to a 5 mL bead tube together. All the following extraction procedure was carried out according to the instructions of the RNeasy PowerWater Kit. The final extracted RNA volume of the 50 mL wastewater or biofilm sample was 50 μL.

2.5. RT-qPCR Assay

The CCDC-N and E-Sarbeco primer-probe sets were adopted according to our previous comparative study on the analytical performance towards wastewater samples [27]. The information on primer-probe sets used in this study and the RT-qPCR conditions were listed in Table S1 in the supplementary material. RT-qPCR assay targeting the N gene was conducted by using a CCDC-N primer-probe set under the conditions [34]. The copy number of the N gene was calculated based on the spiked mocks calibration generated by using 10-fold serial dilutions of a known concentration of 2019-nCoV_N Positive Control (Catalogue No. 10006625) [27].

RT-qPCR assay targeting the E gene was conducted by using an E-Sarbeco primer-probe set under the conditions listed in Table S1 [35]. A standard curve of the E Sarbeco assay was generated by using 10-fold serial dilutions of a known concentration of 2019-nCoV_E Positive Control (Catalogue No. 10006896). Characteristics of the standard curves including the efficiency, the linearity (coefficient of determination, R^2), slope and Y-intercept were listed in Table 1.

Table 1. Characteristics of RT-qPCR calibrations for CCDC-N and E-Sarbeco assay.

Primer-Probe Sets	Efficiency (%)	Linearity (R^2)	Slope (Mean ± SD)	Y-Intercept (Mean ± SD)
CCDC-N	108.2	0.999	−3.139 ± 0.12	40.39 ± 0.18
E-Sarbeco	96.36	0.999	−3.412 ± 0.09	41.87 ± 0.26

All RT-qPCR reactions were conducted in duplicate. For each RT-qPCR run, a series of positive and no template controls were tested to eliminate the false positive and false negative results induced by potential contamination. The RT-qPCR assays were carried out by using the Bio-Rad CFX96 thermal cycler (Bio-Rad Laboratories). The thermocycler conditions were set as reverse transcription at 50 °C for 10 min and initial denaturation at 95 °C for 3 min, followed by 45 cycles of 15 s at 95 °C and 30 s at 58 °C. Data were collected only from the operation of the instrument when the positive control was positive, and the non-template control was negative.

2.6. Data Analysis

The experimental data of duplicate batch tests were integrated to generate a combined decay constant. The relationship between the SARS-CoV-2 RNA concentration in control or sewer reactors and time was, respectively, evaluated by adopting monophasic and biphasic first-order decay kinetics in this study due to its wide suitability for analyzing the viral RNA decay in wastewater [13,15,26]. Hence, the concentration variation of SARS-CoV-2 RNA was linearized by using the natural log (ln)-transformation of the calculated RNA concentration of each sampling time point as shown in Equation (1), where C_t and C_0 are the concentrations (based on gene copies/50 mL) of SARS-CoV-2 N or E gene copies at time t and time 0, respectively. k is the reduction rate constant. The monophasic first-order decay rate constant with the associated 95% confident interval (CI) was estimated by using GraphPad Prism Version 9.0.0 (GraphPad Software, La Jolla, CA, USA). The fit was assessed by the coefficient of determination (R^2) and root mean square error (RMSE). The time required to achieve 90% (T_{90}) reduction of the viral RNA based on the monophasic first-order decay kinetics was further calculated using k values according to Equation (2).

$$\ln\left(\frac{C_t}{C_0}\right) = -kt \quad (1)$$

$$T_{90} = \frac{ln(0.1)}{k} \quad (2)$$

3. Results

3.1. Reduction of SARS-CoV-2 RNA in Wastewater

The initial SARS-CoV-2 RNA concentration of the wastewater used in this study was between 10^4–10^5 gene copies/L (2.1×10^4 and 4.2×10^4 gene copies/L of the N gene; 3.6×10^4 and 2.3×10^5 gene copies/L of E gene, respectively), which was high enough for RNA decay/decrease batch test. In addition, the RNA extraction method adopted in this study was confirmed to have an RT-qPCR-based recovery of 60.5 ± 22.2% for murine hepatitis virus (MHV), a surrogate for SARS-CoV-2 [33]. The matrix spike recoveries of N (CCDC-N) and E (E-Sarbeco) genes were 3.63 ± 1.8% and 41.2 ± 18.7%, respectively, which might be due to the detection limit and efficiency of the different RT-qPCR assay [36]. The 50% detection of CCDC-N and E-Sarbeco assay against the wastewater sample is around 3 to 4 log/L [36]. It indicates that the negative samples tested during the reduction evaluation theoretically had up to about 2 log reduction/L. Thus, we regarded the $\ln(\frac{C_t}{C_0})$ value of all negative results as −4.61 (ln (0.01)) in all the following figures.

The variation of SARS-CoV-2 N and E gene concentrations (mean ± SD) in gravity (GS) and rising main (RM) control and sewer reactors (RM) was shown in Figure S2, respectively. Significant variations of the RNA concentrations were observed, although different reduction trends were generally observed between control and biofilm reactors. Considering the limited positive data obtained in each scenario (we used raw wastewater without spiking), data acquired for N and E genes in the RM and GS reactors were combined to compare the SARS-CoV-2 RNA concentration reduction in the presence and absence of biofilms. As shown in Figure 2A, 6/8 tests of both the 6 h and 12 h wastewater samples collected from the control reactors were positive, and 5/6 positive tests were above the 90% reduction line. This reveals limited decay of SARS-CoV-2 RNA in wastewater within 12 h. It is worth noticing that in the 1 h and 3 h samples of control reactors, only 2 out of the total 8 tests were positive which is abnormal compared to the high RNA concentration and high positive test numbers (6/8 positive) in the 6 h and 12 h samples. Although it seems that no differences between control and biofilm reactors appeared in the 3 h sample, the trend between 1 h to 6 h in control and biofilm reactors was the opposite. Thus, we infer these negative results might be induced by the low initial concentration and the varied RNA recovery of each extraction. Considering the high positive rates and concentrations of the 6 and 12 h samples, these negative results were thus excluded from further data analysis. In addition, there is an abnormal increase between the results of the 0 h and 0.5 h samples. Most of the increased concentrations were obtained by E gene assay (empty triangle). A similar increase was also observed by Hokajarvi et al. [15] for the persistence of the SARS-CoV-2 RNA in wastewater influent at different temperatures. This abnormal increase might be due to the larger variation of the E gene recovery caused by the low sensitivity of the E gene targeting RT-qPCR assay [36]. Meanwhile, a smaller abnormal increase of the N gene than the E gene was observed, which was consistent with the smaller recovery variation of the N gene (SD = 1.8%) than the E gene (SD = 18.7%) [36]. This behavior was also observed in biofilm reactors but was not as high as in control reactors. The RNA desorption from sewer biofilms occurred due to the disturbance of feeding wastewater to the sewer reactors. The biofilm sloughing, together with attached viral RNA on biofilms, might have reduced the initial variations.

In comparison to the low decay in wastewater in control reactors, a significant concentration reduction of SARS-CoV-2 RNA was observed in the wastewater phase of sewer reactors. There are only 4/8, 2/8 and 1/8 tests of the 1, 3 and 6 h samples, respectively, were positive (Figure 2B). In addition, in the sewer biofilm reactor, the decrease in RNA concentration in the wastewater phase was consistent with the decrease in the positive test numbers. The overall results, including the positive ratio and the reduction of concentrations, indicated that the concentration reduction of SARS-CoV-2 RNA in the wastewater phase was enhanced with the sewer biofilms.

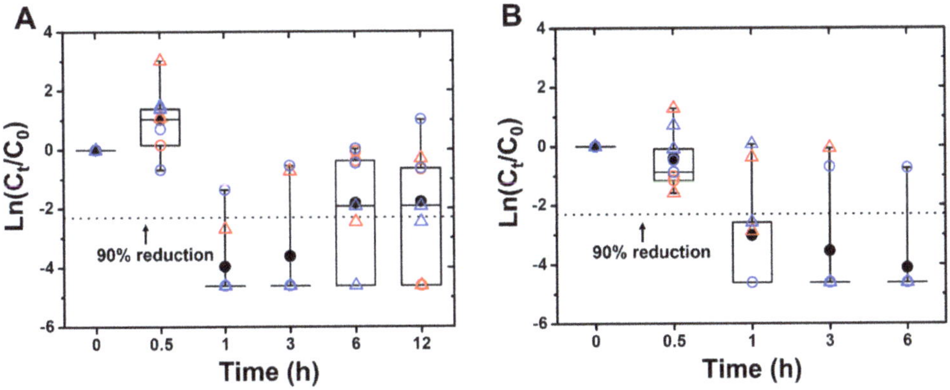

Figure 2. Concentration variation of SARS-CoV-2 RNA (E gene and N gene) in the wastewater phase of (**A**) control (RMC and GSC) and (**B**) biofilm reactors (RM−1st and GS−1st). The red symbols represent the results of the RM reactors. The blue symbols represent the results of the GS reactors. Empty circles represent the results of the N gene. Empty triangles represent the results of the E gene. Solid circles represent the average of the total eight tests. The middle solid lines inside the box represent medians. The top and bottom borders of the box represent the 75%ile and 25%ile, respectively. The top and bottom whiskers represent the 95%ile and 5%ile, respectively.

3.2. In-Sewer Reduction Kinetics Analysis of SARS-CoV-2 RNA in the Wastewater Phase of Sewer Reactors

To further analyze the effect of sewer biofilm on the reduction of SARS-CoV-2 RNA concentration in wastewater, we combined the results of the N and E genes to compare the difference between the RM and GS reactors. We also combined the results of the RM and GS reactors to compare the difference between the SARS-CoV-2 N and E genes. Monophasic and biphasic first-order decay kinetics were adopted to generate the reduction rate k (h^{-1}) and the T_{90} values (h) of SARS-CoV-2 RNA in different types of sewer reactors. The results of the kinetics analysis were provided in Table 2 and Figure S3.

Table 2. Reduction rate k/k_1 (h^{-1}) and k_2 (h^{-1}), transiting time point t_1 (h) and T_{90} values (h) of SARS-CoV-2 RNA (N and E genes) in different types of sewer reactors (RM and GS) with and without biofilms based on the monophasic (Control) or biphasic (Biofilm) first-order decay models.

Targets/Reactors		k/k_1 (h^{-1}) [95% CI]	k_2 (h^{-1}) [95% CI]	t_1 (h) [95% CI]	T_{90} (h) [95% CI]	R^2	RMSE
Control (Monophasic)	RM	0.23 [0.02 to 0.44]	-	-	10.07 [4.52 to 21.63]	0.75	0.89
	GS	0.12 [0 * to 0.31]	-	-	18.77 [0 * to 227.8]	0.48	0.79
	N gene	0.09 [0.02 to 0.16]	-	-	26.19 [0 * to 113.5]	0.81	0.28
	E gene	0.26 [0 * to 0.64]	-	-	8.74 [4.35 to 16.49]	0.53	1.59
	Total	0.18 [0 * to 0.37]	-	-	13.11 [7.38 to 24.14]	0.67	0.8
Biofilm (Biphasic)	RM	2.74 [1.38 to 4.69]	0.37 [0 * to 0.78]	1 [0 * to 2.7]	0.84 [0.56 to 3.41]	0.63	1.36
	GS	2.42 [1 to 4.41]	0.004 [0 * to 0.66]	1.5 [0.5 to NA]	0.95 [0.59 to 3.79]	0.52	1.53
	N gene	3.92 [2.78 to 5.44]	0.07 [0 * to 0.3]	1 [0.72 to 1.64]	0.59 [0.45 to 0.86]	0.66	1.25
	E gene	1.15 [0.73 to 2.35]	0.38 [NA]	NA	2.0 [1.45 to 6]	0.7	1.24
	Total	2.58 [1.63 to 3.71]	0.19 [0 * to 0.58]	1.25 [0.69 to 2.33]	0.89 [0.65 to 2.67]	0.57	1.45

Note(s): * 95% CI truncated at 0 in the case of negative values for k and T_{90}. NA: Unreliable results.

All T_{90} values from the biofilm reactor were less than or equal to 2 h, while all T_{90} values from the control reactor were between 8 to 27 h. The total T_{90} was reduced from 13.11 h in the control reactor to 0.89 h in the biofilm reactor. T_{90} values of the RM and GS reactors were reduced from 10.07 h and 18.77 h in the control reactor to 0.84 h and 0.95 h in the biofilm reactor, respectively. T_{90} values of the SARS-CoV-2 N and E genes were reduced from 26.19 h and 8.74 h in the control reactor to 0.59 h and 2.0 h in the biofilm reactor, respectively. The k values of the control reactor were between 0.09 to 0.26 h^{-1} and the total k value was 0.18 h^{-1}. For the biofilm reactor, the k_1 values were between 1.15

to 3.93 h^{-1} and the total k_1 value was 2.58 h^{-1}, which was around 14.3 times that of the control reactor. The k_2 values of the biofilm reactors were between 0.004 to 0.38 h^{-1} which were very similar to the k value range of the control reactors. In addition, all the valid transiting time point t_1 values of the biofilm reactors were less than or equal to 1.5 h.

3.3. Accumulation of SARS-CoV-2 RNA in Sewer Biofilms

SARS-CoV-2 RNA concentration in sewer biofilms of the RM and GS reactors was evaluated by targeting both the N gene and E gene. The results of gene copies per cm^2 of biofilm surface area were shown in Table 3. SARS-COV-2 N gene was detected in 5 out of 10 biofilm samples, and the average concentration of the positive RM and GS biofilm sample was 258.4 ± 72.7 GC/cm^2 (4 positive samples, mean ± SD (standard deviation)) and 487 GC/cm^2 (1 positive sample), respectively. SARS-COV-2 E gene was detected in 8 of 10 biofilm samples, and the average concentration of positive RM and GS biofilm was 704.7 ± 155.7 GC/cm^2 (3 positive samples) and 1828.7 ± 1180.6 GC/cm^2 (5 positive samples), respectively. In addition, during the 12 h of the second test, an accumulation of 260.9 GC/cm^2 (from 654.5 GC/cm^2 to 915.4 GC/cm^2) and 113.7 GC/cm^2 (from 935.9 GC/cm^2 to 1049.6 GC/cm^2) of the SARS-COV-2 E gene was observed in the RM-1st and GS-1st sewer biofilm reactor, respectively. For the N gene, an accumulation of 365.5 GC/cm^2 was observed in the RM-1st sewer reactor. However, no N gene accumulation was observed in the GS-1st sewer reactor.

Table 3. Accumulation of SARS-CoV-2 RNA in rising main and gravity sewer biofilms.

Type	Batch Test	N Gene (GC/cm^2) Reactor-1st		N Gene (GC/cm^2) Reactor-2nd	E Gene (GC/cm^2) Reactor-1st		E Gene (GC/cm^2) Reactor-2nd
		0 h	12 h	12 h	0 h	12 h	12 h
RM	Test 1	196.1 *	-	285.1 *	ND	-	ND
	Test 2	ND	365.5	187.0	654.5 *	915.4	544.1
GS	Test 1	ND	-	ND	686.1 *	-	3622.4
	Test 2	ND	ND	487.0 *	935.9	1049.6 *	2849.4 *

Note(s): * Duplicate RT-qPCR tests only yielded one positive result. Sample was not acquired. ND: not detected.

Assuming that the wastewater is in perfect plug flow in sewer reactors and since the duplicate tests were conducted on two consecutive days and started at the same time of the day, thus, the 0 h sample of the second test could be regarded as the 24 h sample of the first test, and the 12 h sample of the second test could be regarded as the 36 h sample of the first test. Based on this timeline, the variation of the E and N gene concentration in biofilm samples during 36 h was shown in Figure S4. The results indicated that, in biofilm samples, the concentration of the SARS-CoV-2 E gene was increasing in both the RM-1st (from ND to 654.5 GC/cm^2 to 915.4 GC/cm^2) and the GS-1st (from 686.1 GC/cm^2 to 935.9 GC/cm^2 to 1049.6 GC/cm^2) sewer reactors during two batch tests. The concentration of the SARS-CoV-2 N gene was increasing in the biofilm of the RM-1st (from 196.1 GC/cm^2 to ND to 365.5 GC/cm^2) reactor. However, no SARS-COV-2 N gene was detected in the biofilm samples of the GS-1st reactor over 36 h. In addition, an increase in gene concentration was also observed in the biofilm of the RM-2nd (from ND to 544.1 GC/cm^2) and the GS-2nd (from ND to 487 GC/cm^2) reactors by targeting the E gene and N gene, respectively. However, in cases of results of the RM-2nd biofilm targeting the N gene and the GS-2nd biofilm targeting the E gene, a concentration decline was observed.

Collectively speaking, the SARS-CoV-2 RNA accumulation was observed in both the RM-1st and GS-1st sewer biofilm reactors during two consecutive days of sampling, suggesting that the concentration reduction of RNA in wastewater may be caused by not only the RNA decay in wastewater, but also more so by the RNA attachment on sewer biofilms.

4. Discussion

The above results support a higher reduction of SARS-CoV-2 RNA in sewer reactors, possibly due to biofilms, suggesting that the adsorption of viral particles in sewer pipes may be an important factor affecting the amount of virus in sewage influent. The possible biofilm formation in distribution pipes can enhance such virus adsorption and the age of networks can determine possible biofilm formation in sewage distribution networks. This study evaluated the fate or reduction of SARS CoV 2 RNA in wastewater and the adsorption by sewer biofilms at the same time in a simulated sewer reactor system. A previous study reported an accumulation of ~700 genome copies/cm^2 SARS-CoV-2 RNA on the sewer biofilm of a simulated annular biofilm reactor during a 4-week operation with high COVID-19 incidence in the community [10]. The biofilm was formed over the 4-week operation and, subsequently, sampled for viral RNA detection. However, sewer biofilm samples from a university campus were reported to be negative. In this study, we evaluated the SARS-CoV-2 RNA adsorption by mature sewer biofilms which were cultivated over two years in a simulated sewer reactor system. The accumulation of 260.9 GC/cm^2 and 113.7 GC/cm^2 of SARS-COV-2 E gene on the first RM and GS sewer biofilms in 12 h confirmed the RNA adsorption by biofilm and revealed the importance of sewer biofilms for reducing the concentration of SARS-CoV-2 RNA in influent wastewater. A few earlier studies compared the temperature effects on the decay of SARS-CoV-2 RNA in various types of water and wastewater. The information on the testing conditions, temperature (T), k value, R^2 and T_{90} was listed and compared in Table 4.

Ahmed et al. investigated the decay of SARS-CoV-2 RNA and murine hepatitis virus (MHV) RNA in untreated wastewater at different temperatures [13]. This study determined T_{90} values at 4 °C, 15 °C, 25 °C and 37 °C as 27.8 ± 4.45, 20.4 ± 2.13, 12.6 ± 0.59 and 8.04 ± 0.23 days, respectively, by using a CCDC-N1 RT-qPCR assay. In cases of MHV, the T_{90} values at 4 °C, 15 °C, 25 °C and 37 °C as 56.6 ± 14.2, 28.5 ± 4.43, 17.3 ± 2.46 and 7.44 ± 0.61 days, respectively. The T_{90} value demonstrated a declining trend with the rise in experimental temperature. The less sensitivity to temperatures than murine hepatitis virus (MHV) RNA in untreated wastewater indicated that SARS-CoV-2 RNA is likely to persist long enough in untreated wastewater during in-sewer transportation. Hokajarvi's study [15] also evaluated the stability of SARS-CoV-2 RNA in influent wastewater at 4 °C by targeting both the N and E genes. The T_{90} values of the two gene biomarkers were similar and were comparable to Ahmed's study at 4 °C. However, the T_{90} value of Bivins's study by targeting the SARS-CoV-2 E gene at 20 °C was 3.3 days and 26 days at high titer (10^5 $TCID_{50}$ /mL) and low titer (10^3 $TCID_{50}$ /mL), respectively [14]. These values were inconsistent with the T_{90} value (12.6 ± 0.59 days) in Ahmed's study at 25 °C with an initial SARS-CoV-2 E gene concentration of $10^{7.03 \pm 0.19}$/mL. Caution should be exerted when comparing RNA decay rates, which may be largely affected by the initial concentration and type of SARS-CoV-2 RNA biomarkers at room temperature. In addition to the decay of the SARS-CoV-2 virus, many previous studies have investigated the persistence or decay of other enveloped viruses/bacteriophage. According to the results of Silverman and Boehm's literature review based on 46 and 27 identified decay rate constants of animal coronaviruses and the enveloped bacteriophage Phi6, the average k value of these two types of surrogate was 1.78 ± 0.12 d^{-1} and 7.59 d^{-1} in raw wastewater at 22−25 °C, respectively [37]. These results are closer to the k values of the control reactor in our study (Table 4). They also found that the k values obtained in wastewater were substantially higher than those in freshwater but not statistically different from those observed in estuarine and marine natural waters, and the k value measured using molecular methods were less than those measured by culture methods which cannot directly indicate virus viability [38]. These findings highlighted the effect of the water matrix and the detection methods on the evaluation of enveloped virus decay. In conclusion, previous studies demonstrated that, at room temperature, 90% of the SARS-CoV-2 N gene decayed within 13 days in wastewater, and 90% of the SARS-CoV-2 E gene decayed within 3–26 days, majorly affected by sewage temperature and the targeted assay gene.

Table 4. Comparison of SARS-CoV-2 RNA persistence in wastewater of different studies.

RT-qPCR Assay	Initial Concentration	Virus Types	Testing Method	T (°C)	k (Mean ± SD) [95% CI] (d^{-1})	R^2	T_{90} (d)	Reference
CDC-N1	$10^{7.03 \pm 0.19}$ GC/mL	Gamma-irradiated SARS-CoV-2	4.99 mL of wastewater and 5 mL of gamma-irradiated SARS-CoV-2 were mixed in 15 mL sterile conical tubes. 140 μL aliquot was sampled from each tube at each sampling time point for extraction and RT-qPCR analysis.	4	0.084 ± 0.013 [0.103 to 0.064]	0.79	27.8 ± 4.45 [22.4 to 50.1]	[13]
				15	0.114 ± 0.012 [0.144 to 0.083]	0.71	20.4 ± 2.13 [16.0 to 27.7]	
				25	0.183 ± 0.008 [0.219 to 0.149]	0.87	12.6 ± 0.59 [10.5 to 15.5]	
				37	0.286 ± 0.008 [0.370 to 0.202]	0.74	8.04 ± 0.23 [6.22 to 11.4]	
E-Sarbeco	$10^{5.4}$ GC/mL	1:1000 dilution of SARS-CoV-2 inoculum	Spiked wastewater influent was packed separately into 57 microcentrifuge tubes in portions of 300 μL. Triplicate tubes were stored in the dark and extracted for RT-qPCR analysis.	4	0.04 ± 0.2	0.59	52	[15]
CDC-N2	$10^{6.1}$ GC/mL				0.06 ± 0.0	0.99	36	
E-Sarbeco	High titer (10^5 $TCID_{50}$/mL)	SARS-CoV-2 nCoV WA1-2020 (MN985325.1), isolated from a clinical patient	SARS-CoV-2 nCoV-WA1-2020 (MN985325.1) was diluted 1:10 to wastewater. 1 mL aliquots were pipetted into 2 mL screw-top vials with 3 replicates for each time point.	20	0.67 [0.54 to 0.86]	0.27	3.3 [2.7 to 4.3]	
	Low titer (10^3 $TCID_{50}$/mL)				0.09 [0.00 * to 0.23]	−0.01	26 [9.8 to ∞]	
CDC-N1/N2	135–953 GC/mL	SARS-CoV-2 RNA positive wastewater samples without seeding		4	2.16	-	-	[39]
				10	0.96	-	-	
				35	4.31	-	-	
CDC-N	20.9–41.8 GC/mL	SARS-CoV-2 RNA positive wastewater samples without seeding	50 mL of wastewater was sampled from each reactor at each time point for RNA extraction and RT-qPCR analysis.	22	Control: 2.1 [0.49 to 3.73]	0.81	1.1 [0.00 * to 4.73]	This study
					Biofilm k_1: 94.32 [66.72 to 132.96] Biofilm k_2: 1.68 [0.00 * to 7.2]	0.66	0.03 [0.02 to 0.04]	
E-Sarbeco	36.4–230 GC/mL				Control: 6.32 [0.00 * to 15.35]	0.53	0.36 [0.18 to 0.69]	
					Biofilm:27.6 [17.52 to 56.4]	0.7	0.08 [0.06 to 0.25]	

Note(s): * 95% CI truncated at 0 in the case of negative values for k and T_{90}.

In our study, the estimated T_{90} value was 1.1 day in the control reactor (wastewater only) by targeting the N gene, and was 0.36 day by targeting the E gene, which was shorter than most of the previous studies. However, previous studies usually use wastewater spiked with high SARS-CoV-2 concentration, and the total volume of testing systems was under 5 mL. In addition, the sampled wastewater at different time points was usually at microliter level. These are quite different from the initial RNA concentration and testing conditions used in our study, which may explain the difference between T_{90} values in our study and those in previous reports. In our study, intrinsic SARS-CoV-2 RNA positive wastewater without seeding any type of biomarkers was used, thus the initial concentration was around 1–5 log lower than in previous studies. Furthermore, the total wastewater volume in the lab-scale sewer reactors was 0.75 L, and 50 mL wastewater was sampled at each time point for filtration and RNA extraction. This workflow was consistent with the normal sampling and processing producers for detecting SARS-CoV-2 RNA in wastewater. Weidhaas et al. used a similar large volume of raw wastewater (without seeding) to conduct the decay test, the k value of N gene decay was 2.16 d^{-1}, 0.96 d^{-1} and 4.31 d^{-1} at 4°C, 10°C and 35°C, respectively. These results were consistent with the k value of (2.1 d^{-1} at 22 °C) N gene decay in the control reactors acquired in our study [39].

Fu et al. [11] recently investigated the decay of avian infectious bronchitis virus (IBV) in sewage pipes. The D_{90} (distance required for one log reduction) value of this study was 12.51 km and 8.34 km at 12 °C for the non-biofilm and mature biofilm pipelines, and 8.32 km and 4.54 km at 28 °C, respectively, by assuming the flow rate of wastewater at 0.4 km/h. This study reported the enhanced reduction of viral RNA in wastewater in the presence of sewer biofilm which was cultured for 45 days. In our study, the biofilm of the simulated sewer system had been cultivated for more than two years. In addition, Ahmed et al. [40] reported that a sewer pipe length of over 400 m can produce a SARS-CoV-2 RNA decline in the sewer network. Through the quantification of SARS-CoV-2 RNA for wastewater influent and primary settled solids samples, Graham demonstrated that measuring SARS-CoV-2 RNA concentrations in settled solids is more sensitive than measuring SARS-CoV-2 in influent [41]. In another study about *Campylobacter* spp., the overall reduction followed a biphasic first-order decay model, and the faster reduction rate in the first phase was due to the adsorption of *C. jejuni* and *C. coli* onto biofilm [42]. The sensitivity analysis revealed that the reduction leads to significant differences in WBE back-estimation of *Campylobacter* spp. prevalence, especially for sewer catchments with long hydraulic retention times. All the above studies showed that the virus/RNA attachment on biofilms and solid fractions in wastewater warranted a correction for the WBE applications.

In this study, the 95% CI of reduction parameters is high, with low R^2 and high RSME, which might be difficult to fully interpret the impact of biofilm on the reduction of SARS-CoV-2 in the reactors. However, the 95% CI of the estimated trend lines of the control and biofilm reactors do not overlap with each other in most scenarios (Figure S3). This reflects the different reduction trends of SARS-CoV-2 RNA due to the presence of biofilms. At the 6-h time point, all four samples from the control reactor tested positive for SARS-CoV-2 RNA (Figure 2). In contrast, in the biofilm reactor, only one sample from the GS reactor was detected as positive for the N gene. The results indicate that attachment to sewer biofilms facilitated the reduction of SARS-CoV-2 RNA concentration in the wastewater phase. This is further confirmed by the detection of SARS-CoV-2 RNA accumulation in sewer biofilms over 36 h. The increase of RNA concentration in biofilm samples, measured by the N or E genes, was consistent with the reduction of RNA concentration in wastewater samples in both the RM and GS sewer reactors. The detection of SARS-CoV-2 RNA in wastewater is challenging due to the complex matrix, which contains various components that can affect the performance of RT-qPCR methods differently. In this study, a higher concentration of the E gene was observed in both wastewater and biofilm samples. This result may be attributed to the higher recovery of the E gene compared to the N gene and the variable detection efficiency of RT-qPCR assays for wastewater samples [36]. Similar findings have been reported in previous studies that targeted the N and E genes for SARS-CoV-2 RNA detection [27,36,43,44]. Moreover, we demonstrated that, during in-sewer transportation, the attachment of SARS-CoV-2 virus/RNA on sewer biofilm (the reduction observed in wastewater of sewer biofilm reactors) might be more significant than the virus/RNA decay itself (the reduction observed in the control reactors). According to the results of this study towards the in-sewer decay of *Campylobacter* spp. [42], we supposed that the variation of SARS-CoV-2 RNA concentration of wastewater in sewer reactors was a result of the combined effects of RNA decay, biofilm adsorption, desorption and biofilm sloughing (especially at the beginning of the batch tests due to strong disturbances by feeding wastewater). In Table 2, all the valid T_{90} values (RM, GS and N gene) of the biofilm reactor were shorter than the t_1 value (transiting time point t_1). After t_1, the k_2 value of biofilm reactors were close to the k value of the control reactor. Based on the author's suppose, this behavior revealed that after t_1, there might be a balance between the RNA adsorption and desorption on sewer biofilm and biofilm sloughing. After achieving this balance, the reduction of RNA concentration in wastewater in the sewer reactor will mainly be due to RNA decay, just like that in the control reactor.

Overall, the results showed both RM and GS sewer biofilms can induce SARS-CoV-2 RNA attachment within typical wastewater residence time in sewers. Furthermore, considering the lower decay of SARS-CoV-2 RNA in wastewater, identified in this study, and reported by previous studies, the RNA loss caused by sewer biofilm attachment is more significant than RNA decay itself. However, several methodological limitations should be noticed. The sewer reactor used in this study has a limited total wastewater volume (0.75 L), which could not support multiple samples (50 mL each) at each time point. The biofilm samples were only collected at 0 h and 12 h, which were insufficient to detect the dynamic concentration changes in sewer biofilm.

The initial RNA concentration in non-spiked, intrinsic SARS-CoV-2 positive wastewater was low. This makes it difficult to obtain enough data points to support comprehensive kinetic analysis. In addition, although studies have confirmed that different sewer conditions could result in the different stability of various drugs and chemicals in wastewater, no significant difference in the SARS-CoV-2 RNA reduction was observed between the RM and GS sewer reactors in this study. This might be due to the low initial SARS-CoV-2 RNA concentration of the wastewater used in this study. Further studies should be conducted under various conditions, such as various initial SARS-CoV-2 RNA concentrations and temperatures. Moreover, the potential detachment of SARS-CoV-2 RNA from sewer biofilms should be studied to delineate the full picture of in-sewer dynamics. Other factors such as the higher concentration of RNase (ribonucleases) and other enzymes in wastewater and different biofilms and the different characteristics of biofilms with diverse characteristics of extracellular polymeric substance (EPS) and microbial community under different sewer conditions are worth investigating to analyze the partition of viral RNA into the sewer biofilms. Concentration variance caused by the SARS-CoV-2 mutants should also be considered in future in-sewer decay investigations.

5. Conclusions

This study investigated the impacts of different sewer biofilms on the reduction of SARS-CoV-2 RNA concentration in intrinsically positive wastewater in laboratory-scale sewer reactors. The SARS-CoV-2 RNA concentration in wastewater and biofilms was determined to delineate the fate of viral RNA during in-sewer transportation.

The concentration reduction of SARS-CoV-2 RNA in the wastewater phase was significantly higher in sewer biofilms and a 90% reduction of RNA concentration in wastewater was achieved within 2 h in both rising main and gravity sewer biofilm reactors. Therefore, biofilm grown in sewer networks can affect the fate of SARS-Co-2 during in-sewer transportation.

Accumulation of SARS-CoV-2 RNA in the biofilm of sewer reactors was observed for up to 36 h, which was consistent with the increased RNA reduction in wastewater. Therefore, the adsorption of virus RNA/particles in sewer biofilms is a non-negligible mechanism for the fate of SARS-Co-2 in sewer systems.

The concentration reduction of SARS-CoV-2 RNA in wastewater during in-sewer transportation may be more induced by the biofilm attachment rather than the RNA decay itself under the tested conditions.

Supplementary Materials: The following supporting information can be downloaded at: https://www.mdpi.com/article/10.3390/w15112132/s1, Figure S1: Experimental procedures of duplicate batch tests; Figure S2: Variation of SARS-CoV-2 E and N gene concentration (mean ± SD) in gravity (GS) and rising main (RM) control and sewer reactors, respectively; Figure S3: The in-sewer reduction kinetics of SARS-CoV-2 RNA (N and E gene) in different sewer reactors (RM and GS) with and without biofilms; Figure S4: Short-term SARS-CoV-2 RNA accumulation observed in biofilm reactors during duplicate sampling days. Table S1: RT-qPCR primer-probe sets used in this study [34,35]; Table S2: Characteristics of the influent wastewater used as the feed for the sewer reactors; Table S3: Characteristics of wastewater in sewer reactors over a cycle of 6 h.

Author Contributions: Conceptualization, S.Z. and G.J.; methodology, S.Z.; formal analysis, S.Z.; investigation, S.Z.; data curation, S.Z.; writing—original draft preparation, S.Z.; writing—review and editing, S.Z., E.S., A.T., Y.C., S.P.S., S.G., X.Z., J.S. and G.J.; visualization, S.Z.; supervision, G.J.; funding acquisition, G.J. All authors have read and agreed to the published version of the manuscript.

Funding: This research was supported by the ARC Discovery project (DP190100385). Shuxin Zhang and Elipsha Sharma received support from a University of Wollongong scholarship. NIH grant R21AI157434 to Samendra Sherchan.

Conflicts of Interest: The authors declare no conflict of interest.

References

1. Peccia, J.; Zulli, A.; Brackney, D.E.; Grubaugh, N.D.; Kaplan, E.H.; Casanovas-Massana, A.; Ko, A.I.; Malik, A.A.; Wang, D.; Wang, M.; et al. Measurement of SARS-CoV-2 RNA in wastewater tracks community infection dynamics. *Nat. Biotechnol.* **2020**, *38*, 1164–1167. [CrossRef] [PubMed]
2. Galani, A.; Aalizadeh, R.; Kostakis, M.; Markou, A.; Alygizakis, N.; Lytras, T.; Adamopoulos, P.G.; Peccia, J.; Thompson, D.C.; Kontou, A.; et al. SARS-CoV-2 wastewater surveillance data can predict hospitalizations and ICU admissions. *Sci. Total Environ.* **2022**, *804*, 150151. [CrossRef] [PubMed]
3. Tiwari, A.; Lipponen, A.; Hokajärvi, A.-M.; Luomala, O.; Sarekoski, A.; Rytkönen, A.; Österlund, P.; Al-Hello, H.; Juutinen, A.; Miettinen, I.T.; et al. Detection and quantification of SARS-CoV-2 RNA in wastewater influent in relation to reported COVID-19 incidence in Finland. *Water Res.* **2022**, *215*, 118220. [CrossRef] [PubMed]
4. Sherchan, S.P.; Shahin, S.; Ward, L.M.; Tandukar, S.; Aw, T.G.; Schmitz, B.; Ahmed, W.; Kitajima, M. First detection of SARS-CoV-2 RNA in wastewater in North America: A study in Louisiana, USA. *Sci. Total Environ.* **2020**, *743*, 140621. [CrossRef] [PubMed]
5. Li, J.; Gao, J.; Zheng, Q.; Thai, P.K.; Duan, H.; Mueller, J.F.; Yuan, Z.; Jiang, G. Effects of pH, Temperature, Suspended Solids, and Biological Activity on Transformation of Illicit Drug and Pharmaceutical Biomarkers in Sewers. *Environ. Sci. Technol.* **2021**, *55*, 8771–8782. [CrossRef]
6. Proverbio, D.; Kemp, F.; Magni, S.; Ogorzaly, L.; Cauchie, H.-M.; Gonçalves, J.; Skupin, A.; Aalto, A. Model-based assessment of COVID-19 epidemic dynamics by wastewater analysis. *Sci. Total Environ.* **2022**, *827*, 154235. [CrossRef]
7. Vallejo, J.A.; Trigo-Tasende, N.; Rumbo-Feal, S.; Conde-Pérez, K.; López-Oriona, Á.; Barbeito, I.; Vaamonde, M.; Tarrío-Saavedra, J.; Reif, R.; Ladra, S.; et al. Modeling the number of people infected with SARS-COV-2 from wastewater viral load in Northwest Spain. *Sci. Total Environ.* **2022**, *811*, 152334. [CrossRef]
8. Jiang, G.; Wu, J.; Weidhaas, J.; Li, X.; Chen, Y.; Mueller, J.; Li, J.; Kumar, M.; Zhou, X.; Arora, S.; et al. Artificial neural network-based estimation of COVID-19 case numbers and effective reproduction rate using wastewater-based epidemiology. *Water Res.* **2022**, *218*, 118451. [CrossRef]
9. Li, X.; Zhang, S.; Shi, J.; Luby, S.P.; Jiang, G. Uncertainties in estimating SARS-CoV-2 prevalence by wastewater-based epidemiology. *Chem. Eng. J.* **2021**, *415*, 129039. [CrossRef]
10. Morales Medina, W.R.; D'Elia, S.; Fahrenfeld, N.L. Accumulation of SARS-CoV-2 RNA in Sewer Biofilms. *ACS EST Water* **2022**, *2*, 1844–1851. [CrossRef]
11. Fu, S.; Wang, Q.; He, F.; Zhou, C.; Zhang, J.; Xia, W. The decay of coronavirus in sewage pipes and the development of a predictive model for the estimation of SARS-CoV-2 infection cases based on wastewater surveillance. *medRxiv* 2022. [CrossRef]
12. Bertels, X.; Demeyer, P.; Van den Bogaert, S.; Boogaerts, T.; van Nuijs, A.L.N.; Delputte, P.; Lahousse, L. Factors influencing SARS-CoV-2 RNA concentrations in wastewater up to the sampling stage: A systematic review. *Sci. Total Environ.* **2022**, *820*, 153290. [CrossRef] [PubMed]
13. Ahmed, W.; Bertsch, P.M.; Bibby, K.; Haramoto, E.; Hewitt, J.; Huygens, F.; Gyawali, P.; Korajkic, A.; Riddell, S.; Sherchan, S.P.; et al. Decay of SARS-CoV-2 and surrogate murine hepatitis virus RNA in untreated wastewater to inform application in wastewater-based epidemiology. *Environ. Res.* **2020**, *191*, 110092. [CrossRef] [PubMed]
14. Bivins, A.; Greaves, J.; Fischer, R.; Yinda, K.C.; Ahmed, W.; Kitajima, M.; Munster, V.J.; Bibby, K. Persistence of SARS-CoV-2 in Water and Wastewater. *Environ. Sci. Technol. Lett.* **2020**, *7*, 937–942. [CrossRef]

15. Hokajarvi, A.M.; Rytkonen, A.; Tiwari, A.; Kauppinen, A.; Oikarinen, S.; Lehto, K.M.; Kankaanpaa, A.; Gunnar, T.; Al-Hello, H.; Blomqvist, S.; et al. The detection and stability of the SARS-CoV-2 RNA biomarkers in wastewater influent in Helsinki, Finland. *Sci. Total Environ.* **2021**, *770*, 145274. [CrossRef] [PubMed]
16. O'Brien, J.W.; Banks, A.P.W.; Novic, A.J.; Mueller, J.F.; Jiang, G.; Ort, C.; Eaglesham, G.; Yuan, Z.; Thai, P.K. Impact of in-Sewer Degradation of Pharmaceutical and Personal Care Products (PPCPs) Population Markers on a Population Model. *Environ. Sci. Technol.* **2017**, *51*, 3816–3823. [CrossRef]
17. Ort, C.; van Nuijs, A.L.N.; Berset, J.-D.; Bijlsma, L.; Castiglioni, S.; Covaci, A.; de Voogt, P.; Emke, E.; Fatta-Kassinos, D.; Griffiths, P.; et al. Spatial differences and temporal changes in illicit drug use in Europe quantified by wastewater analysis. *Addiction* **2014**, *109*, 1338–1352. [CrossRef]
18. Hart, O.E.; Halden, R.U. Modeling wastewater temperature and attenuation of sewage-borne biomarkers globally. *Water Res.* **2020**, *172*, 115473. [CrossRef]
19. McCall, C.; Fang, Z.N.; Li, D.; Czubai, A.J.; Juan, A.; LaTurner, Z.W.; Ensor, K.; Hopkins, L.; Bedient, P.B.; Stadler, L.B. Modeling SARS-CoV-2 RNA degradation in small and large sewersheds. *Environ. Sci. Water Res. Technol.* **2022**, *8*, 290–300. [CrossRef]
20. Hvitved-Jacobsen, T.; Vollertsen, J.; Nielsen, A.H. *Sewer Processes: Microbial and Chemical Process Engineering of Sewer Networks*; CRC Press: Boca Raton, FL, USA, 2013.
21. He, C.; Li, J.; Jiang, G.; Chen, S.; Niel, C.; Yuan, Z.; Mueller, J.F.; Thai, P. Transformation of phthalates and their metabolites in wastewater under different sewer conditions. *Water Res.* **2021**, *190*, 116754. [CrossRef]
22. Li, J.; Gao, J.; Thai, P.K.; Mueller, J.F.; Yuan, Z.; Jiang, G. Transformation of Illicit Drugs and Pharmaceuticals in Sewer Sediments. *Environ. Sci. Technol.* **2020**, *54*, 13056–13065. [CrossRef] [PubMed]
23. Thai, P.K.; O'Brien, J.; Jiang, G.; Gernjak, W.; Yuan, Z.; Eaglesham, G.; Mueller, J.F. Degradability of creatinine under sewer conditions affects its potential to be used as biomarker in sewage epidemiology. *Water Res.* **2014**, *55*, 272–279. [CrossRef] [PubMed]
24. Choi, P.M.; O'Brien, J.W.; Li, J.; Jiang, G.; Thomas, K.V.; Mueller, J.F. Population histamine burden assessed using wastewater-based epidemiology: The association of 1,4-methylimidazole acetic acid and fexofenadine. *Environ. Int.* **2018**, *120*, 172–180. [CrossRef] [PubMed]
25. O'Brien, J.W.; Choi, P.M.; Li, J.; Thai, P.K.; Jiang, G.; Tscharke, B.J.; Mueller, J.F.; Thomas, K.V. Evaluating the stability of three oxidative stress biomarkers under sewer conditions and potential impact for use in wastewater-based epidemiology. *Water Res.* **2019**, *166*, 115068. [CrossRef] [PubMed]
26. Shi, J.; Li, X.; Zhang, S.; Sharma, E.; Sivakumar, M.; Sherchan, S.P.; Jiang, G. Enhanced decay of coronaviruses in sewers with domestic wastewater. *Sci. Total Environ.* **2022**, *813*, 151919. [CrossRef]
27. Zhang, S.; Li, X.; Shi, J.; Sivakumar, M.; Luby, S.; O'Brien, J.; Jiang, G. Analytical performance comparison of four SARS-CoV-2 RT-qPCR primer-probe sets for wastewater samples. *Sci. Total Environ.* **2022**, *806*, 150572. [CrossRef]
28. Choi, P.M.; Li, J.; Gao, J.; O'Brien, J.W.; Thomas, K.V.; Thai, P.K.; Jiang, G.; Mueller, J.F. Considerations for assessing stability of wastewater-based epidemiology biomarkers using biofilm-free and sewer reactor tests. *Sci. Total Environ.* **2020**, *709*, 136228. [CrossRef]
29. Li, J.; Gao, J.; Thai, P.K.; Sun, X.; Mueller, J.F.; Yuan, Z.; Jiang, G. Stability of Illicit Drugs as Biomarkers in Sewers: From Lab to Reality. *Environ. Sci. Technol.* **2018**, *52*, 1561–1570. [CrossRef]
30. Kulandaivelu, J.; Gao, J.; Song, Y.; Shrestha, S.; Li, X.; Li, J.; Doederer, K.; Keller, J.; Yuan, Z.; Mueller, J.F.; et al. Removal of Pharmaceuticals and Illicit Drugs from Wastewater Due to Ferric Dosing in Sewers. *Environ. Sci. Technol.* **2019**, *53*, 6245–6254. [CrossRef]
31. Jiang, G.; Sharma, K.R.; Guisasola, A.; Keller, J.; Yuan, Z. Sulfur transformation in rising main sewers receiving nitrate dosage. *Water Res.* **2009**, *43*, 4430–4440. [CrossRef]
32. Li, W.; Zheng, T.; Ma, Y.; Liu, J. Current status and future prospects of sewer biofilms: Their structure, influencing factors, and substance transformations. *Sci. Total Environ.* **2019**, *695*, 133815. [CrossRef] [PubMed]
33. Ahmed, W.; Bertsch, P.M.; Bivins, A.; Bibby, K.; Farkas, K.; Gathercole, A.; Haramoto, E.; Gyawali, P.; Korajkic, A.; McMinn, B.R.; et al. Comparison of virus concentration methods for the RT-qPCR-based recovery of murine hepatitis virus, a surrogate for SARS-CoV-2 from untreated wastewater. *Sci. Total Environ.* **2020**, *739*, 139960. [CrossRef] [PubMed]
34. Vogels, C.B.F.; Brito, A.F.; Wyllie, A.L.; Fauver, J.R.; Ott, I.M.; Kalinich, C.C.; Petrone, M.E.; Casanovas-Massana, A.; Catherine Muenker, M.; Moore, A.J.; et al. Analytical sensitivity and efficiency comparisons of SARS-CoV-2 RT-qPCR primer-probe sets. *Nat. Microbiol.* **2020**, *5*, 1299–1305. [CrossRef] [PubMed]
35. Corman, V.M.; Landt, O.; Kaiser, M.; Molenkamp, R.; Meijer, A.; Chu, D.K.W.; Bleicker, T.; Brünink, S.; Schneider, J.; Schmidt, M.L. Detection of 2019 novel coronavirus (2019-nCoV) by real-time RT-PCR. *Eurosurveillance* **2020**, *25*, 2000045. [CrossRef] [PubMed]
36. Ahmed, W.; Bivins, A.; Metcalfe, S.; Smith, W.J.M.; Verbyla, M.E.; Symonds, E.M.; Simpson, S.L. Evaluation of process limit of detection and quantification variation of SARS-CoV-2 RT-qPCR and RT-dPCR assays for wastewater surveillance. *Water Res.* **2022**, *213*, 118132. [CrossRef] [PubMed]
37. Silverman, A.I.; Boehm, A.B. Systematic Review and Meta-Analysis of the Persistence and Disinfection of Human Coronaviruses and Their Viral Surrogates in Water and Wastewater. *Environ. Sci. Technol. Lett.* **2020**, *7*, 544–553. [CrossRef]
38. Silverman, A.I.; Boehm, A.B. Systematic Review and Meta-Analysis of the Persistence of Enveloped Viruses in Environmental Waters and Wastewater in the Absence of Disinfectants. *Environ. Sci. Technol.* **2021**, *55*, 14480–14493. [CrossRef]

39. Weidhaas, J.; Aanderud, Z.T.; Roper, D.K.; VanDerslice, J.; Gaddis, E.B.; Ostermiller, J.; Hoffman, K.; Jamal, R.; Heck, P.; Zhang, Y.; et al. Correlation of SARS-CoV-2 RNA in wastewater with COVID-19 disease burden in sewersheds. *Sci. Total Environ.* **2021**, *775*, 145790. [CrossRef]
40. Ahmed, F.; Islam, M.A.; Kumar, M.; Hossain, M.; Bhattacharya, P.; Islam, M.T.; Hossen, F.; Hossain, M.S.; Islam, M.S.; Uddin, M.M.; et al. First detection of SARS-CoV-2 genetic material in the vicinity of COVID-19 isolation Centre in Bangladesh: Variation along the sewer network. *Sci. Total Environ.* **2021**, *776*, 145724. [CrossRef]
41. Graham, K.E.; Loeb, S.K.; Wolfe, M.K.; Catoe, D.; Sinnott-Armstrong, N.; Kim, S.; Yamahara, K.M.; Sassoubre, L.M.; Mendoza Grijalva, L.M.; Roldan-Hernandez, L.; et al. SARS-CoV-2 RNA in Wastewater Settled Solids Is Associated with COVID-19 Cases in a Large Urban Sewershed. *Environ. Sci. Technol.* **2021**, *55*, 488–498. [CrossRef]
42. Zhang, S.; Shi, J.; Sharma, E.; Li, X.; Gao, S.; Zhou, X.; O'Brien, J.; Coin, L.; Liu, Y.; Sivakumar, M.; et al. In-sewer decay and partitioning of Campylobacter jejuni and Campylobacter coli and implications for their wastewater surveillance. *Water Res.* **2023**, *233*, 119737. [CrossRef] [PubMed]
43. Ahmed, W.; Bertsch, P.M.; Angel, N.; Bibby, K.; Bivins, A.; Dierens, L.; Edson, J.; Ehret, J.; Gyawali, P.; Hamilton, K.A.; et al. Detection of SARS-CoV-2 RNA in commercial passenger aircraft and cruise ship wastewater: A surveillance tool for assessing the presence of COVID-19 infected travellers. *J. Travel Med.* **2020**, *27*, taaa116. [CrossRef] [PubMed]
44. Zheng, X.; Li, S.; Deng, Y.; Xu, X.; Ding, J.; Lau, F.T.K.; In Yau, C.; Poon, L.L.M.; Tun, H.M.; Zhang, T. Quantification of SARS-CoV-2 RNA in wastewater treatment plants mirrors the pandemic trend in Hong Kong. *Sci. Total Environ.* **2022**, *844*, 157121. [CrossRef] [PubMed]

water

MDPI

Article

Extensive Wastewater-Based Epidemiology as a Resourceful Tool for SARS-CoV-2 Surveillance in a Low-to-Middle-Income Country through a Successful Collaborative Quest: WBE, Mobility, and Clinical Tests

Juan Eduardo Sosa-Hernández [1,†], Mariel Araceli Oyervides-Muñoz [1,†], Elda M. Melchor-Martínez [1], Erin M. Driver [2], Devin A. Bowes [2,3], Simona Kraberger [4], Sofia Liliana Lucero-Saucedo [1], Rafaela S. Fontenele [4,5], Lizeth Parra-Arroyo [1], LaRinda A. Holland [4], Samantha Ayde Peña-Benavides [1], Melanie Engstrom Newell [2,3], Manuel Martínez-Ruiz [1], Sangeet Adhikari [2,6], Laura Isabel Rodas-Zuluaga [1], Rahul Kumar [2], Itzel Y. López-Pacheco [1], Carlos Castillo-Zacarias [7], Hafiz M. N. Iqbal [1], Efrem S. Lim [4,5], Daniel Salas-Limón [8], Arvind Varsani [4,5], Rolf U. Halden [2,6,9,*] and Roberto Parra-Saldívar [1,*]

1 Tecnologico de Monterrey, School of Engineering and Sciences, Monterrey 64849, Mexico; eduardo.sosa@tec.mx (J.E.S.-H.); mariel.oyervides@tec.mx (M.A.O.-M.); elda.melchor@tec.mx (E.M.M.-M.); sofia_liliana@hotmail.es (S.L.L.-S.); a01036078@tec.mx (L.P.-A.); a00823349@tec.mx (S.A.P.-B.); mmruiz@tec.mx (M.M.-R.); a01674156@tec.mx (L.I.R.-Z.); a00824134@tec.mx (I.Y.L.-P.); hafiz.iqbal@tec.mx (H.M.N.I.)
2 Biodesign Center for Environmental Health Engineering, The Biodesign Institute, Arizona State University, Tempe, AZ 85287, USA; erin.driver@asu.edu (E.M.D.); dbowes@asu.edu (D.A.B.); mengstro@asu.edu (M.E.N.); sadhik14@asu.edu (S.A.); rkumar87@asu.edu (R.K.)
3 School for Engineering of Matter, Transport, and Energy, Arizona State University, Tempe, AZ 85287, USA
4 The Biodesign Center for Fundamental and Applied Microbiomics, Arizona State University, Tempe, AZ 85281, USA; simona.kraberger@asu.edu (S.K.); rafasfontenele@asu.edu (R.S.F.); larinda.holland@asu.edu (L.A.H.); efrem.lim@asu.edu (E.S.L.); arvind.varsani@asu.edu (A.V.)
5 School of Life Sciences, Arizona State University, Tempe, AZ 85281, USA
6 School of Sustainable Engineering and the Built Environment, Arizona State University, Tempe, AZ 85281, USA
7 Facultad de Ingenieria Civil, Departamento de Ingenieria Ambiental, Universidad Autonoma de Nuevo Leon, Ciudad Universitaria S/N, San Nicolás de los Garza 66455, Mexico; carlos.castillozcr@uanl.edu.mx
8 Facultad de Ingeniería Civil, Universidad Autonoma de Nuevo Leon, UANL, Av. Universidad s/n. CD. Universitaria, San Nicolás de los Garza 66455, Mexico; salas.daniel@gmail.com
9 OneWaterOneHealth, Arizona State University Foundation, Tempe, AZ 85287, USA
* Correspondence: rolf.halden@asu.edu (R.U.H.); r.parra@tec.mx (R.P.-S.)
† These authors contributed equally to this work.

check for updates

Citation: Sosa-Hernández, J.E.; Oyervides-Muñoz, M.A.; Melchor-Martínez, E.M.; Driver, E.M.; Bowes, D.A.; Kraberger, S.; Lucero-Saucedo, S.L.; Fontenele, R.S.; Parra-Arroyo, L.; Holland, L.A.; et al. Extensive Wastewater-Based Epidemiology as a Resourceful Tool for SARS-CoV-2 Surveillance in a Low-to-Middle-Income Country through a Successful Collaborative Quest: WBE, Mobility, and Clinical Tests. *Water* **2022**, *14*, 1842. https://doi.org/10.3390/w14121842

Academic Editors: Guangming Jiang, Ryo Honda and Sudipti Arora

Received: 5 May 2022
Accepted: 31 May 2022
Published: 8 June 2022

Abstract: The COVID-19 pandemic has challenged healthcare systems worldwide. Efforts in low-to-middle-income countries (LMICs) cannot keep stride with infection rates, especially during peaks. A strong international collaboration between Arizona State University (ASU), Tec de Monterrey (TEC), and Servicios de Agua y Drenaje de Monterrey (Local Water Utilities) is acting to integrate wastewater-based epidemiology (WBE) of SARS-CoV-2 in the region as a complementary approach to aid the healthcare system. Wastewater was collected from four sewer catchments in the Monterrey Metropolitan area in Mexico (pop. 4,643,232) from mid-April 2020 to February 2021 (44 weeks, $n = 644$). Raw wastewater was filtered and filter-concentrated, the RNA was extracted using columns, and the Charité/Berlin protocol was used for the RT-qPCR. The viral loads obtained between the first (June 2020) and second waves (February 2021) of the pandemic were similar; in contrast, the clinical cases were fewer during the first wave, indicating poor coverage. During the second wave of the pandemic, the SARS-CoV-2 quantification in wastewater increased 14 days earlier than the COVID-19 clinical cases reported. This is the first long-term WBE study in Mexico and demonstrates its value in pandemic management.

Keywords: SARS-CoV-2 surveillance; wastewater-based epidemiology; SARS-CoV-2 in wastewater; public health

1. Introduction

The COVID-19/SARS-CoV-2 pandemic is responsible for more than 5 million deaths worldwide according to WHO data [1]. In Mexico, more than 3.8 million cases were reported, with 289,811 deaths since the first case was confirmed in the country on 27 February 2020 [1]. To access a confirmed diagnostic test by public health services in Mexico, it is mandatory to have multiple symptoms. However, according to the CDC, symptoms of COVID-19 may appear 2–14 days after exposure to the virus, which represents a non-diagnostic period for the infected population, increasing the risk of spreading the virus [2]. Additionally, less affluent communities often do not have many testing sites or easy access to those sites in their area or do not have the option to work from home to feed their families. In low-to-middle-income countries (LMICs), access to any COVID-19 test, including PCR testing, is lower and does not cover most of the population besides healthcare institutions' efforts. Mexico has 0.09 clinical tests per thousand people compared with 2.44 in the US, which is replicated in similar economies. Therefore, lower accessibility results in a higher clinical test positivity, which indicates that the authorities are only testing the sickest patients and/or people who seek out medical attention [3].

This year, the National Institute of Statistics and Geography (INEGI) [4–6] reported that 31 million of the working-age Mexican population are engaged in occupations in the informal employment sector without options to stay at home for quarantine. Additionally, more than 50% do not have access to health services, which can increase the risk for rapid viral transmission within and between communities. Furthermore, the described situation poses major challenges for epidemiologists who are trying to understand the dynamics of transmission. Individualized testing provides insight into who is currently infected but does not always capture people who are asymptomatic or still developing symptoms.

The Public Education Secretary set forward a vacation period of one month, beginning 14 March 2020, at the beginning of the pandemic, based on the number of increasing clinical cases reported across the country. The federal government established a mandatory quarantine period on 26 March 2020, where all non-essential activities were suspended. The State's Health Secretary of Nuevo Leon, along with the Nuevo Leon government, applied different restrictions to mobility to reduce the increase in COVID-19 cases in the state of Nuevo Leon at that time. On 1 July, the first reopening of the economy occurred, which led to an increase in the number of cases in Monterrey and its metropolitan area. As a result, the State's Health Secretary and Nuevo Leon's government applied new mobility restrictions, which included reducing the capacity of all businesses, scheduled restrictions, and weekend closures. The presented actions were attempts to mitigate the viral spread and provide a buffer for overwhelmed local healthcare systems.

A late response to assessing, mitigating, and preventing the community spread of the virus could be avoided by integrating alternative epidemiological surveillance strategies, including the use of wastewater-based epidemiology. This approach is based on the chemicals and biomarkers present in the urine and feces in raw wastewater to obtain qualitative and quantitative data on the activity of inhabitants within a given wastewater catchment. This technology was applied for norovirus, hepatitis A virus, and poliovirus detection [7] Water contaminated by nonenveloped viruses could be a potential fomite when a poor plumbing system easily generates aerosols and could represent a public health issue [8]. However, SARS-CoV-2-positive feces from COVID-19 patients were used to infect Vero E6 cells to determine the presence of infectious SARS-CoV-2 particles. The isolated viral particles showed no cytopathic effect on these cells, and SARS-CoV-2 was not detected in the supernatants, suggesting the absence of infective viral particles [9].

Recently, it was used for countries such as Italy [10–12], Australia [7], Germany [13], Spain [14–16], India [17], the Netherlands [18] USA [19–23], France [24,25], Ecuador [26], Israel [27], Turkey [28], Japan [29], and Brazil [30] to detect SARS-CoV-2 in wastewater.

In this study, a previously established collaboration between Arizona State University and Tech de Monterrey [31] was rapidly adapted to implement wastewater surveillance to monitor SARS-CoV-2 in wastewater throughout Nuevo León state, the third-largest metropolitan area in Mexico (Figure 1). In 2010, Nuevo León was reported as a predominantly low-to-middle-income city with a total of 4,689,601 inhabitants in an area of 6357 people per km^2 (INEGI). The aims of this study were to (i) determine the feasibility of monitoring the virus in wastewater, (ii) assess spatial and temporal wastewater trends throughout the area, and (iii) compare these trends with available clinical data and mobility rates in other low-to-middle-income country (LMIC) populations.

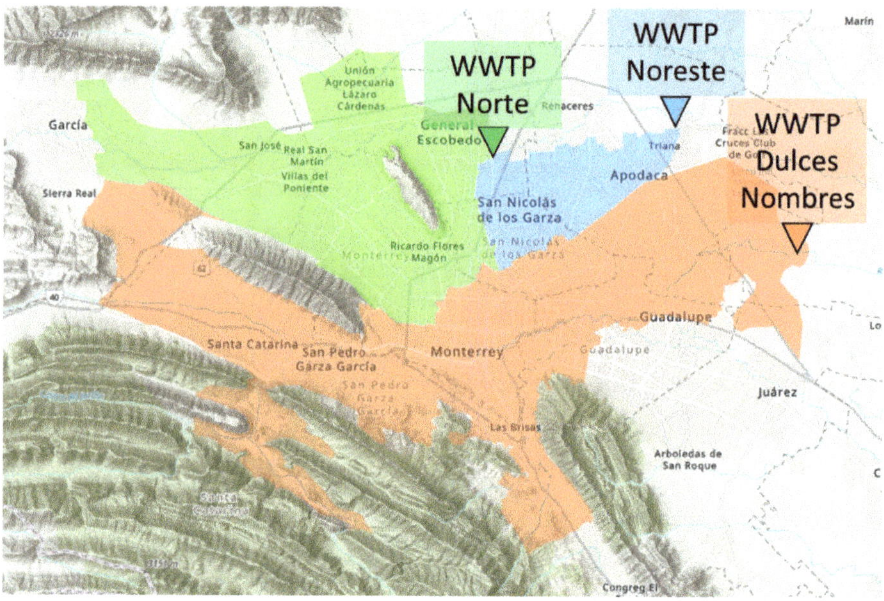

Figure 1. Wastewater catchment area of each wastewater treatment plant, namely, Dulces Nombres (Orange), Norte plant (Green), and Noreste Plant (Blue). The inverted triangles correspond to the WWTP locations. Cadereyta (Cad) WWTP is not shown. Source: Servicios de Agua y Drenaje de Monterrey, 2020.

2. Materials and Methods

2.1. Study Area

The study area covered part of the Monterrey Metropolitan Area (MMA), which is located in the northeast of México in the state of Nuevo León. The municipalities investigated in this study were Apodaca, Garcia, San Pedro Garza García, General Escobedo, Guadalupe, Juárez, San Nicolás de los Garza, Monterrey, El Carmen, Pesquería, and Santa Catarina. This total area had a population of 4.6 M inhabitants and a population density of 6357 inhabitants per km2. Four wastewater treatment plants were sampled; they had the following served populations: Norte with 1,139,123 people (25°47.860′ N, 100°17.483′ W), Noreste with 476,695 (25°48′28.3″ N, 100°09′58.2″ W), Dulces Nombres with 1,695,589 (25°44′16.3″ N, 100°04′09.5″ W), and Cadereyta with 122,337 (25°35′10.0″ N, 99°58′28.8″ W). The population in each catchment area per WWTP does not correspond with the municipality area (as shown in Figure 1). The program ArcGis Pro (version 2.8) was used to determine the exact population in these areas. In short, the shapefiles (geographic

feature format) of the municipalities were downloaded from INEGI (2016) [4] at the level of residential blocks, which was the smallest unit for mapping found in the INEGI data. This data was uploaded to ArcGis Pro and matched with the corresponding WWTP area. Afterward, the population data were updated with the most recent census, which was published this year (INEGI, 2021) [4]. Average daily wastewater flows discharged to each WWTP range were between 17,492 and 605,290 m^3/d. Data obtained from the catchment area of each WWTP were provided by Servicios de Agua y Drenaje de Monterrey.

2.2. *Sample Collection, Storage, and Shipment*

Samples of wastewater influent were collected at each wastewater treatment plant (WWTP) after its primary screening. The sampling method consisted of 1 L grab samples collected each hour, which were used to create a 24 hr composite sample with added volumes weighted according to the corresponding flow rate; the final sample volume was 2 L. Samples were collected in high-density polyethylene (HDPE) bottles and stored on ice at 4 °C until subsequent compositing. Sample collection occurred on Sundays, after which, samples were kept at 4 °C and subsequently shipped to ASU on Tuesday in a cold chain. Shipment was made with an express (next-day) service. The disposable insulator box included ice and gel bags to maintain appropriate temperature and prevent degradation.

2.3. *Sample Processing and Analysis*

The methodological details were previously published [32,33]. Briefly, raw wastewater was filtered using a two-step process: 0.45 µm polyethersulfone (PES) membrane filter unit (Fisher Scientific, Lenexa, KS, USA), then a 0.22 µm Minisart® PES syringe filter (Sartorius, Bohemia, NY, USA). The filtrate was concentrated on Amicon® ultra 15 centrifugal filters with a 10,000 molecular weight cutoff (MWCO) (Millipore Sigma, Burlington, MA, USA) and extracted using a Qiagen Rneasy Mini Kit (Qiagen, Germantown, MD, USA) following a modified version of the RNA purification. Reverse transcriptase–quantitative polymerase chain reaction (RT-qPCR) was performed on an Applied Biosystems QuantStudioTM 3 Real-Time PCR System with the QuantStudio Design and Analysis Software 1.3 (Thermo Fisher Scientific, Waltham, MA, USA) using a SuperScript™ III One-Step RT-PCR System with Platinum™ Taq DNA Polymerase (Invitrogen, Carlsbad, CA, USA) and the Charité/Berlin (World Health Organization) protocol primer and probe E (envelope) gene target (Integrated DNA Technologies, Coralville, IA, USA).

3. Results

The goal of this study was to use an existing international collaborative network in the US and Mexico to implement SARS-CoV-2 monitoring in wastewater in predominantly low-to-middle-income cities to aid in the public health response.

3.1. *SARS-CoV-2 in Wastewater in the Monterrey Metropolitan Area*

The sewer network dedicated to the Monterrey metropolitan area was composed of three catchment areas across several municipalities presented in Figure 1; from largest to smallest catchment size, this involved Dulces Nombres, Norte, and Noreste, serving a total of 4.34 million people, while the fourth location corresponds to Cadereyta catchment (not included in the figure, located at coordinates: 25.585683, −99.974523). Wastewater samples were analyzed from 16 April 2020 to 31 January 2021, with the first SARS-CoV-2 genome copies detected on 27 June from three WWTPs including: "Dulces Nombres" ("DNo", which partially includes the Apodaca, Monterrey, Juárez, Santa Catarina, San Pedro Garza García, Pesquería, and Guadalupe municipalities), "Noreste" ("Nes", which partially includes the Apodaca, General Escobedo, and San Nicolás de los Garza municipalities), and "Cadereyta" ("Cad", which includes the Cadereyta municipality). The first SARS-CoV-2 detection at "Norte" ("Nte", which partially includes the El Carmen, García, General Escobedo, Monterrey, and San Nicolás de los Garza municipalities) occurred the following day on 28 June. At this time, there were 213 confirmed new COVID-19 cases reported by the State's Health Secretary, with a seven-day moving average of 303 new clinical cases. The observed SARS-CoV-2 quantification in genome copies per day (calculated according to the WW flow per day) is shown with the daily number of cases reported by the State's Health Secretary confirmed with the clinical PCR testing (Figure 2).

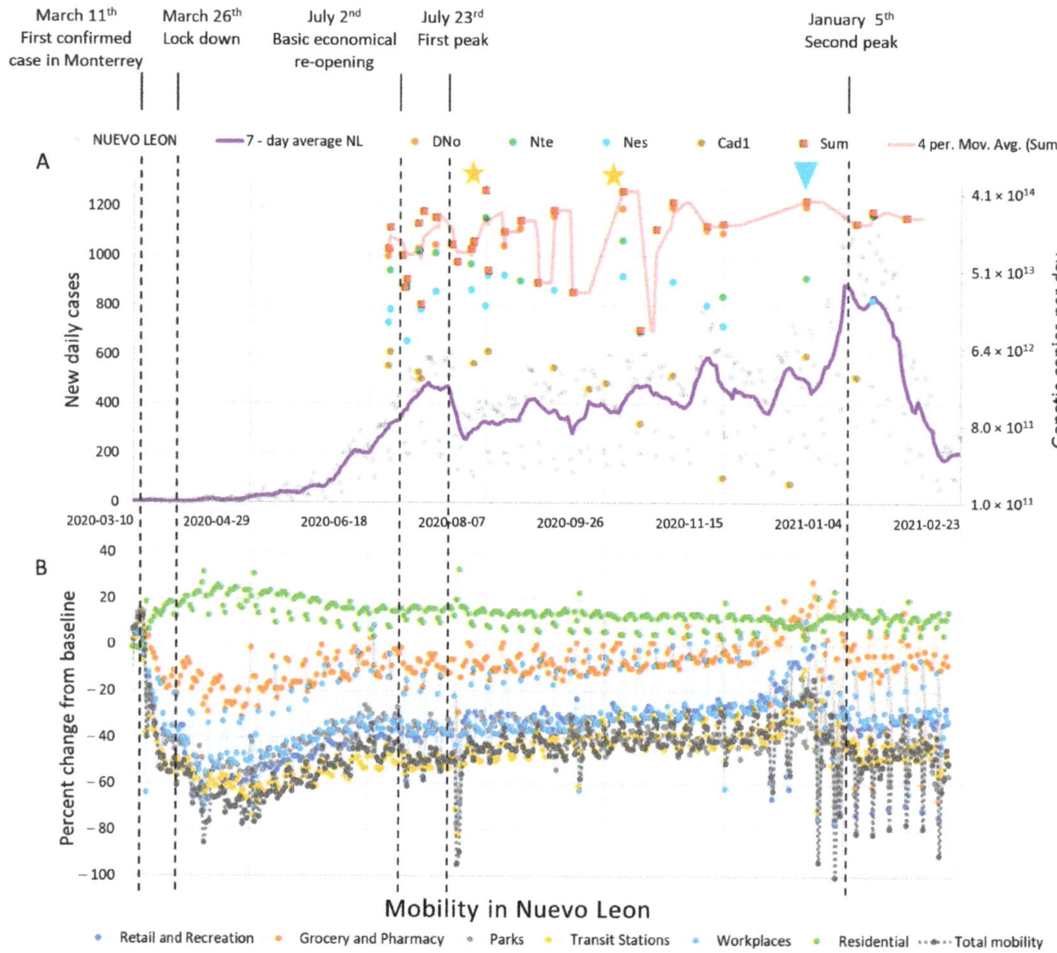

Figure 2. Wastewater monitoring (**A**) and mobility (**B**). The first graph (**A**) shows the official new COVID-19 cases per day (light gray circles) and its seven-day moving window average (purple continuous line) in Nuevo Leon state, México. The data were obtained from COVID-19 México from https://datos.covid-19.conacyt.mx/#DownZCSV (accessed on 16 September 2021) [34], along with the genome copies per day sum for all WWTPs in red squares with a four-day moving window average shown by a light red line. At the top, important events are highlighted with gray dashed lines, where two yellow stars mark the higher quantifications and an inverted blue triangle marks the higher quantification before the second wave peak. (**B**) Mobility reports from Nuevo Leon reported by the Google platform are plotted in the same time frame. The mobility report was separated into the following categories: retail and recreation, grocery and pharmacy, workplaces, residential, and total mobility. Each category was based on the time spent per day in each location type. This information was compared with the new cases and WBE information corresponding to the Monterrey metropolitan area since more than 80% of the population of Nuevo León was located in this area.

In Figure 3, SARS-CoV-2 wastewater-derived results are shown individually for each WWTP with the corresponding catchments' new clinical cases. The Dulces Nombres WWTP had peak genome copies per L of measurements of 4.05×10^5 on 12 July, and the next day, a high in new daily clinical cases was reported (~270 new cases), followed ten days later by another peak in cases (~280 new cases), which was the highest in the first wave of the pandemic. In each case, the highest measured genome copies of SARS-CoV-2 detected in wastewater would correspond to an increase in the number of cases, anywhere from 1 to 10 days after the measured peak in wastewater. Similar results were seen for Norte and Cadereyta WWTPs. The Noreste WWTP showed that the measured SARS-CoV-2 genome copies per L were a leading indicator of increasing clinical caseloads during the first wave of the pandemic but not in the second.

3.2. *COVID-19 Confirmed Clinical Cases vs. WW SARS-CoV-2 Viral Copies in the Monterrey Metropolitan Area*

According to the SARS-CoV-2 Mexican Daily Technical Report, on 9 February 2021, a total of 1,995,892 cases and 174,657 deaths from SARS-CoV-2 were confirmed. In Nuevo Leon, Mexico, 159,814 SARS-CoV-2 cases and 8207 deaths were confirmed from 10 March 2020 to 9 February 2021, with 145,153 confirmed cases in the study area, or 91% of total cases reported in COVID-19 Tablero México—CONACYT (accessed on 15 September 2021) [34]. As described in previous works, wastewater-derived SARS-CoV-2 genome copies in wastewater correlate with the number of active patients contributing to the sewershed [17]. In this study, the same result was observed (Figure 2). During the first wave of increasing clinical cases, SARS-CoV-2 genome copies per L in wastewater samples from the four WWTPs also increased. Similarly, from 12 to 26 July, a decrease in the average of genome copies/L in wastewater samples from all WWTPs occurred with a subsequent reduction in the reported clinical cases (from 612 to 355 new cases) from 23 to 30 July.

Peaks of viral genome copies per L were detected in the four WWTPs during the first wave on 7 August and on 4 October 2020, after the first wave (1.6×10^6 and 1.5×10^6, respectively) (see Figure 2, yellow stars). Two weeks before an increase of 780 clinical cases on 6 September, an increase in genome copies/L was observed before the second wave peak, which was as high as 1.0×10^{13} to 3.3×10^{14} copies. On 20 December, before the second wave peak (5 January 2021), 3.5×10^{14} genome copies were detected, as high as the previously detected peaks of viral genome copies in wastewater samples. This result suggests less clinical testing and an underestimation of active COVID-19 cases in the Monterrey Metropolitan area during the plateau of the first wave (from August to October).

During the second wave, we found an increase in the genome copies in wastewater samples on 20 December, just when the holidays started (blue triangle) (Figure 2). On 5 January, the clinical testing reports rose to the maximum number of cases observed so far since the beginning of the pandemic, reporting 1162 new COVID-19 cases that day, with an average of 889 new cases daily. As we present in this work, the WBE results predicted this maximum peak two weeks after the increased number of genome copies in the WW samples (Figure 2). Our data suggests a peak of cases predicted during this second wave, with a robust two weeks notice.

There are several reports where the RNA from SARS-CoV-2 was monitored for over a year. In Munich [35], around 1.5 million habitants were monitored through wastewater weekly for a year, sampling in six areas, where the sewage viral load preceded the incidence numbers by three weeks. This timelapse of the leading indicator was variable between the six areas. Averages of 4 days to 3 weeks were identified in several studies [23,36]. In San Diego, CA, US, 2.3 million residents were monitored for a period of 3 months (July to October 2020), where the peaks of viral load quantification predicted the increase in COVID-19 cases in the area up to three weeks in advance in most cases. This study suggested that this system is helpful for community surveillance in vulnerable populations [36]. In Massachusetts, from January to May 2020, the viral load trends in the sewage appeared between 4–10 days earlier than the clinical data [23].

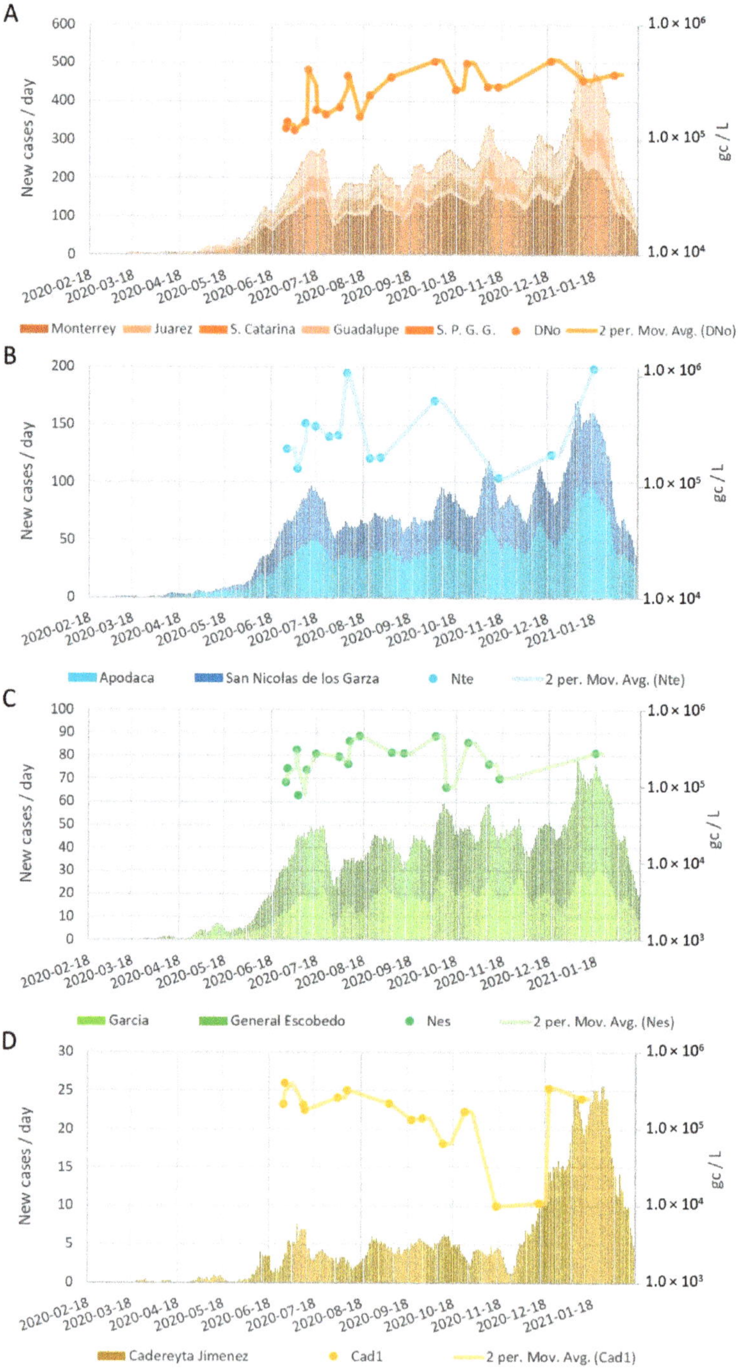

Figure 3. New cases per municipality associated with WWTPs that serve the Monterrey metropolitan area vs. genetic copies per liter in each WWTP: (**A**) Dulces Nombres, (**B**) Norte, (**C**) Noreste, and (**D**) Cadereyta 1.

In Latin America, an epidemiological dynamics report of a 1.2 million population in Argentina from July 2020 to January 2021 showed that increases in SARS-CoV-2 genome copies detection in sewage samples were 3–6 days ahead of the clinical COVID-19 weekly confirmed cases [37]. The only study realized in Mexico was in Queretaro city, where they correlated the accumulated number of new COVID-19 cases with the number of viral genome copies/mL detected in wastewater samples [38].

3.3. SARS-CoV-2 in WW and Mobility Info in the Monterrey Metropolitan Area

Mexico is currently among the 15 countries with the highest confirmed cases of SARS-CoV-2 despite the efforts made by the government. The Mexican government implemented a series of measures beginning on 26 March 2020, such as the suspension until 30 April of non-essential activities in the public, private, and social sectors (all activities except construction, mining, transport equipment development, as well as occupations related to the issuance of health and safety at the workplace); the banning of meetings of more than 50 people in essential sectors; the implementation of basic hygiene, prevention, and healthy distance measures; and urging the population to take co-responsible domiciliary guard, which involved strictly providing home protection to anyone over 60 years of age, pregnant women, or people suffering from chronic or autoimmune diseases [39]. These measures were taken to reduce the number of people infected; the contagion curve showed a lower trend than in other countries for the days at the beginning of infection monitoring (Figure 2).

From 30 March to 18 May, when there was a partial reopening of activities in the country, there was an increase of 46,296 confirmed cases. After 18 May, in the same period length (50 days), there was an increase in confirmed cases of 205,021, meaning that SARS-CoV-2 cases increased four-fold. It is not strange to observe that from 18 May, the exponential phase of contagion and hospitalized SARS-CoV-2 cases in the country approximately began in Nuevo Leon and the study area. Although economic reactivation was necessary, this reactivation caused a significant increase in the number of people infected and hospitalized with SARS-CoV-2. Despite this increase, the number of total confirmed cases in Mexico was lower in comparison to other countries. It can be considered that the actions of the Mexican government were correct, and yet these actions could have been even better if the WBE methodology had been implemented to know the estimated cases that were not reported, such as the asymptomatic cases or cases that lacked diagnosis. Measures such as lockdowns and mobility restrictions were applied by the government to limit the virus from becoming widespread around the world. In France, the COVID-19 epidemic's dynamics were evaluated through SARS-CoV-2 detection in wastewater samples from March to April during the lockdown period. The lockdown effect showed a decrease in the quantities of viral genome copies 29 days after, simultaneously with the reduction in the number of new clinically reported cases [25].

In this study, a mobility data report in Local mobility reports on COVID-19 (accessed on 26 September 2021) [40] was separated into the following categories: retail and recreation, grocery and pharmacy, workplaces, residential, and total mobility. This information included more than 80% of the population of Nuevo Leon since it is located in the metropolitan area. These data were used to compare the mobility and the rise of COVID-19 clinical cases, to monitor the government's mobility restrictions and the pandemic behavior, and compare with our data.

During the first wave of infections, the economy reopening activities began on 2 July (Figure 2B), where all the essential activities were allowed to operate full-time, while non-essential activities were allowed to operate during an established schedule at 50% capacity. Mobility was reduced to 25% capacity in the commercial sector, and services (including restaurants) with an established schedule during weekdays closed on the weekends [41]. After this announcement, there was an average increase of measured SARS-CoV-2 genome copies detected in the four WWTPs, especially in August, after which, there was an increase in the number of clinical cases. At this time there was a greater need for clinical testing, and the health system started to limit the testing to only symptomatic people. It was assumed that the rise in the number of cases was limited due to testing; therefore, a real comparison was not observed in this first wave. To achieve a direct comparison, the testing strategy should have included a higher number of random clinical testing samples since Mexico had the lowest rate of clinical COVID-19 testing (0.6 tests per 1000 population) in the Organization for Economic Co-operation and Development (OECD) countries, where the average was 27.7 [38,42].

When the first COVID-19 case was reported, retail and recreation mobility decreased, as observed in Figure 2B (blue dots), as well as total mobility (gray dots). Jobs mobility was restricted to only essential jobs, and it was observed that the total mobility decreased to 30% of the baseline in May compared with the last year 2019 [5]. Public transportation served 21.5 million people per month, where 1.99 million people were active workers [6]. After the mandatory lockdown, retail and recreation mobility increased considerably from 30 April to when the economy reopened on 2 July. After this period this mobility was maintained until December when the total mobility and the retail and recreation mobility started to rise again and it coincided with the job mobility reduction (Figure 2B). With this, an upswing in the genome copies/L was observed in wastewater and clinical testing, where its maximum number of cases was reported on January 5 (1162) and the wastewater measurement increase was a leading indicator for the rise in the number of cases with a two-week advance (Figure 2).

In addition to the analysis of raw wastewater samples, treated effluent was also examined. It was found that SARS-CoV-2 was not detected in treated wastewater. It is believed that treatment steps in the WWTP acted as a retainer of SARS-CoV-2 genetic material, especially in the activated sludge step [43].

3.4. Limitations of the WBE Methodology in SARS-CoV-2 Detection

As an international collaborative study, samples were shipped from Mexico to the US, presenting a lag time of ~5.5 days between sample collection and sample receipt. Samples were kept at 4 °C before shipment to avoid excess temperature increases. However, longer hold times were shown to reduce the recoverable virus in wastewater samples [7]. Consequently, the detected concentrations of SARS-CoV-2 (genome copies/L) may have been higher than the measured values, indicating a higher estimated disease burden in the communities than originally thought. Unfortunately, the international shipment was the only option since no local laboratories were capable of measuring SARS-CoV-2 in wastewater at that time and it would not be possible until 2021.

4. Conclusions

This work highlighted the importance of the role WBE can play in providing information regarding COVID-19 community infections in LMIC populations. It was demonstrated that the wastewater-based epidemiology methodology could be useful in disease monitoring and the necessity of strategically placing WBE control centers in target communities.

LMIC strategies to assess the pandemic through clinical testing lagged regarding the representation of the actual number of COVID-19 clinical cases. The first detection of genome copies in wastewater occurred on 27–28 June when the average clinical cases were 303 new daily cases. A correlation was demonstrated between the wastewater genomic copies and the clinical case trends. During the first wave, the increase in clinical cases agreed with the wastewater genomic copies increase, and similarly with the decrease. Moreover, during the second wave, an increase in genomic copies was detected around 15 days ahead of the highest peak of clinical cases. Wastewater-based epidemiology is a synergic tool that is used for clinical testing to accurately describe viral circulation in the general population regardless of the access to clinical testing. The difference between reported clinical cases in the first and second peaks contrasted with the same level of genomic copies in wastewater demonstrated its relevance regarding overcoming the gaps in clinical testing generated by infrastructure capacity, economic challenges, and cultural adaptation.

Finally, the use of WBE as a tool to follow up and evaluate emerging health policies on the pandemic control is effective. Additionally, it can be complemented with free available tools on mobility categorized by activity. A complete system requires a set of tools from which this work has showcased three synergic parts that may be further explored.

Author Contributions: Conceptualization, J.E.S.-H., M.A.O.-M., E.M.M.-M. and M.M.-R.; methodology, M.A.O.-M., E.M.D., D.A.B., R.S.F., L.A.H., M.E.N., S.A., R.K. and S.K.; validation, E.M.D. and D.A.B.; formal analysis, J.E.S.-H., M.A.O.-M., E.M.D., D.A.B. and S.A.P.-B.; investigation, J.E.S.-H., M.A.O.-M., E.M.D., D.A.B., S.K., R.S.F., L.A.H., M.E.N., S.A., R.K., R.U.H. and R.P.-S.; data curation, J.E.S.-H., E.M.M.-M., E.M.D., D.A.B., S.L.L.-S. and S.A.P.-B.; writing—original draft preparation, J.E.S.-H., M.A.O.-M., E.M.M.-M., S.L.L.-S., L.P.-A., S.A.P.-B., M.M.-R., L.I.R.-Z., I.Y.L.-P. and C.C.-Z.; writing—review and editing, E.M.D., D.A.B., M.E.N., S.A., R.K., H.M.N.I., S.K., R.S.F., L.A.H., E.S.L., A.V., D.S.-L., R.U.H. and R.P.-S.; visualization, J.E.S.-H., M.A.O.-M., S.L.L.-S., L.P.-A., S.A.P.-B., L.I.R.-Z., I.Y.L.-P. and C.C.-Z.; supervision, J.E.S.-H., E.M.D., H.M.N.I., E.S.L., A.V., D.S.-L., R.U.H. and R.P.-S.; project administration, J.E.S.-H., E.M.D., R.U.H. and R.P.-S.; funding acquisition, J.E.S.-H., R.U.H. and R.P.-S. All authors have read and agreed to the published version of the manuscript.

Funding: We are thankful to Tecnologico de Monterrey for providing financial support to conduct this research.

Institutional Review Board Statement: Not applicable.

Informed Consent Statement: Not applicable.

Data Availability Statement: Not applicable.

Acknowledgments: This project acknowledges the Biotechnology Center—FEMSA and the recently created ad hoc laboratory MARTEC from Tecnológico de Monterey for the development of the project. We also acknowledge the Arizona State University and Biodesign Institute for the fast shift in the previous collaboration supported by the Global Consortium for Sustainability Outcomes with the project title "Facilitating Evidence-based Decision-making for Global Health Out-comes" during the COVID-19 pandemic to increase the capacity for the sample processing to carry out this study. We are thankful to the GIECC Research Group from Tecnológico de Monterey for providing us with the ArcGis Pro license and Duvan Alfonso Alvarado Pachon for the guidance to use the program. CONACYT is thankfully acknowledged for partially supporting this work under the Sistema Nacional de Investigadores (SNI) program awarded to Juan Eduardo Sosa-Hernández (CVU: 375202), Mariel A. Oyervides-Muñoz (CVU: 422778), Elda M. Melchor-Martínez (CVU: 230784), Hafiz M.N. Iqbal (CVU: 735340), Roberto Parra-Saldívar (CVU: 35753), Manuel Martínez-Ruiz (CVU: 418151), and Carlos Castillo-Zacarías (CVU: 359310). Finally, we are thankful to Servicios de Agua y Drenaje de Monterrey for the collaboration in this project.

Conflicts of Interest: R.U.H. and E.M.D. are cofounders of AquaVitas, LLC, 9260 E. Raintree, Ste 130, Scottsdale, AZ 85260, USA, which is an Arizona State University startup company providing commercial services in wastewater based epidemiology. R.U.H. is also the founder of OneWaterOneHealth, which is a non-profit project of the Arizona State University Foundation.

References

1. Mexico: WHO Coronavirus Disease (COVID-19) Dashboard With Vaccination Data | WHO Coronavirus (COVID-19) Dashboard With Vaccination Data n.d. Available online: https://covid19.who.int/region/amro/country/mx (accessed on 31 May 2022).
2. Symptoms of COVID-19 | CDC n.d. Available online: https://www.cdc.gov/coronavirus/2019-ncov/symptoms-testing/symptoms.html (accessed on 31 May 2022).
3. Coronavirus Pandemic (COVID-19)—Our World in Data n.d. Available online: https://ourworldindata.org/coronavirus#citation (accessed on 31 May 2022).
4. SCITEL n.d. Available online: https://www.inegi.org.mx/app/scitel/Default?ev=10 (accessed on 31 May 2022).
5. Transporte de Pasajeros n.d. Available online: https://www.inegi.org.mx/temas/transporteurb/ (accessed on 31 May 2022).
6. Resultados De La Encuesta Nacional De Ocupación Y Empleo. Nueva Edición (Enoen) Cifras Durante El Cuarto Trimestre De 2020 n.d. Available online: https://www.inegi.org.mx/contenidos/saladeprensa/boletines/2021/enoe_ie/enoe_ie2021_02.pdf (accessed on 31 May 2022).

7. Ahmed, W.; Angel, N.; Edson, J.; Bibby, K.; Bivins, A.; O'Brien, J.W.; Choi, P.M.; Kitajima, M.; Simpson, S.L.; Li, J.; et al. First confirmed detection of SARS-CoV-2 in untreated wastewater in Australia: A proof of concept for the wastewater surveillance of COVID-19 in the community. *Sci. Total Environ.* **2020**, *728*, 138764. [CrossRef] [PubMed]
8. Panchal, D.; Prakash, O.; Bobde, P.; Pal, S. SARS-CoV-2: Sewage surveillance as an early warning system and challenges in developing countries. *Environ. Sci. Pollut. Res.* **2021**, *28*, 22221–22240. [CrossRef] [PubMed]
9. Albert, S.; Ruíz, A.; Pemán, J.; Salavert, M.; Domingo-Calap, P. Lack of evidence for infectious SARS-CoV-2 in feces and sewage. *Eur. J. Clin. Microbiol.* **2021**, *40*, 2665–2667. [CrossRef]
10. La Rosa, G.; Iaconelli, M.; Mancini, P.; Ferraro, G.B.; Veneri, C.; Bonadonna, L.; Lucentini, L.; Suffredini, E. First detection of SARS-CoV-2 in untreated wastewaters in Italy. *Sci. Total Environ.* **2020**, *736*, 139652. [CrossRef] [PubMed]
11. Rimoldi, S.G.; Stefani, F.; Gigantiello, A.; Polesello, S.; Comandatore, F.; Mileto, D.; Maresca, M.; Longobardi, C.; Mancon, A.; Romeri, F.; et al. Presence and infectivity of SARS-CoV-2 virus in wastewaters and rivers. *Sci. Total Environ.* **2020**, *744*, 140911. [CrossRef]
12. La Rosa, G.; Mancini, P.; Ferraro, G.B.; Veneri, C.; Iaconelli, M.; Bonadonna, L.; Lucentini, L.; Suffredini, E. SARS-CoV-2 has been circulating in northern Italy since December 2019: Evidence from environmental monitoring. *Sci. Total Environ.* **2021**, *750*, 141711. [CrossRef]
13. Westhaus, S.; Weber, F.-A.; Schiwy, S.; Linnemann, V.; Brinkmann, M.; Widera, M.; Greve, C.; Janke, A.; Hollert, H.; Wintgens, T.; et al. Detection of SARS-CoV-2 in raw and treated wastewater in Germany—Suitability for COVID-19 surveillance and potential transmission risks. *Sci. Total Environ.* **2020**, *751*, 141750. [CrossRef]
14. Chavarria-miró, G.; Anfruns-Estrada, E.; Guix, S.; Paraira, M.; Galofré, B.; Sánchez, G.; Pintó, R.M.; Bosch, A. Sentinel Surveillance of SARS-CoV-2 in Wastewater Anticipates the Occurrence of SARS-CoV-2 Cases. *medRxiv* **2020**. [CrossRef]
15. Randazzo, W.; Cuevas-Ferrando, E.; Sanjuán, R.; Domingo-Calap, P.; Sánchez, G. Metropolitan wastewater analysis for COVID-19 epidemiological surveillance. *Int. J. Hyg. Environ. Health* **2020**, *230*, 113621. [CrossRef]
16. Randazzo, W.; Truchado, P.; Cuevas-Ferrando, E.; Simón, P.; Allende, A.; Sánchez, G. SARS-CoV-2 RNA in wastewater anticipated COVID-19 occurrence in a low prevalence area. *Water Res.* **2020**, *181*, 115942. [CrossRef]
17. Kumar, M.; Patel, A.K.; Shah, A.V.; Raval, J.; Rajpara, N.; Joshi, M.; Joshi, C.G. First proof of the capability of wastewater surveillance for COVID-19 in India through detection of genetic material of SARS-CoV-2. *Sci. Total Environ.* **2020**, *746*, 141326. [CrossRef] [PubMed]
18. Medema, G.; Heijnen, L.; Elsinga, G.; Italiaander, R.; Brouwer, A. Presence of SARS-Coronavirus-2 RNA in Sewage and Correlation with Reported COVID-19 Prevalence in the Early Stage of the Epidemic in The Netherlands. *Environ. Sci. Technol. Lett.* **2020**, *7*, 511–516. [CrossRef]
19. Green, H.; Wilder, M.; Collins, M.; Fenty, A.; Gentile, K.; Kmush, B.L.; Zeng, T.; Middleton, F.A.; Larsen, D.A. Quantification of SARS-CoV-2 and cross-assembly phage (crAssphage) from wastewater to monitor coronavirus transmission within communities. *medRxiv* **2020**, *2*, 1–18. [CrossRef]
20. Nemudryi, A.; Nemudraia, A.; Wiegand, T.; Surya, K.; Buyukyoruk, M.; Cicha, C.; Vanderwood, K.K.; Wilkinson, R.; Wiedenheft, B. Temporal Detection and Phylogenetic Assessment of SARS-CoV-2 in Municipal Wastewater. *Cell Rep. Med.* **2020**, *1*, 100098. [CrossRef]
21. Peccia, J.; Zulli, A.; Brackney, D.E.; Grubaugh, N.D.; Kaplan, E.H.; Casanovas-Massana, A.; Ko, A.I.; Malik, A.A.; Wang, D.; Wang, M.; et al. SARS-CoV-2 RNA concentrations in primary municipal sewage sludge as a leading indicator of COVID-19 outbreak dynamics. *medRxiv* **2020**. [CrossRef]
22. Sherchan, S.P.; Shahin, S.; Ward, L.M.; Tandukar, S.; Aw, T.G.; Schmitz, B.; Ahmed, W.; Kitajima, M. First detection of SARS-CoV-2 RNA in wastewater in North America: A study in Louisiana, USA. *Sci. Total. Environ.* **2020**, *743*, 140621. [CrossRef]
23. Wu, F.; Zhang, J.; Xiao, A.; Gu, X.; Lee, W.L.; Armas, F.; Kauffman, K.; Hanage, W.; Matus, M.; Ghaeli, N.; et al. SARS-CoV-2 Titers in Wastewater Are Higher than Expected from Clinically Confirmed Cases. *mSystems* **2020**, *5*, e00614-20. [CrossRef]
24. Trottier, J.; Darques, R.; Mouheb, N.A.; Partiot, E.; Bakhache, W.; Deffieu, M.S.; Gaudin, R. Post-lockdown detection of SARS-CoV-2 RNA in the wastewater of Montpellier, France. *One Health* **2020**, *10*, 100157. [CrossRef]
25. Wurtzer, S.; Marechal, V.; Mouchel, J.M.; Maday, Y.; Teyssou, R.; Richard, E.; Almayrac, J.L.; Moulin, L. Evaluation of lockdown impact on SARS-CoV-2 dynamics through viral genome quantification in Paris wastewaters. *medRxiv* **2020**. [CrossRef]
26. Guerrero-Latorre, L.; Ballesteros, I.; Villacrés-Granda, I.; Granda, M.G.; Freire-Paspuel, B.; Ríos-Touma, B. SARS-CoV-2 in river water: Implications in low sanitation countries. *Sci. Total Environ.* **2020**, *743*, 140832. [CrossRef]
27. Bar-Or, I.; Yaniv, K.; Shagan, M.; Ozer, E.; Weil, M.; Indenbaum, V.; Elul, M.; Erster, O.; Mendelson, E.; Mannasse, B.; et al. Regressing SARS-CoV-2 Sewage Measurements Onto COVID-19 Burden in the Population: A Proof-of-Concept for Quantitative Environmental Surveillance. *Front. Public Health* **2022**, *9*, 561710. [CrossRef] [PubMed]
28. Kocamemi, B.A.; Kurt, H.; Hacıoglu, S.; Yaralı, C.; Saatci, A.M.; Pakdemirli, B. First data-set on SARS-CoV-2 detection for Istanbul wastewaters in Turkey. *medRxiv* **2020**. [CrossRef]
29. Haramoto, E.; Malla, B.; Thakali, O.; Kitajima, M. First environmental surveillance for the presence of SARS-CoV-2 RNA in wastewater and river water in Japan. *Sci. Total Environ.* **2020**, *737*, 140405. [CrossRef]
30. Fongaro, G.; Stoco, P.H.; Souza, D.S.M.; Grisard, E.C.; Magri, M.E.; Rogovski, P.; Schörner, M.A.; Barazzetti, F.H.; Christoff, A.P.; de Oliveira, L.F.V.; et al. The presence of SARS-CoV-2 RNA in human sewage in Santa Catarina, Brazil, November 2019. *Sci. Total Environ.* **2021**, *778*, 146198. [CrossRef] [PubMed]

31. Ng, K.T.; Rapp-Wright, H.; Egli, M.; Hartmann, A.; Steele, J.C.; Sosa-Hernández, J.E.; Melchor-Martínez, E.M.; Jacobs, M.; White, B.; Regan, F.; et al. High-throughput multi-residue quantification of contaminants of emerging concern in wastewaters enabled using direct injection liquid chromatography-tandem mass spectrometry. *J. Hazard. Mater.* **2020**, *398*, 122933. [CrossRef] [PubMed]
32. Holland, L.A.; Kaelin, E.A.; Maqsood, R.; Estifanos, B.; Wu, L.I.; Varsani, A.; Halden, R.U.; Hogue, B.G.; Scotch, M.; Lim, E.S. An 81-Nucleotide Deletion in SARS-CoV-2 ORF7a Identified from Sentinel Surveillance in Arizona (January to March 2020). *J. Virol.* **2020**, *94*, e00711-20. [CrossRef] [PubMed]
33. Bowes, D.A.; Driver, E.M.; Kraberger, S.; Fontenele, R.S.; Holland, L.A.; Wright, J.; Johnston, B.; Savic, S.; Newell, M.E.; Adhikari, S.; et al. Unrestricted Online Sharing of High-frequency, High-resolution Data on SARS-CoV-2 in Wastewater to Inform the COVID-19 Public Health Response in Greater Tempe, Arizona. *medRxiv* **2021**. [CrossRef]
34. COVID-19 Tablero México—CONACYT—CentroGeo—GeoInt—DataLab n.d. Available online: https://datos.covid-19.conacyt.mx/#DownZCSV (accessed on 31 May 2022).
35. Rubio-Acero, R.; Beyerl, J.; Muenchhoff, M.; Roth, M.S.; Castelletti, N.; Paunovic, I.; Radon, K.; Springer, B.; Nagel, C.; Boehm, B.; et al. Spatially resolved qualified sewage spot sampling to track SARS-CoV-2 dynamics in Munich—One year of experience. *Sci. Total Environ.* **2021**, *797*, 149031. [CrossRef]
36. Karthikeyan, S.; Ronquillo, N.; Belda-Ferre, P.; Alvarado, D.; Javidi, T.; Longhurst, C.A.; Knight, R. High-Throughput Wastewater SARS-CoV-2 Detection Enables Forecasting of Community Infection Dynamics in San Diego County. *mSystems* **2021**, *6*, e00045-21. [CrossRef]
37. Giraud-Billoud, M.; Cuervo, P.; Altamirano, J.C.; Pizarro, M.; Aranibar, J.N.; Catapano, A.; Cuello, H.; Masachessi, G.; Vega, I.A. Monitoring of SARS-CoV-2 RNA in wastewater as an epidemiological surveillance tool in Mendoza, Argentina. *Sci. Total Environ.* **2021**, *796*, 148887. [CrossRef]
38. Carrillo-Reyes, J.; Barragán-Trinidad, M.; Buitrón, G. Surveillance of SARS-CoV-2 in sewage and wastewater treatment plants in Mexico. *J. Water Process Eng.* **2020**, *40*, 101815. [CrossRef] [PubMed]
39. Sala de Prensa IMSS. Gobierno de Mexico n.d. Available online: http://www.imss.gob.mx/prensa/archivo/202003 (accessed on 1 June 2022).
40. Local mobility reports on COVID-19 n.d. Available online: https://www.google.com/covid19/mobility/ (accessed on 1 June 2022).
41. Presentación de la "Actualización de medidas de mitigación contra COVID-19" | 02-07-2020 | Gobierno del Estado de Nuevo León n.d. Available online: https://www.nl.gob.mx/publicaciones/presentacion-de-la-actualizacion-de-medidas-de-mitigacion-contra-covid-19-02-07-2020 (accessed on 1 June 2022).
42. Hasell, J.; Mathieu, E.; Beltekian, D.; Macdonald, B.; Giattino, C.; Ortiz-Ospina, E.; Roser, M.; Ritchie, H. A cross-country database of COVID-19 testing. *Sci. Data* **2020**, *7*, 345. [CrossRef] [PubMed]
43. Balboa, S.; Mauricio-Iglesias, M.; Rodriguez, S.; Martínez-Lamas, L.; Vasallo, F.J.; Regueiro, B.; Lema, J.M. The fate of SARS-COV-2 in WWTPS points out the sludge line as a suitable spot for detection of COVID-19. *Sci. Total Environ.* **2021**, *772*, 145268. [CrossRef] [PubMed]

water

Article

Imprints of Lockdown and Treatment Processes on the Wastewater Surveillance of SARS-CoV-2: A Curious Case of Fourteen Plants in Northern India

Sudipti Arora [1,*,†], Aditi Nag [1,†], Ankur Rajpal [2], Vinay Kumar Tyagi [2,*], Satya Brat Tiwari [3], Jasmine Sethi [1], Devanshi Sutaria [1], Jayana Rajvanshi [1], Sonika Saxena [1], Sandeep Kumar Shrivastava [4], Vaibhav Srivastava [5], Akhilendra Bhushan Gupta [6], Absar Ahmed Kazmi [2] and Manish Kumar [7]

1 Dr. B. Lal Institute of Biotechnology, 6-E, Malaviya Industrial Area, Malviya Nagar, Jaipur 302017, India; aditinag1@gmail.com (A.N.); sethi.28.jasmine@gmail.com (J.S.); devanshisnehal@gmail.com (D.S.); rajvanshi97@gmail.com (J.R.); sonika@blalbiotech.com (S.S.)
2 Environmental Biotechnology Group (EBiTG), Department of Civil Engineering, Indian Institute of Technology, Roorkee 247667, India; ankur.envt@gmail.com (A.R.); kazmifce@iitr.ac.in (A.A.K.)
3 School of Civil and Environmental Engineering, Nanyang Technological University, Singapore 637141, Singapore; sbtiwarigroup@gmail.com
4 Centre for Innovation, Research & Development (CIRD), Dr. B. Lal Clinical Laboratory Pvt. Ltd., Jaipur 302017, India; sandeepshrivastava@blallab.com
5 Discipline of Earth Science, Indian Institute of Technology Gandhinagar, Gujarat 382-355, India; vaibhavsri921@gmail.com
6 Department of Civil Engineering, Malaviya National Institute of Technology, Jaipur 302017, India; abgupta.ce@mnit.ac.in
7 School of Engineering, University of Petroleum & Energy Studies, Dehradun 248007, India; manish.kumar@ddn.upsc.ac.in
* Correspondence: sudiptiarora@gmail.com or sudiptiarora@blalbiotech.com (S.A.); vinayiitrp@gmail.com or vinay.tyagi@ce.iitr.ac.in (V.K.T.); Tel.: +91-9829675677 (S.A.)
† Sudipti Arora and Aditi Nag are equal contributors and thus shared the first authorship.

Citation: Arora, S.; Nag, A.; Rajpal, A.; Tyagi, V.K.; Tiwari, S.B.; Sethi, J.; Sutaria, D.; Rajvanshi, J.; Saxena, S.; Shrivastava, S.K.; et al. Imprints of Lockdown and Treatment Processes on the Wastewater Surveillance of SARS-CoV-2: A Curious Case of Fourteen Plants in Northern India. *Water* **2021**, *13*, 2265. https://doi.org/10.3390/w13162265

Academic Editor: William Frederick Ritter

Received: 1 July 2021
Accepted: 15 August 2021
Published: 19 August 2021

Abstract: The present study investigated the detection of severe acute respiratory syndrome–coronavirus 2 (SARS-CoV-2) genomes at each treatment stage of 14 aerobic wastewater treatment plants (WWTPs) serving the major municipalities in two states of Rajasthan and Uttarakhand in Northern India. The untreated, primary, secondary and tertiary treated wastewater samples were collected over a time frame ranging from under-lockdown to post-lockdown conditions. The results showed that SARS-CoV-2 RNA was detected in 13 out of 40 wastewater samples in Jaipur district, Rajasthan and in 5 out of 14 wastewater samples in the Haridwar District, Uttarakhand with the E gene predominantly observed as compared to the N and RdRp target genes in later time-points of sampling. The Ct values of genes present in wastewater samples were correlated with the incidence of patient and community cases of COVID-19. This study further indicates that the viral RNA could be detected after the primary treatment but was not present in secondary or tertiary treated samples. This study implies that aerobic biological wastewater treatment systems such as moving bed biofilm reactor (MBBR) technology and sequencing batch reactor (SBR) are effective in virus removal from the wastewater. This work might present a new indication that there is little to no risk in relation to SARS-CoV-2 while reusing the treated wastewater for non-potable applications. In contrast, untreated wastewater might present a potential route of viral transmission through WWTPs to sanitation workers and the public. However, there is a need to investigate the survival and infection rates of SARS-CoV-2 in wastewater.

Keywords: aerobic wastewater treatment; COVID-19; RT-qPCR based detection; SARS-CoV-2; sewage surveillance; wastewater based epidemiology

1. Introduction

The coronavirus disease (COVID-19) caused by Severe Acute Respiratory Syndrome Coronavirus 2 (SARS-CoV-2) emerged as a worldwide public health emergency within a few months of its outbreak in Wuhan, China in 2019. The extent of the pandemic COVID-19 is widespread and is currently confirmed to be present in more than 213 countries/regions worldwide [1,2]. SARS-CoV-2 spreads through air droplets and physical contact [3]. Hence, early detection and rapid containment protocols are crucial for its control and elimination. It is challenging to check and control the disease spread in developing countries like India because of the dense population [3]. These challenges are evidenced by an initial gradual increase in infection rates during the lockdown, followed by a sharp increase in the number of positive cases alongside the lifting of lockdown restrictions in India. In India, the number of COVID-19 positive cases increased from 1993 in May 2020 to 78,512 in August 2020, as of the last day of August 2020 [4] due to the lifting of the lockdown in early June 2020, signifying the immediate need of attention. Wastewater-based epidemiology (WBE) is a useful tool for community-wide detection of epidemics and pandemics in a given community.

The combination of clinical and environmental surveillance has been useful for public health practices [5]. Relying on clinical testing alone for detection and control is insufficient due to the scale of the spread and the existence of many asymptomatic and pauci-symptomatic cases. However, current evidence indicates that there is a need for better understanding of the role of wastewater as a potential source of epidemiological data and as a factor of public health risk. A well-validated WBE is imperative for viral surveillance, while appropriate sampling, the concentration of the virus in wastewater, population dynamics, and ethical concerns are crucial factors for a reliable WBE approach, particularly in regard to its utility as an early warning system [6].

The third quarter of 2020 has been exceptional in terms of discovering new knowledge pertaining to the WBE of SARS-CoV-2 genetic material loads, its testing methods, and various strong implications emerging from its use around the world. SARS-CoV-2 RNA was reported in wastewater in Brazil even before the first clinically confirmed cases [7], and various other studies detected SARS-CoV-2 RNA in wastewater across the globe in untreated wastewater [8–12]. However, there are very few recent studies on the occurrence of SARS-CoV-2 throughout the entire process of wastewater treatment in wastewater treatment plants (WWTP); which highlights the removal of the SARS-CoV-2 genome around the globe, such as in Spain [13,14], USA [15], Japan [16], Italy [17], Germany [18], Paris [19], China [20] and India [6,21–25]. Some of these studies have reported the presence of SARS-CoV-2 even in treated wastewater, and the main findings are highlighted in Table 1. In particular, in the study undertaken by Wurtzer et al. [19], six samples of treated wastewater were found to be positive for SARS-CoV-2 [19]. Randazzo et al. [14] found secondary treated samples to be positive, while none of the tertiary treated effluents was positive. Westhaus et al. [18] reported positive samples even after tertiary treatment (ozonation). It is important to investigate the presence of the virus in wastewater, as the treated effluent is to be utilized for irrigation purposes.

The study from India by Kumar et al. [7] reported two-point samplings, taken on 8 May 2020 and 27 May 2020, regarding anaerobic treatment systems (UASB). Thus, there remain questions pertaining to the capability of aerobic WWTPs from India at each stage of the treatment to achieve the decay of SARS-CoV-2 RNA in wastewater. Although WBE-based surveillance is a systematically integrated technique used for detection and diagnostics in many countries worldwide [26], India still lacks awareness about the far-reaching benefits of this surveillance. While infrastructure development is quintessential and a long-drawn process, sufficient good quality data on WBE from these regions can be useful for future planning and computational modeling. Furthermore, there are challenges and apprehensions about implementing WBE in developing countries, such as India, due to the poor water supply network and sewerage system. Thus, the present study is important in the context of WBE from an Indian perspective around its capability, owing to unplanned and incomplete sewer systems, sewer overflows and leakage situations, and strong seasonal fluctuations. These considerations warranted a study that can track the presence of SARS-CoV-2 viral RNA after each of the wastewater treatment stages in Indian settings in order to understand the effects of wastewater treatment on RNA decay. Hence, this study will help to allay the commonly-perceived fear of the commons pertaining to the effectiveness of WWTPs.

Table 1. Review on the efficacy of treatment processes in different stages of wastewater treatment plants (WWTPs) for the removal of SARS-CoV-2 RNA in different studies.

Country	No. of WWTP	Treatment Stages	Process Used	Result	References
Spain	6	Untreated	-	Positive samples: 35/42	[14]
		Secondary treatment	Activated sludge	Positive samples: 2/18	
		Tertiary treatment	Coagulation, Flocculation, Sand filtration, Disinfection (UV, NaClO)	Positive samples: 0/12	
Spain	1	Primary treated	Primary settler	Positive samples: 1/4	[13]
		Secondary treatment and Tertiary treatment	SBR and Chemical removal, Microfiltration	Positive samples: 0/5	
Southern Louisiana, USA	2	Untreated	-	Positive samples: 2/7	[15]
		Secondary treated	Activated sludge	Positive samples: 0/4	
		Tertiary treated	Chlorination	Positive samples: 0/4	
Japan	1	Untreated	-	Positive samples: 0/5	[16]
		Secondary treated	Activated sludge and aeration	Positive samples: 1/5	
		River water	-	Positive samples: 0/3	
Italy	3	Untreated	-	Positive samples: 4/8	[17]
		Tertiary treated	Peracetic acid or high-intensity UV lamps	Positive samples: 0/4	
Germany	9	Untreated	Activated sludge	Positive samples: 9/9	[19]
		Tertiary treated	Ozonation	Positive samples: 4/4	
France	3	Untreated wastewater	-	Positive samples: 23/23	[18]
		Tertiary treated	Data not available	Positive samples: 6/8	
China	1	Untreated	-	Positive samples: 0/4	[20]
		Tertiary treated	Septic tank	Positive samples: 7/9	
India	1	Untreated wastewater	-	Positive samples: 1/2	[22]
		Primary treated	UASB Inlet	Positive samples: 1/2	
		Secondary treated	UASB Outlet	Positive samples: 1/2	
		Tertiary treated	Aeration	Positive samples: 1/2	
		Final Effluent	-	Positive samples: 0/2	
India	14	Untreated and primary treated	-	Positive samples: 12/33	Present study
		Secondary treated	MBBR, SBR	Positive samples: 0/7	
		Tertiary treated	UV Chlorination	Positive samples: 0/14	

Note: - = Information not available, MBBR= Moving Bed biofilm reactor, NaClO = Sodium hypochlorite, SBR = Sequencing Batch Reactor, Cl_2 = Chlorine disinfection, UV—Ultra violet disinfection, MLD = million liters per day, UASB = Up-flow Anaerobic Sludge Blanket.

Recently, the world Sustainable Development Summit 2021 highlighted the importance of WBE for SARS-CoV-2 monitoring and discussed the various challenges involved in implementing WBE in India, in order to create policy making decisions. In this context, the main objective of this study is to track the aerobic biological wastewater treatment system for the decay of SARS-CoV-2 and its genomic RNA along the treatment process and evaluate its effectiveness. This study was conducted to detect SARS-CoV-2 RNA in both untreated and treated wastewater samples collected from multiple locations in order to assess the health risks posed by the reuse of effluents coming from WWTPs. Additionally, an attempt has been made to correlate the detected Threshold Cycle (Ct) values of target genes viz. RNA-dependent polymerase (RdRP) gene, nucleocapsid (N) SARS-CoV-2 specific genes, and gene E, which are characteristic of pan-Sarbecoviruses in the qualitative detection of SARS-CoV-2. This can serve as a crude indication of the genome load over the time-frame which includes lockdown, partial lockdown and no lockdown conditions. The experiments were carried out to detect the presence of SARS-CoV-2 in influent, primary, secondary, and tertiary treated effluent samples from 14 wastewater treatment systems (13 WWTPs and one pump house) in four cities (Roorkee, Rishikesh, Haridwar, Jaipur) in the two North Indian states of Uttarakhand and Rajasthan, and to possibly decipher the potential of current biological treatment systems for removing the virus. This is the study report for a comprehensive data analysis that gives insights into aerobic biological treatment systems' role in decaying the SARS-CoV-2 viral genome.

2. Material and Methods

2.1. Sample Collection and Transportation

The wastewater samples (grab samples) (1 L) were collected during nine different time-points from seven wastewater treatment facilities in Jaipur city (Rajasthan), and grab and composite samples were collected during three different time points from seven wastewater treatment facilities in Uttarakhand state (from the cities of Rishikesh, Haridwar, and Roorkee). The samples were collected during the morning hours, between 6 AM and 10 AM, from different wastewater treatment systems when the sewage flow rate was higher. The samples collected in Uttarakhand were transported through cold chain transportation to Jaipur. All samples were pre-processed in the Environmental Biotechnology Laboratory at Dr. B. Lal Institute of Biotechnology, Jaipur. It was ensured during the packaging that the samples stayed at 4 °C for the whole transit. After sampling, the sampling bottle's surface was disinfected with 90% ethanol, was labeled, and was immediately transported (2–4 °C) to the laboratory. The sampling was carried out during the months of May to August 2020. Detailed information on the numbers of samples, sampling sites, designed capacity, the average flow rate of WWTPs, and the current treatment (secondary and tertiary) technology from the states of Rajasthan and Uttarakhand is provided in Table 2. The locations sites of the different WWTPs are highlighted in Figure 1.

Table 2. Details on the sampling location sites along with treatment characteristics of WWTPs located in Rajasthan and Uttarakhand states.

Site No.	Sampling Location	Type of Secondary Treatment Technology	Type of Tertiary Treatment	Dosage &Contact Time of Tertiary Treatment	Design Capacity (MLD)	Flow Rate (Avg. MLD)	Number of Connected Residents (Approx.)
Site 1	Ramniwas Garden, Jaipur 26.8963° N, 75.8100° E	MBBR	UV	NA	1	~1	>7000
Site 2	Central Park, Jaipur 26.9048° N, 75.8073° E	SBR	Cl_2 (Bleach Powder)	4 ppm by dropping system	1	1	>7000
Site 3	Delawas, Jaipur 27.3735° N, 75.8926° E	ASP	No treatment	3 ppm, 30 min	65	~62.5	>480,000
Site 4	Jawahar Circle, Jaipur 26°50′29″ N,75°48′0″ E	MBBR	UV	NA	1	~1	>7000
Site 5	Brahmpuri, Jaipur 26.9373° N, 75.8250° E	SBR	No treatment	NA	27MLD	~8	>59,000
Site 6	MNIT, Jaipur 26.8640° N, 75.8108° E	MBBR	Cl_2 (Hypochlorite)	2.5–3 ppm, 30 min	1	~1	>7000
Site 7	Dravyavati River Project, Jaipur 26.7980° N, 75.8039° E	SBR	Cl_2 (Hypochlorite)	3–5 ppm, 30 min	65	~65	>480,000
Site 8	IIT Roorkee 29.8649° N, 77.8965° E	SBR	UV	40000 microwatt sec/cm^2	3	~1.5	>7000

Table 2. *Cont.*

Site No.	Sampling Location	Type of Secondary Treatment Technology	Type of Tertiary Treatment	Dosage &Contact Time of Tertiary Treatment	Design Capacity (MLD)	Flow Rate (Avg. MLD)	Number of Connected Residents (Approx.)
Site 9	Muni kiReti, Rishkesh 30.1199° N, 78.3031° E	MBBR	Cl_2 (Hypochlorite)	3 ppm, 30 min	5	~3.5	>15,000
Site 10	Swarg Ashram, Rishikesh 30.1165° N, 78.3131° E	SBR	Cl_2 (Hypochlorite)	3 ppm, 30 min	3	~1.5	>10,000
Site 11	Chandreshwar Nagar, Rishikesh 30.1115° N, 78.3056° E	MBBR	Cl_2 (Hypochlorite)	3 ppm, 30 min	7.5	~2.5	>10,000
Site 12	Sarai, Haridwar 29.9043° N, 78.1080° E	SBR	Cl_2 (Hypochlorite)	3 ppm, 30 min	14	~15	>100,000
Site 13	Jagjeetpur, Haridwar 29.9174° N, 78.1316° E	MBBR	Cl_2 (Hypochlorite)	3 ppm, 30 min	68	~45	>200,000
Site 14	Pump House, Haridwar	NA	NA	NA	NA	NA	>200,000

Note: NA = Information not available, ASP = Activated sludge process, MBBR = Moving Bed biofilm reactor, SBR = Sequencing Batch Reactor, Cl_2 = Chlorine disinfection, UV—Ultra violet disinfection, MLD = million liters per day.

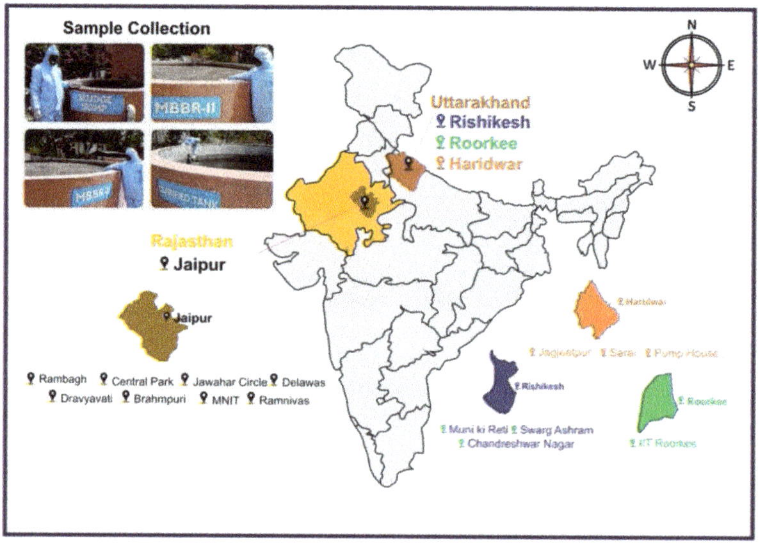

Figure 1. Locations of samples collected from the Uttarakhand and Rajasthan states of India.

2.2. Sample Pre-Processing

The samples were pre-processed using two different methods, as specified in Figure 2. In method A, a 50 mL sample was transferred into sterile falcon tubes in the Biosafety Cabinet (BSL-II), followed by surface sterilization of the falcons using 70% ethanol and UV light exposure for 30 min. The heat inactivation of the virus was then achieved by placing the falcon tubes in a water bath at 60 °C for 90 min. The samples were further filtered through a 0.45μm membrane using a vacuum filter assembly. The filtrate was then transferred to a fresh falcon containing 4 g PEG (Himedia) and 0.9 g NaCl (Himedia). The content was dissolved through manual mixing, followed by centrifugation at 4 °C for 30 min at 7400× *g*. The pellet obtained was then resuspended in 1X Phosphate Buffer Saline (PBS). The method B used for the detection of the SARS-CoV-2 virus was performed by the transfer of a 1 mL sample in a 1.7 mL centrifuge tube, followed by centrifugation at 7400× *g* for 15 min. The supernatant was collected in a fresh tube and was again centrifuged at 7000 rpm for 15 min. The supernatant thus obtained was used for nucleic acid extraction. The samples for the duration of May and June were pre-processed using method A. Because method B requires a shorter duration (2 h 40 min), as compared to method A (as described in Figure 2), and because both methods gave a similar efficiency of detection (as reported in our previous study, Arora et al. [27], all samples from July 2020 onwards were pre-processed using method B.

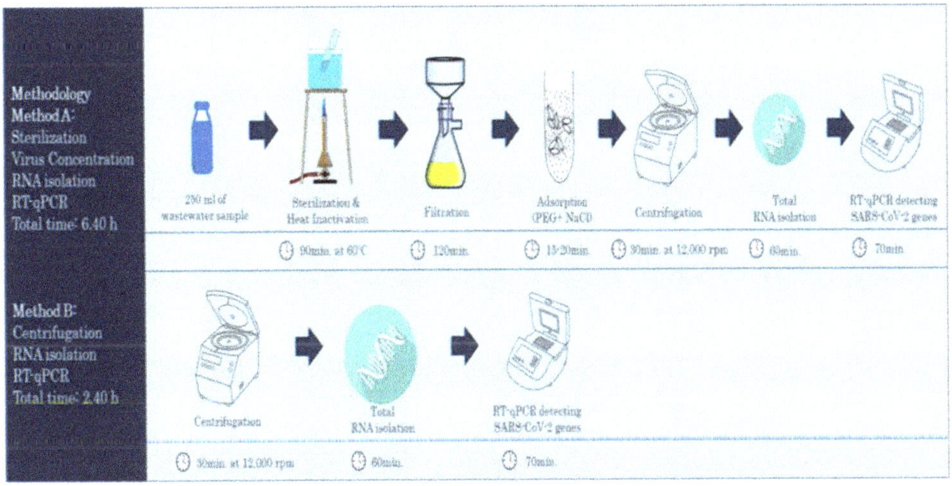

Figure 2. Methodology for the detection of SARS-CoV-2 viral genome through RT-qPCR in wastewater samples.

2.3. RNA Isolation

RNA extraction and the subsequent steps of detection were completed at the Dr. B. Lal Clinical Laboratory Pvt. Ltd., Jaipur (which is authorized by ICMR to conduct COVID-19 testing in humans). The viral RNA molecules present in the wastewater samples from May to early July 2020 were isolated using a Biospin kit (Cat# BSC77M1). As per the vendor's instructions, 10 μL proteinase K and 200 μL of lysis buffer were added to 200 μL of the sample into a 1.5 mL centrifuge tube, followed by vortex mixing and incubation at 56 °C for 15 min in a heating block. 250 μL of ethanol was then added to the sample and mixed by vortexing for 15 s. The mixture was then transferred to the spin column and centrifuged at 10,000× *g*, followed by sequential washing with the three wash buffers provided in the kit, followed by centrifugation at 10,000× *g* for 1 min at each washing step. After the complete drying of the spin column, the RNA was eluted out using a 50–100 μL elution buffer. Centrifugation was done at 12,000× *g* for 1 min. The RNA from the samples collected in late July to August was extracted using the automated Kingfisher Flex System ™ (Cat#5400610).

2.4. Real-Time PCR for Detection of SARS-CoV-2

For the qualitative detection of SARS-CoV-2 in the wastewater samples, RT-qPCR was performed using a commercially available FDA-approved Allplex™ 2019-nCoV Assay kit (Cat# RP10244Y, 208 RP10243X), as per the vendor's instructions for the qualitative detection of SARS-CoV-2 genomic RNA in the sample on Applied Biosystems™ QuantStudio™ 5. The master mix was prepared using the kit content, which was composed of an amplification and detection reagent, enzyme mix for one-time RT-qPCR, buffer containing dNTPs, buffer for one-step PCR and RNase-free water. Each PCR tube contained 8 μL sample RNA, 5 μL 2019-nCoV oligo -mixture, 5 μL Real-time one-step buffer and 2 μL Real-time One-step enzyme and the final volume of the mixture were adjusted to 25 μL using RNase free water. Additionally, as per the kit protocol, "no template" as a negative control, assay target gene plasmids as positive controls, and MS2 phage DNA as internal controls were used to validate each round of reactions. Three genes (E, RdRp, and N) were targeted to detect the presence of the SARS-CoV-2 genome. A list of the different fluorophores used for the detection run is given in Supplementary Materials. Thermal cycling reactions were performed at 50 °C for 20 min, 95 °C for 15 min, 44 cycles at 94 °C for 15 s, and 45 cycles at 58 °C for 30 s, in a thermal cycler. The Allplex™ 2019-nCoV Assay is an in vitro diagnostic real-time reverse transcriptase-polymerase chain reaction (RT-PCR) test used for the qualitative detection of SARS-CoV-2 viral nucleic acids. The kit has reagents RP-V IC (1000 μL), composed of MS2 phage genome as an exogenous internal control. As per the manufacturer's instructions, internal control Ct values above 40 are considered invalid. The PCR controls are provided with the Allplex™ 2019-nCoV Assay to confirm the validity of each PCR run on the same plate. Negative Control (NC) is used as a PCR control to confirm test validity and the absence of any contaminants during testing. The "no template" control is prepared using RNase-free water added to Master Mix prior to PCR. NC was included in each test run.The PC is constructed using plasmids encoding Allplex™ 2019-nCoV Assay target sequences and was included in each test run.

The criteria, of choosing two out of three genes with valid Ct values as criterion for overall positive or not, is based on manufacturer's instruction. This has been extensively discussed in Kumar et al. [21,22], bas well. In addition, it is important to note that, as we detect titer of RNA and talk about the entire SARS-CoV-2 genome, it would not be rational to say that the entire genome exists based on just one gene detection. Therefore, at least two out of three genes need to be present in a sample in order for the sample to be announced as positive.

3. Results and Discussions

3.1. Detection of SARS-CoV-2 RNA in Untreated Wastewater Samples

In India, the country-wide lockdown and prohibitions pertaining to the containment of the pandemic were enforced on 24 March 2020. This was initially declared as a 21-day restriction period, but was extended multiple times until 31 May 2020. With the onset of the post-lockdown period, most of the states lifted restrictions and prohibitions, bar a few precautionary measures. By July 31, while the rate of the country's confirmed positive cases was constantly rising [28], many of the states were completely out of the lockdown conditions. The window of wastewater influent sample collection in the present study started as early as 4 May 2020 and continued throughout the lockdown and until post-lockdown, on 14 August 2020. The observations discussed in Table 3 are taken from sampling completed at various sites in Rajasthan and Uttarakhand and tested for the presence of the viral genome. The early sampling in Rajasthan coincided with the onset of community restrictions and the lockdown duration and continued through the gradual relaxation of the lockdown. The sampling was carried out until August 2020, and coincided with the time window when most establishments, including offices, barbershops, markets, malls, were opened and unrestricted city movements were allowed. In Uttarakhand, sampling was conducted under partial lockdown conditions, with offices closed and restricted activities permitted in some areas.

The wastewater of the entire Jaipur city, and the outskirts area is connected through a sewerage network that joins the main trunk sewer with a 1800 mm diameter and an average flow of 130 MLD, terminating at Delawas Sewage treatment plant (STP) Site 3, based on the activated sludge process (centralized treatment facility at 125 MLD capacity). From this main trunk, settled sewage is withdrawn at a rate of 1 MLD Sites 1, 2, and 4, has been connected to three decentralized treatment plants based on MBBR technology for more than ten years in order to maintain public parks. Besides this, Site 6 has its own WWTP plant at MNIT Jaipur campus area, based on MBBR technology of a capacity of 1 MLD. Site 7 at Dravyavati River WTWP has a capacity of 65 MLD. Site Brahmpuri WWTP (Site 5) has a capacity of 27 MLD. Almost all of the WWTPs are working at full capacity, and hence this can be considered to be their average flow rates. The wastewater from the majority of the Haridwar city gets collected to the pump house (Site 14), and then travels to two major WWTPs at Site 12 and 13, which are SBR and MBBR based, respectively. These

WWTPs are not working in their full capacity, and their average flow rate is provided in Table 2.

Table 3. Results showing Ct values of targeted genes for detection of SARS-CoV-2 RNA in untreated and treated wastewater samples collected from different WWTPs.

(A) Rajasthan							
Site No.	**Sampling Site**	**Type of Sample**	**Sampling Date**	C_{TF}	C_{TR}	C_{TN}	**Final Result Interpretation**
Site 1	MBBR, Ramniwas Garden, Jaipur	Untreated	4 May 2020	33	36	33	Positive
			15 May 2020	-	-	-	Negative
			20 May 2020	31	38	34	Positive
			12 June 2020	32	37	34	Positive
			12 July 2020	36	-	36	Positive
			11 August 2020	35	36	36	Positive
		Secondary treated	11 August 2020	-	-	-	Negative
		Tertiary treated	4 May 2020	-	-	-	Negative
			12 June 2020	35	-	-	Negative
			26 July 2020	-	-	-	Negative
			11 August 2020	-	-	-	Negative
Site 2	SBR, Central Park, Jaipur	Untreated	4 May 2020	-	-	-	Negative
			15 May 2020	-	-	-	Negative
			20 May 2020	-	-	-	Negative
		Secondary Treated	11 August 2020	-	-	-	Negative
		Tertiary treated	4 May 2020	-	-	-	Negative
			11 August 2020	-	-	-	Negative
Site 3	ASP, Delawas, Jaipur	Untreated	4 August 2020	32	36	36	Positive
			8 August 2020	34	35	36	Positive
		Secondary treated	11 August 2020	-	-	-	Negative
Site 4	MBBR, Jawahar Circle, Jaipur	Untreated	4 May 2020	-	-	-	Negative
			12 June 2020	-	-	-	Negative
			4 August 2020	34	-	38	Positive
			8 August 2020	-	-	-	Negative
			11 August 2020	-	-	-	Negative
			29 October 2020	34	-	33	Positive
		Primary treated sample	29 October 2020	31	-	32	Positive
		Secondary treated	8 August 2020	-	-	-	Negative
		Tertiary treated	8 August 2020	-	-	-	Negative
Site 5	SBR, Brahmpuri, Jaipur	Untreated	15 May 2020	-	-	-	Negative
			20 May 2020	36	-	37	Positive
			12 June 2020	36	-	36	Positive
			11 August 2020	33	34	35	Positive
		Secondary treated	12 June 2020				Negative
			11 August 2020	-	-	-	Negative
Site 6	MBBR, MNIT, Jaipur	Untreated	4 May 2020	-	-	-	Inconclusive
			4 August 2020	-	-	-	Negative
		Tertiary treated	4 May 2020	-	-	-	Negative
Site 7	SBR, Dravyavati River Project, Jaipur	Untreated	4 May 2020	–	-	-	Negative
		Tertiary treated	4 May 2020	-	-	-	Negative

Table 3. *Cont.*

(B) Uttarakhand							
Site 8	SBR, IIT Roorkee	Untreated	25 July 2020	33	37	37	Positive
			14 August 2020	36	30	37	Positive
		Secondary treated	11 August 2020	-	-	-	Negative
		Tertiary treated	11 August 2020	-	-	-	Negative
Site 9	MBBR, Muni kiReti, Rishkesh	Untreated	25 July 2020	36	37	37	Positive
		Tertiary treated	25 July 2020	-	-	-	Negative
Site 10	SBR, Swarg Ashram, Rishikesh	Untreated	25 July 2020	-	-	-	Negative
Site 11	MBBR, Chandreshwar Nagar, Rishikesh	Untreated	25 July 2020	-	-	-	Negative
		Tertiary treated	30 July 2020	-	-	-	Negative
Site 12	SBR, Sarai, Haridwar	Untreated	25 July 2020	36	-	37	Positive
		Tertiary treated	25 July 2020	-	-	-	Negative
Site 13	MBBR, Jagjeetpur, Haridwar	Untreated	25 July 2020	33	37	-	Positive
		Tertiary treated	25 July 2020	-	-	-	Negative
Site 14	Sewage, Pump House, Haridwar	Untreated	25 July 2020	-	-	37	Negative

C_{TE} = C_T value of E gene, C_{TR} = C_T value of RdRp gene, C_{TN} = C_T value of N gene, C_{TIC} = C_T value of Internal Control. The value of Ct above 40 indicates that the gene tested is not present in the sample. The presence of at least two out of three positive genes in a sample was ruled to be positive for the SARS-CoV-2 genome.

The observations from the RT-qPCR-based qualitative genome detection showed that the spread of COVID-19 in Northern India is highly extensive. Table 3 shows the results of the untreated and treated wastewater samples, and the Ct values of all three target genes (E, RdRp, and N) were interpreted accordingly. A total of 40 samples from Rajasthan and 14 samples from Uttarakhand collected between May 4 and August 14 (2020) were tested for the presence of SARS-CoV-2 RNA. Some variations in Ct values are probably due to the complexity and variability of the sewage samples and has been described before [29,30]. The interpretation of results in Table 3 (from different sites, as described in Table 1) was based on categorizing a sample as positive when the cycle threshold took place below cycle 40, for either two of the three genes such as RNA-dependent polymerase (RdRP), nucleocapsid (N) SARS-CoV-2 specific genes and gene E, characteristic of pan-Sarbeco viruses detection, as per manufacturer's instructions. The Ct values of positive samples were in the range of 30–38, corresponding to a mild to moderate genome load presence in all of the untreated wastewater samples. It was observed that the areas served by the WWTPs of Jaipur city that showed positive results reported a continuous increase in confirmed positive patients, which corroborated with the Ct values. As observed in Table 2, during May 2020, the samples showed the presence of SARS-CoV-2 RNA in untreated wastewater from Site 1 and Site 5, which corroborated with the positive patients' cases around the area (as reported in [27], while the rest of the sites showed negative samples. With the withdrawal of lockdown restrictions by the end of June 2020, a significant increase in the COVID positive cases was observed. The number of positive cases reached the thousands, as reported by [4]. This can be corroborated with the decrease in the Ct values from >40 to 33, which can be seen for certain sites such as Site 5, Brahmpuri, Jaipur. However, the results obtained from the Site 1 samples are quite captivating. The Ramniwas Garden WWTP (Site 1) is currently serving the walled city area of Jaipur that includes the major hotspot of the city, the Ramganj area, having the maximum reported cases in May 2020. The detected genome load in the sample collected from Site 1 increased during the lockdown, and decreased in the post-lockdown period. One of the key reasons behind this observation is that the SMS Hospital sewage is also collected and treated at Site 1 WWTP. SMS Hospital had a COVID-19 positive patient load in May 2020 (during lockdown in the city), and the load gradually decreased in July and August. This is because patients started to recover, and there was also a decline in positive patient admissions in the hospital. Therefore, this pattern of Ct values and infectivity during and post lockdown periods can be explained.

The Ct values in the wastewater samples collected from Uttarakhand indicate genome loads within the same order as the loads present in Jaipur during June 2020. This was during the time when restrictions were being partially lifted in the cities. The increase in the viral genome load concurs with the gradual rise in the number of infected individuals, which rose from 2138 and 720 active cases on June 1 to 10,260 and 5912 active cases by the end of August 2020 in Jaipur and Haridwar districts, respectively (number of cases for Jaipur district were obtained from the newspaper, the case numbers for Haridwar were reported by the Department of Medical Health and Family Welfare [31] (https://health.uk.gov.in/, accessed on 31 August 2020).

A few studies have also tried to correlate the Ct values with the genome load and the probability of that load being infective [32,33]. Different lab-scale studies were conducted investigating the percentage of the cell cultures turning positive at various Ct levels of the SARS-CoV-2 genome detected in the clinical samples of sputum and nasal swabs [32]. The study shows that a sample with a genome load with Ct values greater or equal to 34 could not infect the cell lines tested and postulates that the patients with higher or equal to 34 Ct values may be discharged [33]. However, these studies have been conducted in labs for clinical samples and might not correspond to the infection probabilities that could occur through the wastewater contamination. Thus, it would be interesting to investigate the possibilities of transmission routes through contaminated wastewater [34].

It was further observed that gene E was frequently detected with the lowest Ct values during the qualitative detection compared to the other two genes (Genes N and RdRp) (as reported in Table 3 and Figure S1). In Uttarakhand, the Ct of E gene in four samples out of five positive samples was the lowest compared to those of N and RdRp genes in the same samples. In Jaipur, all the five-times sampling in August 2020 (overall ten out of twelve samples), showing positive results for the presence of the SARS-CoV-2 genome, had the lowest Ct values for the E gene. Furthermore, it was observed that for the samples collected during the earlier time points from Jaipur, the Ct values of E and N indicated a marginal difference in their respective gene loads (Figure S1a). In contrast, the sample collected later in the time window from July to August clearly showed the prevalence of the E gene over the other two genes (as depicted in Figure S1b–d), which was reflected by the rise in the graph E gene for fewer samples.

Thus, the present study highlights the effect of lifting the lockdown restrictions with the increase in the viral genome load per unit of wastewater. In the post lockdown period (August 2020), the rapid increase in the numbers of COVID-19 patients was corroborated by the decrease in Ct values (Site 5). Additionally, the genes tested for SARS-CoV-2 in the wastewater showed different gene load levels, as indicated by the Ct values. The E gene seems to be present more abundantly than N and RdRp in all of the samples. Either of two reasons can explain this observation. One reason could be related to the host-pathogen interaction, which is different for different populations. Thus, it is possible that a particular gene is more abundantly or stably expressed in a community. Another reason could be that the E gene is responsible for the structural assembly of the viral particles without interacting with N and RdRp genes [35].

Additionally, the self-assembly of SARS-CoV-2 requires the interaction of the N gene with viral RNA for compaction and packaging into the viral capsid [36]. Thus, the observed Ct value trends might be interesting and may indicate shedding of viral capsids before the packaging is complete. Alternatively, the trends of Ct values could indicate the differential expression rates of these genes under different conditions, such as the genetic makeup of a community, geographical, climatic, etc. Therefore, these dynamics in E, N, and RdRp gene detection might prove useful in understanding the viral host interactions and transmission probabilities through wastewater.

The study also highlights the methods of transporting the wastewater samples, and its effect on RNA detection. The samples from the state of Uttarakhand were collected and immediately transported to Jaipur at 4 °C using cold chain transportation. Despite the gap of approximately 3–4 days between sampling and pre-processing, the Ct values observed for these samples indicated a mild to moderate range of genome load, similar to the immediately pre-processed samples. This indicates that transportation before pre-processing did not significantly affect the detection and that transportation of the collected wastewater samples at 4 °C might be a sufficient measure for genome detection. This observation is important in the context of using only qualitative detection of SARS-CoV-2 in wastewater, which was to be transported over a longer distance using cold chain transportation. However, in the case of quantification, more studies are required in order to understand this.

3.2. Secondary Aerobic Biological Treatments Are Sufficient to Decay the SARS-CoV-2 RNA beyond Detection

The presence of the SARS-CoV-2 genome in untreated wastewater is a cause forconcern as the wastewater is a potential route of viral transmission to sanitation workers. Additionally, aerosolization of wastewater during its treatment can promote infection via air, provided that the viral particle is active. The sludge and treated water from these treatment facilities are used for agricultural purposes, which can put end users' health at risk. To investigate the probability of such a transmission route, samples from various stages of aerobic wastewater treatment were collected and checked for the presence of viral RNA.

The samples were collected from primary treatment, secondary treatment, and tertiary treatment stages of wastewater treatment plants. Different treatment methods were available at different sites, which provided comprehensive information on the fate of viral RNA after different methods of treating the wastewater had been implemented. While most of the sites selected in Jaipur have only secondary treatment technology, three of them (viz. Site 1 Ramniwas Garden WWTP, Site 4 Jawahar Circle WWTP, and Site 6 MNIT WWTP) have tertiary treatment technology, with a UV disinfection unit. However, during our test window, only Site 1, Ramniwas Garden WWTP, was connected to a community that was considered to be a hotspot for the pandemic, while the other two sites were not. In order to investigate the effect of each wastewater treatment stage on SARS-CoV-2 decay, samples from each of the stages mentioned above were checked. The sites which were not serving the hotspot communities showed the absence of viral RNA in influent and samples from the subsequent stages. However, it was observed that even where we could detect the presence of viral RNA in the influent, the viral RNA decayed beneath the level of detection immediately after the second stage treatment (Table 3) and remained undetectable at consequent stages as well. In fact, the detectable and intact SARS-CoV-2 viral genome was not observed in any of the post-secondary treated wastewater samples, regardless of the type of biological treatment (i.e., Activated sludge process, MBBR or SBR). The absence of viral RNA in the effluent was consistent between the sampling that was conducted during the lockdown and after the lifting of the lockdown regulations. While it is important to note that there has been a constant increase in the number of COVID-19 cases, and by extension, the SARS-CoV-2 genome load per sample unit, treated samples were still consistently negative for the viral genome presence.

The observations thus far indicated that the viral RNA was decaying between the stages of influent and post-secondary treatment. These observations raised questions about which particular treatment could be directly involved in the decay. Any of three possible reasons could be the source of the decay: firstly, that the viral load was going down during the primary treatment procedures itself due to the settling down of the viral particles; secondly, it was very likely that the secondary treatment procedures, which are effective in virus removal by the biofilm generated by the microbes, were responsible; or thirdly, it was possible that both the primary and secondary treatments together were responsible for the decay with a partial depreciation in load starting at the primary treatment stage itself. Samples were taken and analyzed from untreated wastewater and during all the subsequent stages of treatment in order to delineate between these mechanisms. The Ct values of all target genes obtained between untreated and post-primary treated wastewater samples showed little variation, while most of the viral RNA was decayed during the secondary treatment stage. This indicates the possibility that the second mechanism, where the decay of RNA happens due to the biological treatments, is in action. This evidence thus shows that the wastewater treatment facilities are capable of degrading the viral RNA significantly. The biological treatment stages were capable of completely degrading the intact SARS-CoV-2 genome beyond the detection sensitivity and this did not depend on the tertiary treatment (i.e., disinfection stage). Direct chlorination of untreated sewage might not significantly reduce the detected viral loads if the chlorine demand of the sample is not satisfied. Chlorine demand is directly proportional to the organic waste matter present in the water samples [37]. Since there is a very large quantity of organic matter in the sewage, it is understandable if chlorination alone is not highly effective. Evidently, Zhang et al. [20] found an unexpected occurrence of SARS-CoV-2 viral RNA in aseptic tank even after disinfection with sodium hypochlorite. They suggested reevaluation of the existing disinfection approach (free chlorine: >6.5 mg/L after 1.5-h contact).

In contrast, the absence of any detectable viral genome in wastewater samples collected post-secondary treatment might indicate the efficacy of the biofilm generated by the microflora in the biological reactors in removing the viral genome loads. This hypothesis is based on several studies that have reported biofilms' role in the removal of various types of viruses [38,39]. A biofilm can be defined as a well-organized community consisting of cooperating microorganisms immobilized in an extracellular polysaccharide (EPS) matrix [40,41]. Biofilms can be an association of single or multiple species of bacteria, fungi, algae, protozoans, and rotifers in combinations [41]. Thus, it is possible that these biotic constituents of the biofilm, along with the abiotic components, such aspH, temperature, or minerals present, are an integral part of the SARS-CoV-2 viral decay and removal from the wastewater. Further investigation into the role of biofilm is much needed, as the aerobic biological wastewater treatment process investigated in this study seems to be more efficient in the decay of viral RNA than the anaerobic UASB system, in which the decay is completed only after the post-secondary treatment aeration stage [20].The findings from the present study indicate that secondary aerobic biological WWTPs contribute to reducing the virus concentration due to the adverse environmental conditions (i.e., temperature, solids, pH, or disinfectants) to make the water fit for reuse.

3.3. Comparison of the Efficiency of Studied Treatment Processes

We examined the efficiency of MBBR, SBR, and ASP wastewater treatment processes by comparing the changes in the Ct values of the SARS-CoV-2 E gene, RdRp gene, and N gene before and after the treatment wastewater samples (i.e., influent and effluent). A paired-samples T-test was carried out in order to evaluate the efficacy of MBBR and SBR treatment processes (Figure 3a,b), while a comparison was made between influent and effluent wastewater samples of the ASP treatment process (Figure 3c). The results showed significant removal of all three targeted SARS-CoV-2 genes from the MBBR plant ($p < 0.05$), while a substantial decrease ($p > 0.05$) in E and N genes was noticed in the SBR treatment process depicted by a post treatment increase in Ct values of genes (Figure 3a,b). Likewise, all three genes were successfully removed from the ASP treatment process (Figure 3c).

In addition to this, the paired T-test between the inlet and outlet wastewater samples, taken on the same date during the study, displayed a significant reduction/removal of SARS-CoV-2 genes, except for one occasion in the case of MBBR treatment (Figure 4). Contrary to this, the reduction of SARS-CoV-2 genes in wastewater samples was insignificant in the SBR treatment process. The results suggest that all three treatment processes successfully reduced/removed the virus genetic load in wastewater samples; however, the performance of MBBR was found to be higher than that of the SBR and ASP treatment processes.

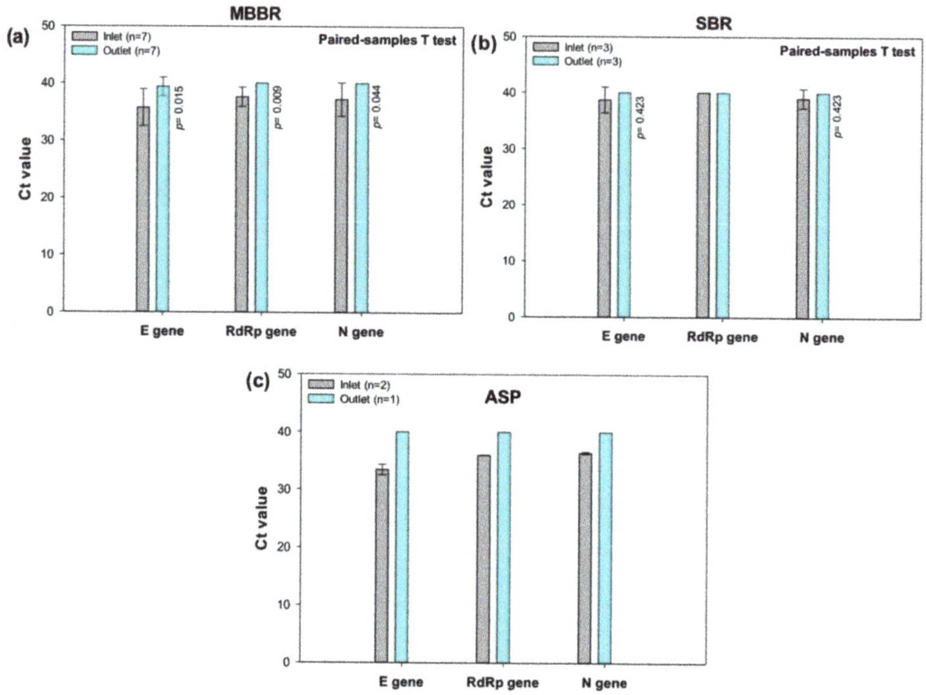

Figure 3. Comparison between the inlet and outlet wastewater samples for SARS-CoV-2 genetic load in (**a**) Moving bed Bioreactor (MBBR) based treatment; (**b**) Sequencing batch reactors (SBR); and (**c**) Activated Sludge Process (ASP).

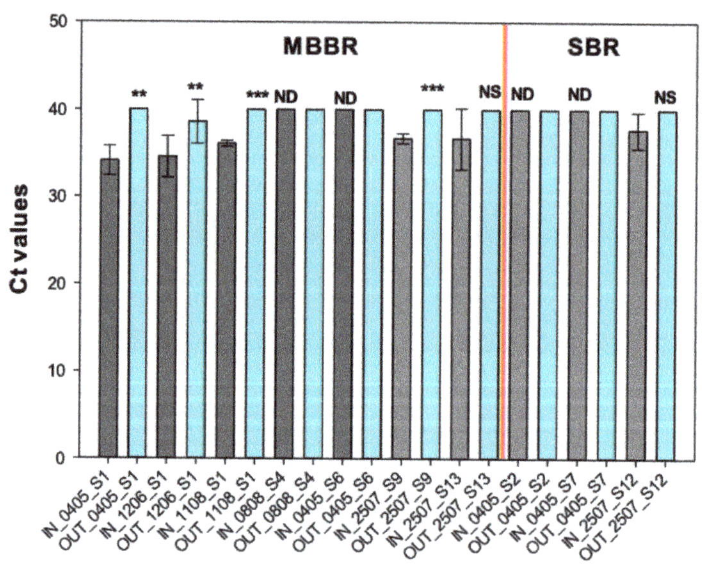

Figure 4. Paired T-test between the inlet and outlet wastewater samples, taken on the same date for SARS-CoV-2 genetic load in Moving bed Bioreactor (MBBR) based treatment and Sequencing batch reactors (SBR). where *** = $p < 0.01$; ** = $p < 0.05$; NS = not significant; ND = not detected; and RT-PCR was run for 40 cycles).

4. Conclusions

The present study reports 33.3 percentage of positive wastewater samples for SARS-CoV-2 from two states of Northern India, with none of the secondary or tertiary treated samples found to be positive. The samples from multiple locations make the study more representative and indicate the potential applications of WBE across diverse geographical and climatic conditions. This highlights the fact that aerobic biological wastewater treatment systems might be sufficient for SARS-CoV-2 removal, diminishing any possibility of the fecal route of disease transmission through treated wastewater. This study also assessed the efficacy of each stage of aerobic wastewater treatment systems and established a surveillance system through sewage monitoring of the potential virus circulation. This study further highlights that Ct values correspond to positive patient cases in the area and illustrated the effect of physical distancing and lockdown regulations, as is evident from several negative samples during the lockdown period. The paired T-test between the inlet and outlet wastewater samples, taken on the same date during the study, displayed a significant reduction/removal of SARS-CoV-2 genes. The results suggest that all three treatment processes successfully reduced/removed the virus genetic load in wastewater samples; however, the performance of MBBR was found to be higher than that of the SBR and ASP treatment processes. This study opens up a new direction of treatment efficacy on SARS-CoV-2 removal and stresses the need to understand the survival of SARS-CoV-2 under natural conditions in various aerobic wastewater treatment systems. The present study aims to add to the existing literature on WBE and contribute to an efficient and resilient public health emergency response mechanism in India for the future. This study provides a comprehensive data analysis and gives insights into the role of aerobic biological treatment systems in decaying the SARS-CoV-2 viral genome.

Supplementary Materials: The following are available online at https://www.mdpi.com/article/10.3390/w13162265/s1, Figure S1: Graphs showing trends in the three genes from the months of May to August for (a) May 20 (b) June 12 (c) July 25 (d) August 4 (e) August 11 (2020).

Author Contributions: Conceptualization, S.A., A.N., A.B.G.; methodology, A.N., J.S., J.R.; validation, V.K.T., M.K., formal analysis, A.B.G., A.A.K., A.R.; investigation, A.N., D.S., J.S., J.R., resources, S.A., S.K.S., S.S.; data curation, S.B.T., V.S., M.K.; writing—original draft preparation, A.N.; writing—review and editing, S.A., V.K.T., S.B.T., V.S., M.K.; visualization, V.K.T., supervision, A.B.G.; project administration, S.A.; funding acquisition, V.K.T., A.A.K. All authors have read and agreed to the published version of the manuscript.

Funding: This research was funded by research grants from the Department of Biotechnology-GoI [Grant No. BT/RLF/Re-entry/12/2016].

Institutional Review Board Statement: Not applicable.

Informed Consent Statement: Not applicable.

Data Availability Statement: Not applicable.

Acknowledgments: We would like to acknowledge the constant support received from B. Lal Gupta (Director) and Aparna Datta (principal) for providing support. We would also like to thank the Centre for Innovation, Research & development (CIRD, Dr. B. Lal Clinical laboratory Pvt. Ltd.) for their support in analysis. We would also thank the Jaipur development authority (JDA) officials & plant operators at WWTPs and all other participating wastewater treatment plants for the collection of sewage samples in this study.

Conflicts of Interest: The authors declare no conflict of interest.

References

1. Johns Hopkins University. 2020. Available online: https://www.arcgis.com/apps/opsdashboard/index.html#/85320e2ea5425604dfaaa75ae62e5c06e61 (accessed on 16 May 2020).
2. Dong, E.; Du, H.; Gardner, L. An interactive web-based dashboard to track COVID-19 in real time. *Lancet Infect. Dis.* **2020**, *20*, 533–534. [CrossRef]

3. World Health Organization (WHO). Poliovirus Detected from Environmental Samples in Israel and West Bank and Gaza Strip. WHO. 2013. Available online: https://www.who.int/csr/don/2013_09_20_polio/en/ (accessed on 21 May 2020).
4. World Health Organization. Coronavirus Disease (COVID-19) Situation Report. Available online: https://www.who.int/emergencies/diseases/novel-coronavirus-2019/situation-reports (accessed on 20 May 2020).
5. Kroiss, S.J.; Ahmadzai, M.; Ahmed, J.; Alam, M.M.; Chabot-Couture, G.; Famulare, M.; Mahamud, A.; McCarthy, K.A.; Mercer, L.D.; Muhammad, S.; et al. Assessing the sensitivity of the polio environmental surveillance system. *PLoS ONE* **2018**, *13*, e0208336. [CrossRef]
6. Kumar, M.; Joshi, M.; Patel, A.K.; Joshi, C.G. Unravelling the early warning capability of wastewater surveillance for COVID-19: A temporal study on SARS-CoV-2 RNA detection and need for the escalation. *Environ. Res.* **2021**, *196*, 110946. [CrossRef]
7. Fongaro, G.; Stoco, P.H.; Souza, D.S.M.; Grisard, E.C.; Magri, M.E.; Rogovski, P.; Schorner, M.A.; Barazzetti, F.H.; Christoff, A.P.; de Oliveira, L.F.V.; et al. SARS-CoV-2 in human sewage in Santa Catalina, Brazil, November 2019. *medRxiv* **2020**. [CrossRef]
8. Ahmed, W.; Angel, N.; Edson, J.; Bibby, K.; Bivins, A.; O'Brien, J.W.; Choi, P.M.; Kitajima, M.; Simpson, S.L.; Li, J.; et al. First confirmed detection of SARS-CoV-2 in untreated wastewater in Australia: A proof of concept for the wastewater surveillance of COVID-19 in the community. *Sci. Total Environ.* **2020**, *728*, 138764. [CrossRef] [PubMed]
9. La Rosa, G.; Mancini, P.; Ferraro, G.B.; Veneri, C.; Iaconelli, M.; Bonadonna, L.; Lucentini, L.; Suffredini, E. SARS-CoV-2 has been circulating in northern Italy since December 2019: Evidence from environmental monitoring. *Sci. Total Environ.* **2020**, *750*, 141711. [CrossRef] [PubMed]
10. Lodder, W.; de RodaHusman, A.M. SARS-CoV-2 in wastewater: Potential health risk, but also data source. *Lancet Gastroenterol. Hepatol.* **2020**, *5*, 533–534. [CrossRef]
11. Medema, G.; Heijnen, L.; Elsinga, G.; Italiaander, R.; Brouwer, A. Presence of SARS-Coronavirus-2 in sewage. *medRxiv* **2020**, *7*, 511–516.
12. Wu, F.; Xiao, A.; Zhang, J.; Gu, X.; Lee, W.L.; Kauffman, K.; Hanage, W.; Matus, M.; Ghaeli, N.; Endo, N.; et al. SARS-CoV-2 titers in wastewater are higher than expected from clinically confirmed cases. *medRxiv* **2020**, *5*, e00614-20.
13. Balboa, S.; Mauricio-Iglesias, M.; Rodríguez, S.; Martínez-Lamas, L.; Vasallo, F.J.; Regueiro, B.; Lema, J.M. The fate of SARS-CoV-2 in wastewater treatment plants points out the sludge line as a suitable spot for incidence monitoring. *medRxiv* **2020**. [CrossRef]
14. Randazzo, W.; Truchado, P.; Cuevas-Ferrando, E.; Simón, P.; Allende, A.; Sánchez, G. SARS-CoV-2 RNA in wastewater anticipated COVID-19 occurrence in a low prevalence area. *Water Res.* **2020**, *181*, 115942. [CrossRef]
15. Sherchan, S.P.; Shahin, S.; Ward, L.M.; Tandukar, S.; Aw, T.G.; Schmitz, B.; Ahmed, W.; Kitajima, M. First detection of SARS-CoV-2 RNA in wastewater in North America: A study in Louisiana, USA. *Sci. Total Environ.* **2020**, *743*, 140621. [CrossRef]
16. Haramoto, E.; Malla, B.; Thakali, O.; Kitajima, M. First environmental surveillance for the presence of SARS-CoV-2 RNA in wastewater and river water in Japan. *medRxiv* **2020**. [CrossRef]
17. Rimoldi, S.G.; Stefani, F.; Gigantiello, A.; Polesello, S.; Comandatore, F.; Mileto, D.; Maresca, M.; Longobardi, C.; Mancon, A.; Romeri, F.; et al. Presence and infectivity of SARS-CoV-2 virus in wastewaters and rivers. *Sci. Total Environ.* **2020**, *744*, 140911. [CrossRef] [PubMed]
18. Westhaus, S.; Weber, F.A.; Schiwy, S.; Linnemann, V.; Brinkmann, M.; Widera, M.; Greve, C.; Janke, A.; Hollert, H.; Wintgens, T.; et al. Detection of SARS-CoV-2 in raw and treated wastewater in Germany–suitability for COVID-19 surveillance and potential transmission risks. *Sci. Total. Environ.* **2020**, *751*, 141750. [CrossRef] [PubMed]
19. Wurtzer, S.; Marechal, V.; Mouchel, J.M.; Maday, Y.; Teyssou, R.; Richard, E.; Almayrac, J.L.; Moulin, L. Evaluation of lockdown impact on SARS-CoV-2 dynamics through viral genome quantification in Paris wastewaters. *medRxiv* **2020**. [CrossRef]
20. Zhang, D.; Ling, H.; Huang, X.; Li, J.; Li, W.; Yi, C.; Zhang, T.; Jiang, Y.; He, Y.; Deng, S.; et al. Potential spreading risks and disinfection challenges of medical wastewater by the presence of Severe Acute Respiratory Syndrome Coronavirus 2 (SARS-CoV-2) viral RNA in septic tanks of Fangcang Hospital. *Sci. Total Environ.* **2020**, *741*, 140445. [CrossRef] [PubMed]
21. Kumar, M.; Patel, A.K.; Shah, A.V.; Raval, J.; Rajpara, N.; Joshi, M.; Joshi, C.G. First proof of the capability of wastewater surveillance for COVID-19 in India through detection of genetic material of SARS-CoV-2. *Sci. Total Environ.* **2020**, *746*, 141326. [CrossRef] [PubMed]
22. Kumar, M.; Kuroda, K.; Patel, A.K.; Patel, N.; Bhattacharya, P.; Joshi, M.; Joshi, C.G. Decay of SARS-CoV-2 RNA along the wastewater treatment outfitted with Upflow Anaerobic Sludge Blanket (UASB) system evaluated through two sample concentration techniques. *Sci. Total Environ.* **2020**, *754*, 142329. [CrossRef]
23. Chakraborty, P.; Pasupuleti, M.; Shankar, M.J.; Bharat, G.K.; Krishnasamy, S.; Dasgupta, S.C.; Jones, K.C. First surveillance of SARS-CoV-2 and organic tracers in community wastewater during post lockdown in Chennai, South India: Methods, occurrence and concurrence. *Sci. Total Environ.* **2021**, *778*, 146252. [CrossRef]
24. Hemalatha, M.; Kiran, U.; Kuncha, S.K.; Kopperi, H.; Gokulan, C.G.; Mohan, S.V.; Mishra, R.K. Surveillance of SARS-CoV-2 spread using wastewater-based epidemiology: Comprehensive study. *Sci. Total Environ.* **2021**, *768*, 144704. [CrossRef]
25. Kopperi, H.; Tharak, A.; Hemalatha, M.; Kiran, C.; Gokulen, C.G.; Mishra, R.K.; Mohan, S.V. Methodological Approach for Wastewater Based Epidemiological Studies for SARS-CoV-2. *medRxiv* **2021**. [CrossRef]
26. World Health Organization. 2020. Available online: https://www.who.int/emergencies/diseases/novel-coronavirus-2019 (accessed on 31 August 2020).

27. Arora, S.; Nag, A.; Sethi, J.; Rajvanshi, J.; Saxena, S.; Shrivastava, S.K.; Gupta, A.B. Sewage surveillance for the presence of SARS-CoV-2 genome as a useful wastewater based epidemiology (WBE) tracking tool in India. *Water Sci. Technol.* **2020**, *82*, 2823–2836. [CrossRef] [PubMed]
28. WHO. Coronavirus Disease (COVID-19) Dashboard. Available online: https://covid19.who.int (accessed on 17 May 2021).
29. Petterson, S.; Grøndahl-Rosado, R.; Nilsen, V.; Myrmel, M.; Robertson, L.J. Variability in the recovery of a virus concentration procedure in water: Implications for QMRA. *Water Res.* **2015**, *87*, 79–86. [CrossRef] [PubMed]
30. Silva-Sales, M.M.; Martínez-Puchol, S.; Gonzales-Gustavson, E.; Hundesa, A.; Gironès, R. High Prevalence of Rotavirus A in Raw Sewage Samples from Northeast Spain. *Viruses* **2020**, *12*, 318. [CrossRef] [PubMed]
31. Department of Medical Health and Family Welfare. Available online: https://health.uk.gov.in/ (accessed on 31 August 2020).
32. Liu, R.; Yi, S.; Zhang, J.; Lv, Z.; Zhu, C.; Zhang, Y. Viral Load Dynamics in Sputum and Nasopharyngeal Swab in Patients with COVID-19. *J. Dent. Res.* **2020**, *99*, 1239–1244. [CrossRef] [PubMed]
33. La Scola, B.; Le Bideau, M.; Andreani, J.; Hoang, V.T.; Grimaldier, C.; Colson, P.; Gautret, P.; Raoult, D. Viral RNA load as determined by cell culture as a management tool for discharge of SARS-CoV-2 patients from infectious disease wards. *Eur. J. Clin. Microbiol. Infect. Dis.* **2020**, *39*, 1059. [CrossRef]
34. Kumar, M.; Alamin, M.; Kuroda, K.; Dhangar, K.; Hata, A.; Yamaguchi, H.; Honda, R. Potential discharge, attenuation and exposure risk of SARS-CoV-2 in natural water bodies receiving treated wastewater. *NPJ Clean Water* **2021**, *4*, 8. [CrossRef]
35. Vennema, H.; Godeke, G.J.; Rossen, J.; Voorhout, W.F.; Horzinek, M.C.; Opstelten, D.J.; Rottier, P.J. Nucleocapsid-independent assembly of coronavirus-like particles by co-expression of viral envelope protein genes. *EMBO J.* **1996**, *15*, 2020–2028. [CrossRef]
36. Ye, Q.; West, A.M.; Silletti, S.; Corbett, K.D. Architecture and self-assembly of the SARS-CoV-2 nucleocapsid protein. *Protein Sci.* **2020**, *29*, 1890–1901. [CrossRef]
37. Rudolfs, W.; Ziemba, J.V. The Efficiency of Chlorine in Sewage Disinfection as Affected by Certain Environmental Factors. *J. Bacteriol.* **1934**, *27*, 419–442. [CrossRef] [PubMed]
38. Quignon, F.; Sardin, M.; Kiene, L.; Schwartzbrod, L. Poliovirus-1 inactivation and interaction with biofilm: A pilot-scale study. *Appl. Environ. Microbiol.* **1997**, *63*, 978–982. [CrossRef] [PubMed]
39. Esfahani, A.R.; Batelaan, O.; Hutson, J.L.; Fallowfield, H. Combined physical, chemical and biological clogging of managed aquifer recharge and the effect of biofilm on virus transport behavior: A column study. *J. Water Process. Eng.* **2019**, *33*, 101115. [CrossRef]
40. Mortensen, B.K. Formation and Detection of Biofilms. Bac to Force. 2014. Available online: https://www.bactoforce.fi/wp-content/uploads/2014/09/Formation-and-detection-of-biofilms.pdf (accessed on 10 September 2020).
41. Maurya, A.; Raj, A. Recent advances in the application of biofilm in bioremediation of industrial wastewater and organic pollutants. *Microorg. Sustain. Environ. Health* **2020**, 81–118. [CrossRef]

water

MDPI

Article

Monitoring of SARS-CoV-2 Variants by Wastewater-Based Surveillance as a Sustainable and Pragmatic Approach—A Case Study of Jaipur (India)

Aditi Nag [1,†], Sudipti Arora [1,*,†], Vikky Sinha [1], Ekta Meena [1], Devanshi Sutaria [1], Akhilendra Bhushan Gupta [2] and Krishna Mohan Medicherla [3]

1 Dr. B. Lal Institute of Biotechnology, 6-E, Malviya Industrial Area, Malviya Nagar, Jaipur 302017, India; aditinag1@gmail.com (A.N.); sinhavikky30@gmail.com (V.S.); ektabibt@gmail.com (E.M.); devanshisnehal@gmail.com (D.S.)
2 Department of Civil Engineering, Malaviya National Institute of Technology, Malviya Nagar, Jaipur 302017, India; abgupta.ce@mnit.ac.in
3 Birla Institute of Scientific Research, Jaipur 302001, India; kmohan@bisr.res.in
* Correspondence: sudiptiarora@gmail.com
† These authors contributed equally to this work and thus shared the first authorship.

Abstract: Wastewater-based surveillance has been emerging as an efficient and advantageous tool to predict COVID-19 prevalence in the population, much earlier (7–28 days) than reported clinical cases, thus providing sufficient time to organize resources and optimize their use in managing COVID-19. Since the commencement of the COVID-19 pandemic, SARS-CoV-2 genetic lineages have emerged and are circulating all over the world. The assessment of SARS-CoV-2 variants of concern (VOCs) in wastewater has recently been proven to be successful. The present research demonstrates a case study utilizing an established approach to perform monitoring of SARS-CoV-2 variants from 11 distinct wastewater treatment plants across Jaipur (India) during the second peak period of COVID-19 (from 19 February 2021 to 8 June 2021). The sequences obtained were analyzed to detect lineage using the Pangolin tool and SNPs using the mpileup utility of Samtools, which reported high genome coverage. The mutation analyses successfully identified the penetration of the B.1. in the first two weeks of sampling (19–26 February), followed by the B.1.617.2 variant into Jaipur in the first week of March 2021. B.1.617.2 was initially discovered in India in October 2020; however, it was not reported until early April 2021.The present study identified the presence of B.1.617.2 in early March, which correlates well with the clinical patient's data (290 cases were reported much later by the government on 10 May 2021). The average total genome coverage of the samples is 94.39% when mapped onto the severe acute respiratory syndrome coronavirus 2 isolate Wuhan-Hu-1; a complete genome (NC_045512.2) sequence and SNP analysis showed that 37–51 SNPs were identified in each sample. The current study demonstrates that sewage surveillance for variant characterization is a reliable and practical method for tracking the diversity of SARS-CoV-2 strains in the community that is considerably faster than clinical genomic surveillance. As a result, this method can predict the advent of epidemiologically or clinically important mutations/variants, which can help with public health decision making.

Keywords: COVID-19; next generation sequencing; SARS-CoV-2; sewage; variants of concerns

Citation: Nag, A.; Arora, S.; Sinha, V.; Meena, E.; Sutaria, D.; Gupta, A.B.; Medicherla, K.M. Monitoring of SARS-CoV-2 Variants by Wastewater-Based Surveillance as a Sustainable and Pragmatic Approach—A Case Study of Jaipur (India). *Water* **2022**, *14*, 297. https://doi.org/10.3390/w14030297

Academic Editor: Laura Bulgariu

Received: 15 December 2021
Accepted: 13 January 2022
Published: 19 January 2022

1. Introduction

SARS-CoV-2 has had an ever-expanding epidemiology since its introduction in December 2019, with an increase in the number of patients and its spread. As of 17 November 2021, there have been 255,252,955 clinically confirmed SARS-CoV-2 cases registered worldwide, with 5,133,224 deaths [1]. Viruses evolve and diversify throughout time, and the introduction of new variants such as the diversity of SARS-CoV-2 is an expected occurrence due to evolution and adaptability, as what has been seen internationally. Upon the suggestion of WHO's Technical Advisory Group on Virus Evolution, WHO recognized the variant B.1.1.529, known as Omicron, as a variant of concern on 26 November 2021 [2]. Most mutations that arise will not provide the virus with a selection advantage. Some alterations or combinations of mutations—such as greater transmissibility via increased receptor binding or the ability to avoid the host immune response by changing surface features recognized by antibodies [3,4]—may do so. Understanding the genetic nature of circulating SARS-CoV-2 in the community is critical in this scenario. Due to limited resources, it is essentially impossible to conduct a genomic surveillance of clinical patients on a large scale as the number of cases and mutations grows. In such cases, the wastewater-based epidemiological paradigm would cover community-wide genetic variant monitoring and would prove to be a practical, cost-effective, and important resource for analyzing SARS-CoV-2 mutations in the population [5]. The World Health Organization (WHO) has also acknowledged environmental sewage surveillance as a method of protecting public health by monitoring and detecting viral infections in circulation [6]. The use of genomic surveillance of wastewater could prove to be a valuable tool for detecting, identifying, predicting, and building an early warning system for the identification of variants of concern (VOCs) in circulation to support public health actions. Table 1 shows the list of currently designated variants of concerns, as defined by WHO [7,8].

Table 1. Currently designated variants of concern by WHO [9].

WHO Label	Pango Lineage	GISAID Clade	Additional Amino Acid Changes Monitored	Earliest Documented Samples	Date of Designation
Alpha	B.1.1.7	GRY	+S: 484K +S: 452R	United Kingdom, Sep-2020	18 December 2020
Beta	B.1.351	GH/501Y.V2	+S: L18F	South Africa, May-2020	18 December 2020
Gamma	P.1	GR/501Y.V3	+S: 681H	Brazil, Nov-2020	11 January 2021
Delta	B.1.617.2	G/478K.V1	+S: 417N +S: 484K	India, Oct-2020	VOI: 4 April 2021 VOC: 11 May 2021
Omicron	B.1.1.529	GRA	+S: R346K	Multiple countries, Nov-2021	VUM: 24 November 2021 VOC: 26 November 2021

The appearance of new SARS-CoV-2 variants with mutations linked to greater transmissibility, a weaker antibody response, or both [10–12] has recently attracted attention. SARS-CoV-2 mutations have been found across the world, raising worries about the efficiency of treatment and vaccinations. Because of its positive-sense single-stranded RNA genetic material, SARS-CoV-2 has a high mutation rate, posing a public health threat. Different highly infectious forms of SARS-CoV-2 that have emerged because of mutations in the SARS-CoV-2 genome have been identified as variants of concerns (VOCs) (Table 1). The B.1.1.7 lineage of SARS-CoV-2, for example, which was discovered in the United Kingdom (UK) in November 2020, is thought to be 40–80% more contagious than the initial strain [13]. Other SARS-CoV-2 lineages from Brazil (P.1), Southern African countries (B.1.351), and India (B.1.617.2) are also more transmissible than the early 2020 variants. Because Omicron has been designated as a variant of concern, WHO recommends that countries improve their surveillance and sequencing of cases, share genome sequences on publicly available databases such as GISAID, report initial cases or clusters to WHO, and conduct field investigations and laboratory assessments to better understand if Omicron has different transmission or disease characteristics, or how it affects the effectiveness of vaccines. In terms of viral pathogenicity, virulence, and transmission, the variants of concern (VOCs) are crucial [2].

The clinical monitoring, which includes whole genome sequencing from infected patients' nasopharyngeal samples, is an effective method, although it is costly, labor-intensive, and time-consuming. It also has serious flaws, such as only including genomes of symptomatic patients, which, according to a recent study [14–16], accounts for only two-thirds of the population, with the remaining one-third of SARS-CoV-2 infections being asymptomatic and not covered by clinical genomic monitoring. As a result, finding variants in wastewater could be a more efficient, quicker, and practical way to track the appearance and spread of new variants in a community. Several research works have attempted to sequence the SARS-CoV-2 genome and detected genetic variations from wastewater samples in several locations, including Montana, USA [16], California, USA [17], Switzerland [18], London [19], Canada [20], and India [21], among others. To effectively control the COVID-19 pandemic in a coordinated manner, it is critical to promote this unique strategy on a much larger scale globally and to build a repository of dominant variants in circulation.

With almost 5000 cases recorded in Jaipur during the second wave of COVID-19, it was vital to analyze the wastewater from the city to understand the infection dynamics and focus on the variations circulating in the community. To highlight the necessity of variant surveillance, we conducted an amplicon-based metagenomic landscape of SARS-CoV-2 in the wastewater of the Jaipur region during the period of the second wave, from 19 February to 8 June 2021.

The study's goal was to see if the SARS-CoV-2 RNA was present in wastewater streams using NGS technology (Illumina NextSeq500) for sequencing, which could be used to look for mutations. The current investigation could provide crucial information about circulating variations in the community far before a clinically confirmed discovery, as well as investigate the possibility of WBE as a cost-effective and practical surveillance technique.

2. Methodology

2.1. Sample Collection and Transportation

For the monitoring of the SARS-CoV-2 variants, influent samples were collected from 11 municipal wastewater treatment plants (WWTPs) located across Jaipur city. Figure 1 depicts the geographic locations of the study's sampling sites, which cover roughly 60–70% of the city's sewerage network. Table 2 summarizes the details of WWTPs. The samples were collected weekly between 19 February and 8 June 2021 in this temporal investigation. All the samples were collected as one-liter grabs in sterile bottles and brought to the Dr. B. Lal Institute of Biotechnology's Environmental Biotechnology Laboratory in Jaipur for further research and analysis. As previously described [22–25], samples were kept at 4 °C for 24 h after sampling. For sample collection, appropriate precautions, including ambient temperatures, were considered. During the whole sample process, the concerned staff wore standard personal protective equipment (PPE).

Table 2. Characteristics of WWTP sampling sites.

Site No.	Sampling Location	Type of Secondary Treatment Technology	Type of Tertiary Treatment	Dosage and Contact Time of Tertiary Treatment	Design Capacity (m^3/Day)	Flow Rate (Average. MLD)	Number of Connected Residents (Approximately)
Site 1	Brahmpuri, Jaipur 26.9373° N, 75.8250° E	SBR	No treatment	NA	27,000 m^3/day	~8	>59,000
Site 2	Central Park Garden, Jaipur 26.9048° N, 75.8073° E	SBR	Cl_2 (Bleach Powder)	4 ppm by dropping system	1000 m^3/day	~1	>7000

Table 2. *Cont.*

Site No.	Sampling Location	Type of Secondary Treatment Technology	Type of Tertiary Treatment	Dosage and Contact Time of Tertiary Treatment	Design Capacity (m^3/Day)	Flow Rate (Average. MLD)	Number of Connected Residents (Approximately)
Site 3	Ramniwas Garden, Jaipur 26.8963° N, 75.8100° E	MBBR	UV	NA	1000 m^3/day	~1	>7000
Site 4	MNIT, Jaipur 26.8640° N, 75.8108° E	MBBR	Cl_2 (Hypochlorite)	2.5–3 ppm, 30 min	1000 m^3/day	~1	>2000
Site 5	Jawahar Circle Garden, Jaipur 26.5029 ^{0}N, 75.480^0 E	MBBR	UV	NA	1000 m^3/day	~1	>7000
Site 6	Dravyavati River, Jaipur 26.7980° N, 75.8039° E	SBR	Cl_2 (Hypochlorite)	3–5 ppm, 30 min	65,000 m^3/day	~65	>480,000
Site 7	Dhelawas, Jaipur 27.3735° N, 75.8926° E	ASP	No treatment	3 ppm, 30 min	65,000 m^3/day	~62.5	>480,000
Site 8	Paldi Meena, Jaipur 26.8759° N, 75.8945° E	SBR	No treatment	NA	3000 m^3/day	0.6–0.7	~5000
Site 9	Ralawata, Jaipur 26.76873° N, 75.93092° E	ASP	Cl_2 (Hypochlorite)	10 kg per hour	30,000 m^3/day	20–22	~170,370
Site 10	Kho Nagorian, Jaipur 26.84063° N, 75.88546° E	SBR	No treatment	NA	50,000 m^3/day	~45.5	>480,000
Site 11	Dr. B. Lal Institute of Biotechnology Institute's Campus WTWP 26.85697° N, 75.82749° E	Biokube^{TM}	No treatment	NA	7.5 m^3/day	7.5 KLD	~500

Note: MNIT = Malaviya National Institute of Technology, MBBR = Moving Bed Biofilm Reactor, SBR = Sequencing Batch Reactor, ASP = Activated Sludge Process-plug flow process, Biokube^{TM} = Plug & Play type packaged sewage treatment plant—Mars 4000, Cl_2 = Chlorine disinfection, UV = Ultraviolet disinfection, MLD = million liters per day, NA = Not applicable.

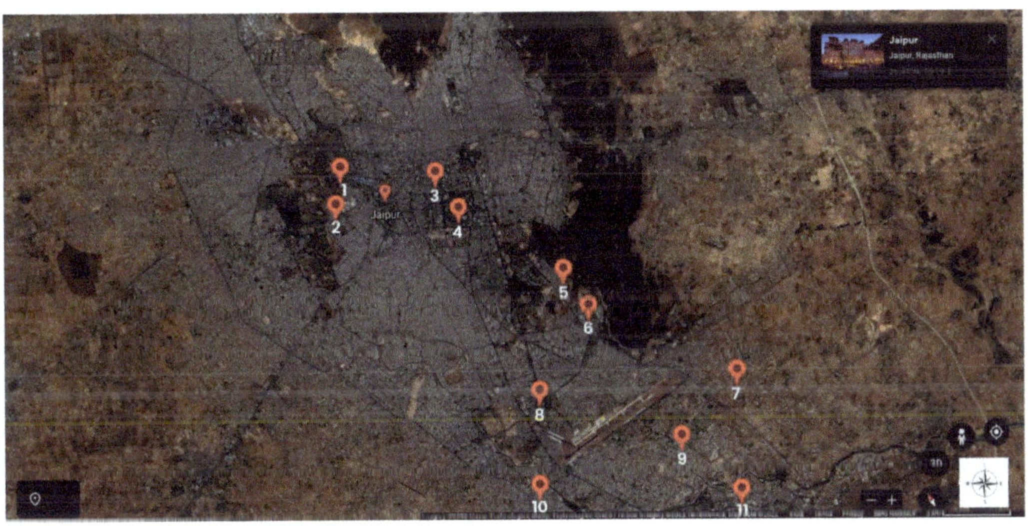

Figure 1. Geographical locations of wastewater treatment plant sampling sites as presented on the map: 1. Brahmpuri, 2. Dravyavati River, 3. Ramniwas Garden, 4. Central Park Garden, 5. MNIT, 6. Dr. B. Lal Institute of Biotechnology, 7. Paldi Meena, 8. Jawahar Circle, 9. Kho Nagorian, 10. Dhelawas, 11. Ralawata.

2.2. Sample Preparation for SARS-CoV-2 Detection Using RT-qPCR

The samples for RNA isolation were produced with minor changes to the methodology previously reported [25,26]. The wastewater samples were exposed to UV light for 30 min to sterilize the surface before being thoroughly mixed. Furthermore, 1 mL of the sample was aliquoted and centrifuged at 7000× *g* for 30 min (to remove debris and undesirable components), with the supernatant being processed for RNA extraction, as described in [27]. The viral RNA was extracted from the treated wastewater samples using the automated KingFisher^{TM} Flex equipment, as reported in [24], using the MagMAX Viral/Pathogen Nucleic Acid Isolation Kit (Applied Biosystems, Waltham, MA USA), according to the manufacturer's instructions. The aliquots of eluted RNA in the plates were then sealed and stored at −20 °C until needed.

2.3. SARS-CoV-2 Quantitative and Qualitative Detection

The presence of SARS-CoV-2 RNA was detected qualitatively and quantitatively in total RNA isolated from wastewater samples using a CFX 96 Thermal Cycler (Bio-Rad, Hercules, CA, USA) RT-PCR machine and two commercially available kits. Kit 1 contained 2019-nCoV MOM (prepared master mix), 5X Real-time One-step Buffer, Real-time One-step Enzyme, and exogenous Internal Control for qualitative detection of SARS-CoV. Kit 2 contained 2019-nCoV MOM (prepared master mix), 5X Real-time One-step Buffer, Real-time One-step Enzyme, and exogenous Internal Control for qualitative detection of SARS-CoV. (IC). Kit 1 was designed to read the E gene, N gene, and RdRp gene on the Cal Red 610 and Quasar 670 fluorophore channels, with FAM and HEX as internal controls. A total of 11 uL of extracted RNA was mixed with 14 uL of RT-PCR master mix to make the PCR reaction. The reaction protocol was as follows: 1 cycle at 50 °C for 20 min, 1 cycle at 95 °C for 15 min, 45 cycles of denaturation at 94 °C for 15 s, combined annealing and extension for 30 s at 58 °C, plate read, and detection. The Bio-Rad CFX Manager software version 3.1 was used to analyze the PCR run (Bio-Rad Laboratories). The identification of a minimum of two genes (out of three) in a sample was judged positive based on Ct values, according to the manufacturer's instructions.

InnoDetect One Step COVID-19 (Kit 2) was also used to further quantify the presence of the SARS-CoV-2 viral genome in wastewater samples, with two different plasmid DNA consisting of the N gene and the ORF1ab gene, separately used to prepare a standard curve (ranging from 10 pg/uL to 0.01 fg/uL), as per the manufacturer's instructions. The quantification of the individual genes in the samples was then based on these standard curves. To make a primary stock with a concentration of 40 ng/uL, RNase-free water was employed. For individual identification, kit 2 included a master mix, primer probes (N gene, ORF1ab, and RNaseP), and three fluorophore channels (HEX/VIC, FAM, and ROX/Texas Red, respectively). SARS-CoV-2 viral RNA was utilized as a positive control, while DNase RNase-free water was used as a negative control. A reverse transcription stage at 42 °C for 15 min and 1 cycle, cDNA initial denaturation at 95 °C for 3 min and 1 cycle, denaturation at 95 °C for 15 s, and combined annealing and extension at 60 °C for 40 s were all followed by plate read and detection. Positive samples had a quantifiable presence of either of the two genes (N or ORF1ab) or both genes. The samples that tested positive for SARS-CoV-2 were chosen for further analysis.

2.4. Sample Pre-Processing for NGS RNA Extraction: Virus Filtration and Concentration

The virus was inactivated by transferring the samples to 50 mL Tarsons falcon (code 546041) tubes in a biosafety cabinet (BSL2), which was followed by a 70% ethanol spray over the surface of the falcon tubes and 30 min of UV light exposure for surface sterilization. After UV irradiation, the samples were placed in a water bath at 60 °C and incubated for 90 min to ensure that the virus was heat-inactivated. After the coronavirus was inactivated, the samples were brought to room temperature and filtered using a vacuum filter assembly via a 0.45 μ membrane. Each sample's filtrate was placed in a new 50 mL falcon containing 0.9 g sodium chloride (NaCl) and 4 g polyethylene glycol (PEG). By gently mixing the ingredients, the contents were dissolved. After that, the PEG and NaCl-containing samples were centrifuged at 4 °C for 30 min at 7000 rpm. After that, the pellet was re-suspended in 1X Phosphate Buffer Saline (PBS) and processed for RNA extraction.

2.5. Total RNA Extraction

ZymoBIOMICS®96Magbead DNA/RNA Kit R2136 was used to extract total RNA according to the manufacturer's instructions. Using the simple MAG extractor, 3 mL of lysis buffer containing concentrated sewage was extracted. Extracted NA were eluted in a 55 L elution buffer and stored at 70 °C until appropriate samples were pooled and transmitted to an NGS service provider (Eurofins, Bangalore, India, Company Headquarters Luxembourg, France) for sequencing.

2.6. Sample Pooling for NGS

The wastewater samples were collected continuously from Sites 1–11 for 11 weeks. The samples found positive each week were then processed to extract total RNA. A total of 51 samples out of more than 110 influent samples collected were found positive for SARS-CoV-2 presence. These 51 RNA samples were pooled week-wise to make composite RNA samples for 11 weeks, starting from 19 February to 10 June, as described in Table 2.

2.7. Qualitative and Quantitative Analysis of RNA Samples

The quality and quantity of the viral RNA samples were checked by Nanodrop, and results are shown in Table S1.

2.8. Preparation of 2 × 150 NextSeq500 ARTIC Library

NGSeq ARTIC SARS-CoV-2 Kit and Illumina TruSeq Nano DNA Library Prep Kit were used to construct paired-end sequencing libraries from the QC-passed and selected viral RNA samples. In a nutshell, isolated viral RNA was reverse transcribed to cDNA before being used in a PCR experiment with the oligo mix pools provided. Both amplicon pools were combined and purified using AMPure XP beads before being fragmented with a Covaris M220. End Repair Mix was used to repair the fragmented amplicons by eliminating the 3′ overhangs and filling in the 5′ overhangs with 5′–3′ polymerase activity, followed by adapter ligation to the fragments. The low rate of chimaera (concatenated template) creation is ensured by this method. AMPure XP beads were used to size the ligated products. As specified in the kit methodology, the size-selected products were PCR amplified using the index primer, and index adapters were ligated to the ends of the DNA fragments, preparing them for hybridization onto a flow cell.

2.9. Quantity and Quality Check (QC) of Library on Agilent 4200 Tape Station

The PCR enriched libraries were purified with AMPure XP beads and evaluated on an Agilent Technologies 4200 Tape Station system with a high-sensitivity D1000 Screen tape, as detailed in Table S2.

2.10. Cluster Generation and Sequencing

The PE illumina libraries were loaded onto NextSeq500 for cluster creation and sequencing after getting the Qubit concentration for the libraries and the mean peak sizes from the Agilent TapeStation profile. On the NextSeq500, paired-end sequencing allowed the template fragments to be sequenced both forward and backward. After re-synthesis of the reverse strand during sequencing, the adapters were designed to facilitate selective cleavage of the forward strands. The reverse strand was then utilized to sequence the fragment from the opposite end.

2.11. Bioinformatic Analysis

The lineage detection was performed by using Pangolin tools. SNP analysis was conducted using the mpileup utility of Samtools (v 0.1.18). Multiple sequence alignment was generated using CLUSTAL W, and the visualization and analysis were completed using Jalview 2.11.1.1.

3. Results

3.1. SARS-CoV-2 Gene Concentrations in Wastewater and the Quality and Depth of the Data Obtained

Three SARS-CoV-2 genes (E, RdRP, and N gene regions) were found in all 110 wastewater influents during this 5-month study, with quantities ranging from 1×10^4 to 1×10^6 gene copies/L of wastewater. The SARS-CoV-2 gene concentrations were relatively steady at the start of the study, grew fast in mid-March, peaked in April and May, then plateaued in June, according to the general observed trend. This pattern matches the trend of the daily COVID-19 new confirmed cases in Jaipur [28,29]. Table S3 shows the results of genome concentrations in several samples across different WWTP concentrations. Sample pooling (from several WWTPs) is described in Table 3, and high-quality clean reads were acquired by processing raw sequences with Trimmomatic reads.

Table 3. Description of samples and raw data reads.

Sample No.	Sample Coding	Sample Collection Date	Type of Sample (Total No. of Samples)	Sample Pooled from WWTPs	Total Number PE of Reads
1	JPR 1	27 February 2021	Individual RNA	3	1,409,611
2	JPR 2	20 February 2021	Pooled RNA (2)	9 and 2	1,552,223
3	JPR 3	12 March 2021	Pooled RNA (2)	5 and 7	1,570,305
4	JPR 4	19 March 2021	Pooled RNA (3)	7, 4, 5	1,579,011
5	JPR 5	1 April 2021	Pooled RNA (7)	4, 2, 3, 1, 5, 8, 4	1,385,609
6	JPR 6	9 April 2021	Pooled RNA (7)	5, 2, 8, 6,4, 1, 3	1,507,711
7	JPR 7	20 April 2021	Pooled RNA (6)	5, 8, 4, 2, 3, 1	1,352,690
8	JPR 8	1 May 2021	Pooled RNA (5)	5, 4, 1, 6, 8	1,588,702
9	JPR 9	8 May 2021	Pooled RNA (5)	4, 6, 5, 11, 1	1,606,125
10	JPR 10	15 May 2021	Pooled RNA (6)	2, 7, 5, 4, 6, 8	1,414,505
11	JPR 11	24 May 2021	Pooled RNA (6)	9, 7, 10, 6	1,194,455

3.2. Lineage Identification and Multiple Sequence Analysis with Other Reference Sequences

The high-quality reads of the samples were subsequently aligned to the severe acute respiratory syndrome coronavirus 2 isolate Wuhan-Hu-1, complete genome (NC_045512.2) reference sequence using BWA MEM (version 0.7.17). Consensus sequence was extracted using Samtools mpileup. The mostly probable lineage of the sample consensus sequence was assigned using the Pangolin tool. The identified lineage for the samples is provided in Table 4.

Table 4. Lineage identification in 11 pooled samples from different WWTPs.

Sample	Lineage
JPR 1	B.1
JPR 2	B.1
JPR 3	B.1.617.2
JPR 4	B.1.617.2
JPR 5	B.1.617.2
JPR 6	B.1.617.2
JPR 7	B.1.617.2
JPR 8	B.1.617.2
JPR 9	B.1.617.2
JPR 10	B.1.617.2
JPR 11	B.1.617.2

3.3. Analysis of Phylogeny and SNPs in the Samples

This analysis was performed by comparing the obtained sequences to a consensus sequence from the ancestral SARS-CoV-2 Wuhan city variant. However, since these samples were not just a single patient sample but rather a pool of viral RNA obtained from various WWTPs of the city at a given time point, it was important to assess if this data obtained could be applied and used to monitor the emerging variants and phylogenies in the whole city and aid in genome surveillance initiatives. To find out if the phylogenetic analysis and ancestry could be studied from the sequence information obtained from these samples, multiple sequence alignment was performed by using Clustal Omega [30], and phylogenetic trees were obtained by the neighbor joining method (Figure 2). The alignments were first completed to check if the sequences obtained could be analyzed amongst themselves without any external reference sequence (Figure 2a). The analysis was repeated using several reference sequences, viz., the Delta variant obtained in January 2021 from India in the state of Tamil Nadu, Delta variant from Netherlands in the month of June 2021, Delta variant sequence obtained from Italy in September 2021, and a completely different variant sequence, the Mu variant, obtained from Columbia, USA in April 2021 (Figure 2b). This analysis demonstrated the usability of the genomic sequence obtained from the wastewater samples, which can be used as effectively in variant surveillance, at par with the patient data.

The sequences obtained from the pooled samples not only gave robust reads but were also aligned amongst themselves with the different variant sequences. Since all the samples were classified as B.1.617.2, except for the first two time point samples, it was a worthy question to look at the genetic variations in these samples over the period of sampling. By comparing them with NC_045512.2, SNP detection was carried out, and the following SNPs could be identified with their genomic coordinates, reference-alternate bases associated genes, and amino acid changes.

Out of these, a total of 423 gene-related SNPs were detected from all the samples, out of which a total sum of 254 belonged to the ORF1ab polyprotein gene. Out of these, 61 were unique SNPs found across 11 of the samples analyzed. In these 11 samples, there were 12 types of SNP mutations which did not change the amino acid phenotype in the sequence (Ser549Ser, Phe924Phe, and Thr5960Thr in all samples; Tyr4227Tyr and Leu6666Leu in ten; Leu4386Leu and Gly334Gly in five; Gly6068Gly, Asn6333Asn, and Ser2500Ser in three; and Pro1692Pro and Thr1773Thr in two samples were detected in the 11 pooled samples). Table 5 describes the SNPs in 11 samples. Amongst the SNPs which led to a change in the amino acid sequences, six SNPs were present in all the samples; these were Ala2529Thr, Ala3209Val, Val3718Ala, Pro4715Leu, Pro5401Leu, and Pro5971Leu. Furthermore, there were also a few SNPs which were not present in all the samples but only in a few out of 11 samples sequenced, i.e., Pro1640Leu (91% of samples), Gly5063Ser (72%), Pro5401Leu (63%), Thr3750Ile (36.4%), Arg6676Gln (36.4%), Thr5036Met (27.4%), Ala385Val (18.2%), Lys798Asn(18.2%), Gln1021His (18.2%), Pro2046Leu (18.2%), Thr3255Ile (18.2%), Thr5477Ile (18.2%), and His6517Tyr (18.2%).

Three SNP variants were found to be linked with ORF3A, out of which Ser26Leu was found in every sample, Gly100Val was found in 18%, and Ala110Ser in only 9% of instances. It was interesting to note that the only three polymorphisms in ORF7a (Val82Ala, Leu116Phe, and Thr120Ile) which were detected in 11 wastewater samples were present in 72% of the samples and were highly correlated and always present together as a set. No other SNPs were detected in the samples. Furthermore, the ORF8 gene was linked with three types of SNPs detected in the study. Out of these, Ala65Ser was uniformly detected in all the samples, while Phe120Leu and Asp119Val were present in only 27% and 18% of the samples, respectively. Unlike ORF7a, these SNPs were correlated with respect to their associated gene. Strangely, only two out of 11 samples showed any SNPs related to gene M during our analysis. Three types of SNPs were detected: Asn66Lys (two samples), Val70Ile (unique detection), and Pro71His (two samples). In the Membrane protein M gene, five types of SNPs were detected. Out of these, Ile82Thr was present in 91% of the samples sequenced but the other four SNPs were unique mutations found only in one of the 11 samples. In the E gene, a total of two samples only showed SNP mutations, two of which were present together, Asn66Lys and Pro71His. In addition to these, only one more uniquely present SNP, Val70Ile, was detected.

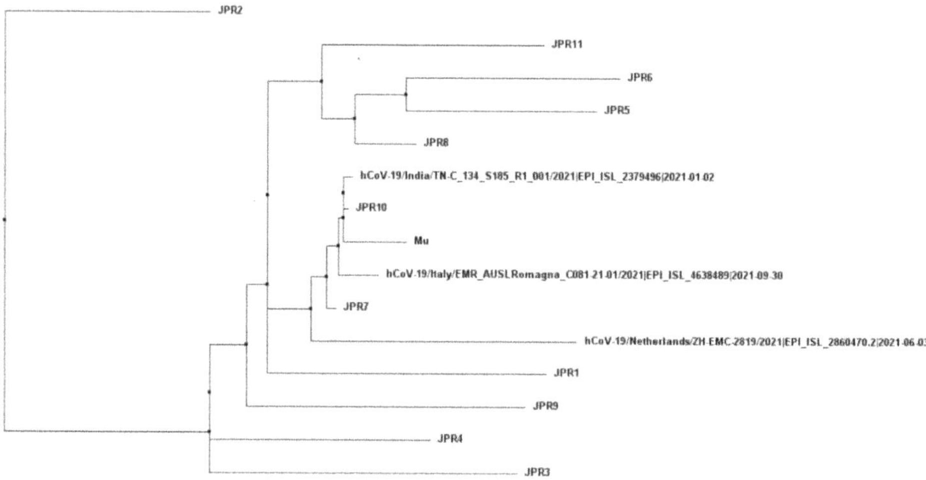

(a)

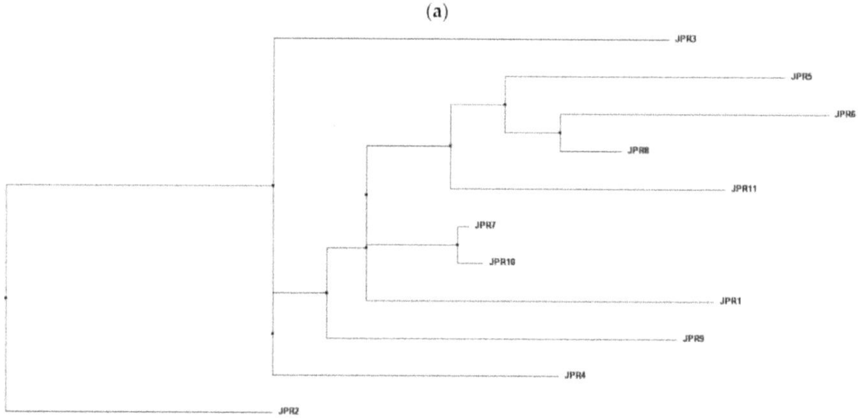

(b)

Figure 2. Phylogenetic analysis with the sequences obtained from wastewater samples. (**a**) Phylogenetic tree of 11 samples collected from Jaipur between 27 February (JPR1) and the end of May 2021 (JPR11). JPR2-4 corresponds to samples collected weekly from March, JPR5-7 from April, and JPR 8-11 from May. (**b**) Shows the phylogeny of the collected samples with GISAID-submitted sequences from different time points as a reference. Mu stands for the sequence collected in April for the Mu variant, observed for the first time in Columbia, USA.

Table 5. SNP identification in samples.

Sample	Number of SNPs Detected	Number of Genic SNPs Detected
JPR 1	39	35
JPR 2	44	41
JPR 3	40	36
JPR 4	51	47
JPR 5	39	36
JPR 6	41	37
JPR 7	42	39
JPR 8	45	42
JPR 9	40	36
JPR 10	44	40
JPR 11	37	34
Total	462	423

Out of 423 instances of gene-related SNPs, the N and S genes had 33 and 65 linked detected SNP sites across all the samples, respectively. The N gene had eight unique SNP mutations across the samples. These were Ile337Ile, Thr379Ile, Arg385Lys, Arg185Leu, Asp63Gly, Gln281Lys, Ser327*, Arg189Ser. Although Asp377Tyr and Arg203Met were present in 91% and 82% of the samples, the presence of both in the same sample was only 72%. On the contrary, two of the SNP types, Asp402Asn and Asp402Val, although present in only about 27%, were always present together. The S gene had 13 different types of SNPs associated with it. Out of these, Ile472Ile, Gln218His, Gln1208His, Pro426Leu, and Gly142Asp were uniquely identified. Furthermore, Thr19Arg, Leu452Arg, Thr478Lys, and Pro681Arg were the ones which were found in all the samples, and Asp614Gly was found in 91%. Ala222Val, Asp950Asn, and Lys1191Asn all occurred approximately in 18% of the samples; there was no correlation in sample overlap amongst them. There were additional 39 SNPs which could be detected in the intergenic regions of various samples.

3.4. Comparisons of SNPs with the Delta and Delta Plus Variants

There are several key mutations in single nucleotide bases that have been reported to characterize the Delta and Delta Plus variants. The most widely inspected probably belong to the surface glycoprotein gene of the Spike protein. The characteristic mutations which are present in Delta strains as compared to the ancestral variant are Thr19Arg, (Val70Phe*), Thr95Ile, Gly142Asp, Glu156-, Phe157-, Arg158G, (Ala222Val*), (Trp258Leu*), (Lys417Asn*), Leu452Arg, Thr478Lys, Asp614Gly, Pro681Arg, and Asp950Asn, out of which the three shown with an "*" here even more prevalent in the Delta plus variant.However, the SNPs found and those overlapping with known SNPs for the S gene were Thr19Arg, Thr478Lys, Leu452Arg, Ala222Val, Gly142Asp, Asp614Gly, Asp950Asn, and Pro681Arg only; other SNPs from the known list and the observed list did not match for the S gene. Two observations worth noting are: Firstly, the Lys417Asn mutation was not observed in any of the samples; however, 18% of the samples showed a Lys1191Asn SNP change. Secondly, the Pro681Arg SNP was observed in 100% of the samples, which is an SNP site present in the Mu variant as Pro681His.

In addition to the S gene, there are other mutations which are reported in non-S loci for the Delta and Delta Plus variants, e.g., ORF1ab, M, N, ORFs 3a, 7a, and 8, E, etc. The signature mutations in these sites were also compared with the SNPs detected in the Jaipur samples and are summarized in the Table 6.

Table 6. Comparison of the SNPs present at non-spike loci in the Delta and Delta plus variants with the Jaipur samples.

Genes	SNPs Overlapping with Delta and Delta Plus	SNPs Which Were Absent in Jaipur Samples	Total Types of Non-Unique Detected SNPs	Total Unique SNPs Detected
ORF1ab	Thr3646Ala	Ala1146Thr	30	61
	Thr3255Ile	Ala3209Val		
	Val3718Ala	Val2930Leu		
	Pro2046Leu	Pro2287Ser		
	Pro1640Leu	Ala1306Ser		
		Pro2046Leu		
		Thr3750Ile		
ORF3a	Ser26Leu		2	1
M	Ile82Thr		1	4
N	Arg203Met	Gly215Cys	4	8
	Asp63Gly			
	Asp377Tyr			
ORF7A	Thr120Ile	Thr40Ile	3	0
	Val82Ala			

3.5. *Variations in Various Genes Identified in the WBE Samples*

Since the SNPs were variably present in the genes of different samples, multiple sequence alignment was performed with the predicted peptide sequences to identify the extent of variations across various sample sequences. Figure 3 depicts the regions of variations and phylogenetic relationships of the S, N, M, and ORF1ab genes. As can be observed from the phylogenetic relationships, the higher number of SNPs was reflected by the higher variability in the alignment, which in turn was associated with the more diversified clades of the tree. The existence of such diversity in the same time frame in a single city might allow for an understanding of the evolution of variants and possibly hold some key insights into the host–pathogen interactions. Figure 3 shows the analysis of gene variations of the S, M, N, and ORF1ab genes.

3.6. *Comparing the Sequences Obtained in Context of Patient Samples of the City Obtained during the Same Time Frame*

Since the wastewater sample genome sequences showed phylogenetic trees that were well-aligned with the different timeline reference sequences, an attempt was made to investigate if the sequences showed any correlation between the community viral variants and the hospital patient viral variants prevalent in the city at the time of sampling. The phylogenetic tree was calculated for the wastewater genome sequences with eight patient samples and a Delta variant reference sequence. These patient samples were collected between the same window of wastewater sampling in the city of Jaipur (from February 2021 to May 2021). The Phylogenetic tree is depicted in Figure 4.

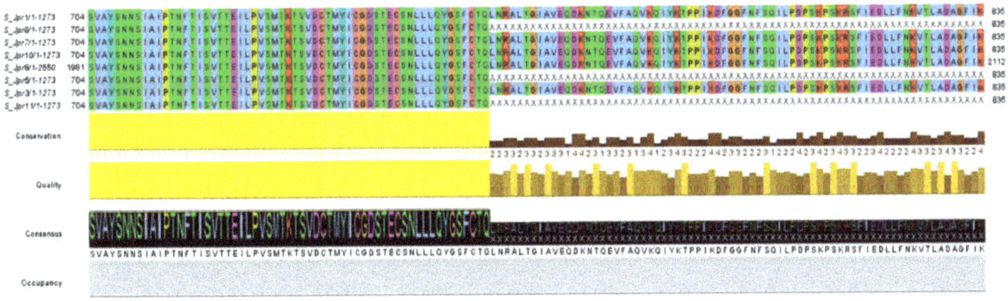

(a)

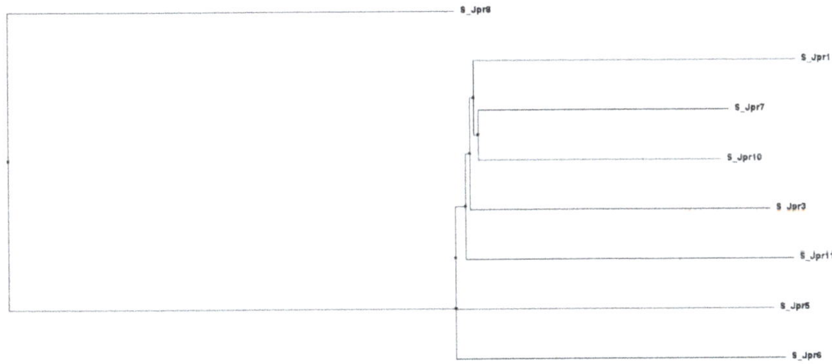

(b)

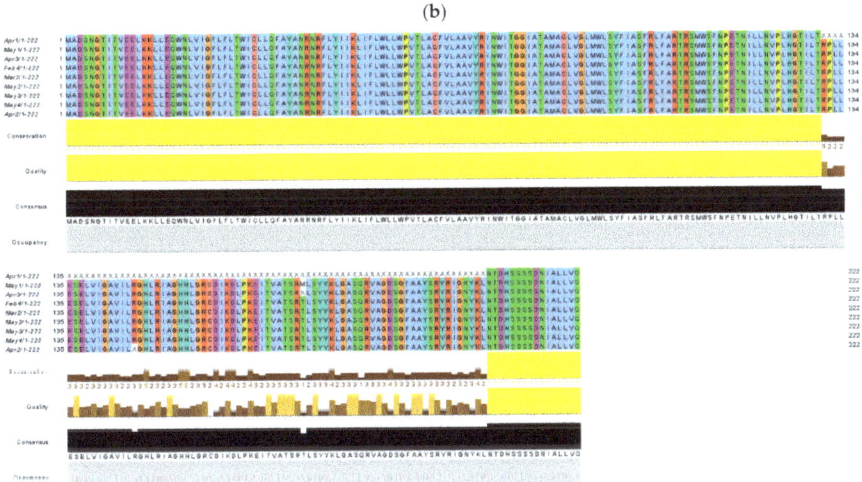

(c)

Figure 3. *Cont.*

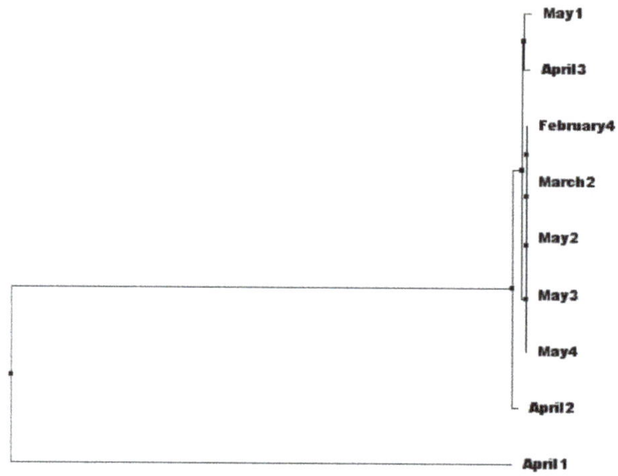

(**d**)

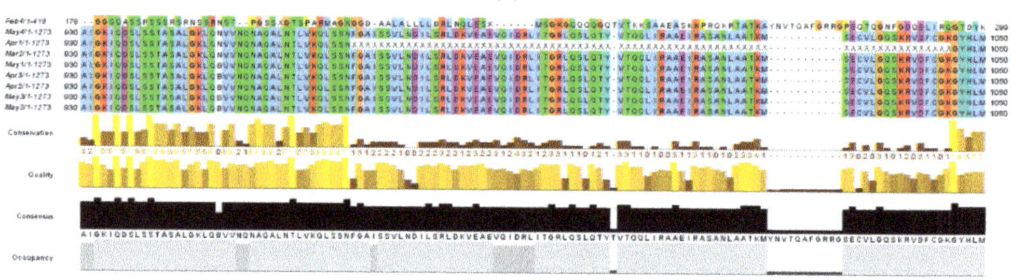

(**e**)

Figure 3. *Cont.*

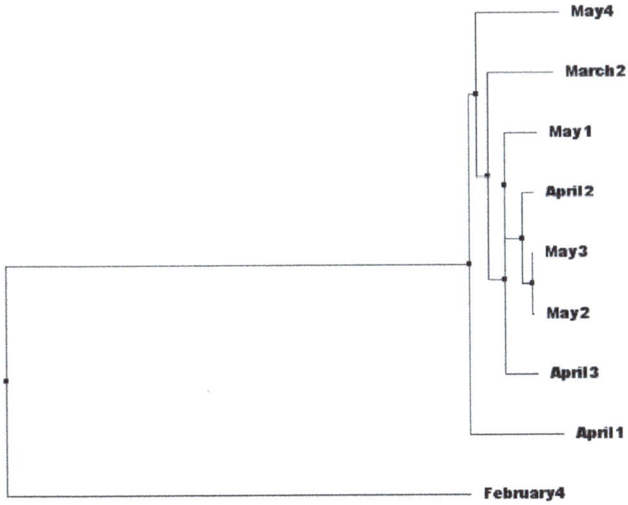

(f)

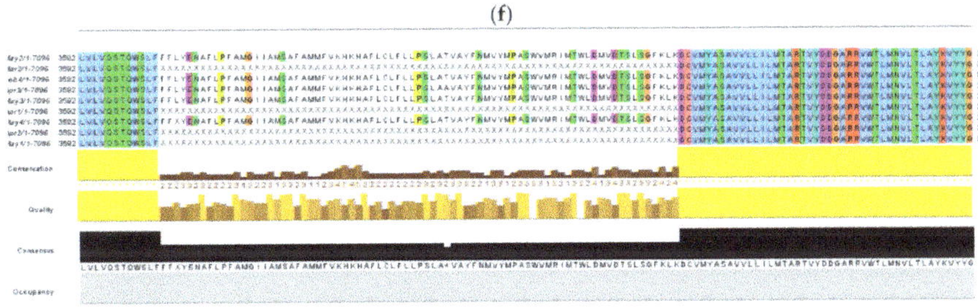

(g)

Figure 3. *Cont.*

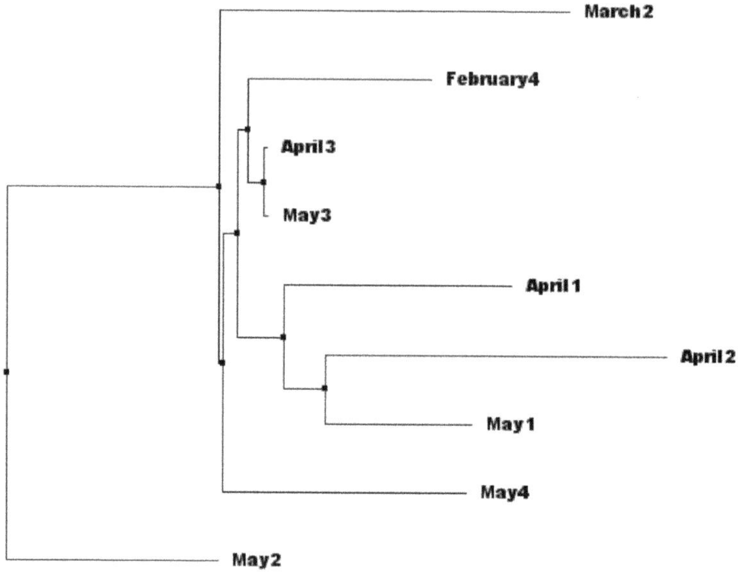

(**h**)

Figure 3. Analysis of gene variations of the S (**a**,**b**), M (**c**,**d**), N (**e**,**f**), and ORF1ab genes (**g**,**h**). (**a**) Spike alignment between 704 and 835 bps, (**b**) Spike Phylogenetic tree, (**c**) M gene alignment at 1–222 bps, (**d**) Phylogenetic tree of the M gene, (**e**) N gene at 930–1050 bps, (**f**) Phylogenetic tree for the N gene, (**g**) ORF1ab polyprotein multiple sequence alignment at 3592–3724 bps, (**h**) Phylogenetic tree of ORF1ab.

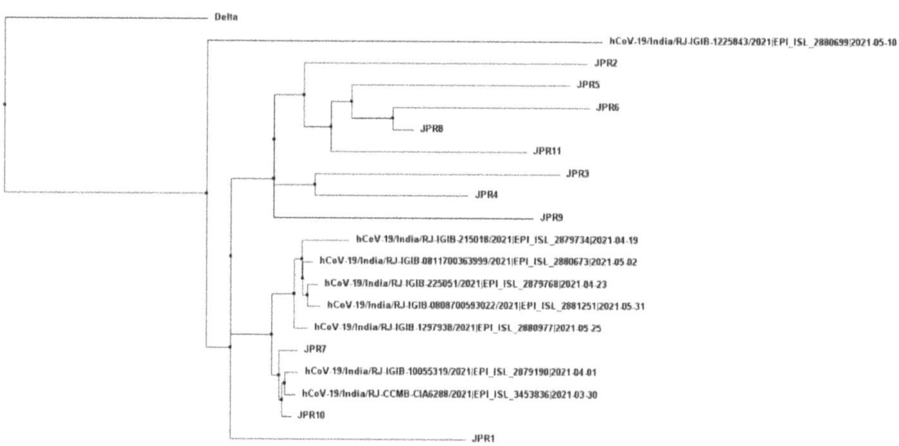

Figure 4. Phylogenetic tree of wastewater sample with clinical patient sample.

3.7. *Sequences on Public Databases*

The obtained sequences have already been submitted to GISAID (EPI_ISL_4106468 EPI_ISL_4198506, EPI_ISL_4198392, EPI_ISL_4198393, EPI_ISL_4198461, EPI_ISL_4197532, EPI_ISL_4198503, EPI_ISL_4198504, EPI_ISL_4198119, EPI_ISL_4198513, EPI_ISL_4198502) and the NCBI gene database (OK090988 and OK342100-OK342107) to further strengthen the research and advancements in combating this pandemic, and are available for various analyses [31–33].

4. Discussion

It has been established that investigating the spread of SARS-CoV-2 and identifying variations in wastewater may reliably determine viral strain prevalence [7]. The method uses pooled samples and can only detect mutations per place in the genome, not specific strains. Nonetheless, it has several advantages over human clinical sampling, primarily in terms of ethical concerns and population capture, as only a few samples of wastewater can provide a comprehensive picture of viral variety as compared to thousands of samples from humans. From February to June 2021, we used a well-established system of wastewater sampling and viral concentration techniques to conduct a large-scale surveillance of SARS-CoV-2 variants using next-generation sequencing of SARS-CoV-2 genomes, which were sampled monthly in 11 different WWTPs across Jaipur.

Attempts have been made across the world to keep track of the pandemic and its causal agents by keeping a constant eye on new varieties. This monitoring is carried out through many national and international programs, such as INSACOG and GISAID [30,33,34], which are connected to the daily samples taken and sequenced from patients. When compared to the viral load circulating in community wastewater samples, which also contains many organic and inorganic contaminants, the viral load seen in a single patient sample proves to be extremely concentrated. The study's first goal was to collect RNA samples of sufficient quantity and quality in order to be processed on next-generation sequencing (NGS) technologies. The Nanodrop quality checks were passed by all eleven samples prepared for sequencing.

Despite the high Ct values that are typical of such materials, the sequencing of SARS-CoV-2 from wastewater was effective, with good coverages of the genome acquired for most samples and sufficient depth to call it a mutation. These findings made it possible to keep a close eye on variations all throughout Jaipur, even in areas where clinical sampling revealed little morbidity [28]. The increased proportions of sequenced viruses belonging to the B.1.617.1 and B.1.617.2 lineages submitted to the GISAID EpiCoV database have been linked to an increase in case numbers in India.

In December 2020, B.1.617.1 was discovered for the first time in India. It first rose in proportion in India, peaking at roughly 50% of weekly reported sequences in the GISAID EpiCoV database in late March 2021, before declining in April 2021 [35]. It has been discovered in India (290), the United Kingdom (247), the United States (137), Singapore (64), Germany (28), Australia (21), Denmark (21), Bahrain (9), Japan (9), Angola (8), Switzerland (8), Hong Kong (7), Ireland (7), Portugal (7), Belgium (5), Luxembourg (5), South Korea (5), Canada (4), the Netherlands (4), New Zealand (4), Sweden (4), France (3), Jordan (3), Czechia (2), Guadeloupe (2), Russia (2), Sint Maarten (2), Spain(2), Cambodia (1), Greece (1), Italy (1), Malaysia (1), and Mexico (1).There are signs that the variant is becoming more common in the EU/EEA, although it is still only found in a small percentage of sequences, with the greatest proportions reported in Luxembourg (0.6%) and Spain (0.3%) [36].

Within B.1.617, there are three separate lineages, each with its own mutation profile, necessitating evaluation at the lineage level rather than as a whole. The spike protein amino acid alterations L452R, E484Q, D614G, P681R, and Q1071H characterize B.1.617.1 (some viruses also carry V382L). The spike protein changes T19R, 157–158, L452R, E484Q, D614G, P681R, and D950N have allowed it to be classified as a VOI by ECDC and WHO, and as a VOC by the UK, indicating that its transmissibility is at least as high as that of VOC B.1.1.7. B.1.617.3 is defined by the spike protein changes T19R, 157–158, L452R, E484. The ECDC and WHO have categorized this lineage as a VOI, but the UK has classified it as a VUI [37]. Below are some of the specific spike protein modifications associated with these lineages that have been described as having an impact on viral properties.

E484Q (only B.1.617.1 and B.1.617.3)—changes at this site are associated with reduced neutralization by convalescent sera and specific therapeutic antibodies. L452R—changes at this site are associated with increased transmissibility and reduced neutralization by convalescent plasma and specific therapeutic antibodies. Although this has not been proven in practice, for the P681R—alteration is positioned exactly near the furin cleave site and might potentially influence S1/S2 cleavage, cell entrance, and infectivity. D614G—this alteration is linked to increased transmissibility with a high degree of certainty. The great majority of currently circulating viruses, however, carry it [38].

To summarize, this study demonstrates the benefits of wastewater sample sequencing as a reliable method of monitoring the diversity of SARS-CoV-2 strains circulating in a community, warning against the emergence of epidemiologically or clinically relevant mutations or variants, and assisting in public health decision making.

The detection of existing circulating variants and dominant mutations in populations using SARS-CoV-2 genome sequencing in wastewater explains the link between dominant variants and the pandemic situation in Jaipur, India. It demonstrates the potential of SARS-CoV-2 genomic surveillance in wastewater as an early warning indicator system and its ability to detect rapidly emerging new variants in any significant epidemic.

5. Significance of the Work and Conclusions

The usefulness of wastewater surveillance in COVID-19 trend tracking in several areas has been demonstrated in this study. The COVID-19 cases were substantially linked to SARS-CoV-2 gene quantities in wastewater. In early March 2021, the B.1.617.2 variant was first detected in Jaipur as part of the national SARS-CoV-2 sequencing program, which was newly established to track VOC in clinical samples. This demonstrates that surveillance by wastewater is a robust strategy that may cover wide areas with few samples, providing early notification as regards the penetrance and spread of VOC in a region where clinical sampling is insufficient. Thus, integrating WBE surveillance with genome sequencing and pathogen variant surveillance might be an effective and efficient step towards monitoring, detecting, and managing new infection waves similar to those shown by infectious SARS-CoV-2 variants early on, and therefore protect the communities from loss of life and economic resources in any future catastrophes similar to this this pandemic.

Supplementary Materials: The following supporting information can be downloaded at: https://www.mdpi.com/article/10.3390/w14030297/s1, Table S1: Details of Samples sent for Sequencing & RT-qPCR Detection; Table S2: Mapping and Consensus Statistics; Table S3: RT qPCR Detection.

Author Contributions: Conceptualization, S.A., A.N. and A.B.G.; methodology, A.N., V.S. and E.M.; validation, A.B.G. and K.M.M.; formal analysis, A.N.; investigation, A.N., V.S., D.S. and E.M.; resources, S.A.; data curation, A.N., S.A., A.B.G. and K.M.M.; writing—original draft preparation, A.N. and S.A.; writing—review and editing, A.B.G. and K.M.M.; visualization, K.M.M.; supervision, A.B.G.; project administration, S.A.; funding acquisition, S.A. and A.N. All authors have read and agreed to the published version of the manuscript.

Funding: This research was funded by research grants from the Institutional Internal Mural Grant (IMG), supported by Institutional Research Scientific Committee (IRSC) (Grant No. BIBT/IRSC/IMG/ 20-21/ 042.

Institutional Review Board Statement: Not applicable.

Informed Consent Statement: Not applicable.

Data Availability Statement: Not applicable.

Acknowledgments: We would like to acknowledge the constant support received from B. Lal Gupta (Director), Saksham Gupta (Deputy Director), and Aparna Datta (Principal). We would also like to thank the Sandeep Shrivastava (Head, Centre for Innovation, Research and Development (CIRD, Dr. B. Lal Clinical laboratory Pvt. Ltd.) for their support in analysis. We would also like to thank the Jaipur development authority (JDA) officials and plant operators at the WWTPs and all other participating wastewater treatment plants for the collection of sewage samples in this study.

Conflicts of Interest: The authors declare no conflict of interest.

References

1. Jin, H.; Wang, H.; Li, X.; Zheng, W.; Ye, S.; Zhang, S.; Zhou, J.; Pennington, M. Economic burden of COVID-19, China, January–March, 2020: A cost-of-illness study. *Bull. World Health Organ.* **2021**, *99*, 112. [CrossRef]
2. Available online: https://www.who.int/news/item/28-11-2021-update-on-omicron (accessed on 14 December 2021).
3. Bakhshandeh, B.; Jahanafrooz, Z.; Abbasi, A.; Goli, M.B.; Sadeghi, M.; Mottaqi, M.S.; Zamani, M. Mutations in SARS-CoV-2; Consequences in structure, function, and pathogenicity of the virus. *Microb. Pathog.* **2021**, *154*, 104831. [CrossRef]
4. Harvey, W.T.; Carabelli, A.M.; Jackson, B.; Gupta, R.K.; Thomson, E.C.; Harrison, E.M.; Ludden, C.; Reeve, R.; Rambaut, A.; Peacock, S.J.; et al. SARS-CoV-2 variants, spike mutations and immune escape. *Nat. Rev. Microbiol.* **2021**, *19*, 409–424. [CrossRef]
5. Amereh, F.; Negahban-Azar, M.; Isazadeh, S.; Dabiri, H.; Masihi, N.; Jahangiri-Rad, M.; Rafiee, M. Sewage systems surveillance for SARS-CoV-2: Identification of knowledge gaps, emerging threats, and future research needs. *Pathogens* **2021**, *10*, 946. [CrossRef] [PubMed]
6. Available online: https://www.who.int/emergencies/diseases/novel-coronavirus-2019/situation-reports (accessed on 14 December 2021).
7. Nemudryi, A.; Nemudraia, A.; Wiegand, T.; Surya, K.; Buyukyoruk, M.; Cicha, C.; Vanderwood, K.K.; Wilkinson, R.; Wiedenheft, B. Temporal detection and phylogenetic assessment of SARS-CoV-2 in municipal wastewater. *Cell Rep. Med.* **2020**, *1*, 100098. [CrossRef] [PubMed]
8. Di Marcantonio, C.; Chiavola, A.; Gioia, V.; Frugis, A.; Cecchini, G.; Ceci, C.; Spizzirri, M.; Boni, M.R. Impact of COVID19 restrictions on organic micropollutants in wastewater treatment plants and human consumption rates. *Sci. Total Environ.* **2021**, *811*, 152327. [CrossRef] [PubMed]
9. Available online: https://www.who.int/en/activities/tracking-SARS-CoV-2-variants/ (accessed on 14 December 2021).
10. Altmann, D.M.; Boyton, R.J.; Beale, R. Immunity to SARS-CoV-2 variants of concern. *Science* **2021**, *371*, 1103–1104. [CrossRef]
11. Challen, R.; Brooks-Pollock, E.; Read, J.M.; Dyson, L.; Tsaneva-Atanasova, K.; Danon, L. Risk of mortality in patients infected with SARS-CoV-2 variant of concern 202012/1: Matched cohort study. *BMJ* **2021**, *372*, n579. [CrossRef]
12. Grint, D.J.; Wing, K.; Williamson, E.; McDonald, H.I.; Bhaskaran, K.; Evans, D.; Evans, S.J.; Walker, A.J.; Hickman, G.; Nightingale, E.; et al. Case fatality risk of the SARS-CoV-2 variant of concern B. 1.1. 7 in England, 16 November to 5 February. *Eurosurveillance* **2021**, *26*, 2100256. [CrossRef]
13. Davies, N.G.; Jarvis, C.I.; Edmunds, W.J.; Jewell, N.P.; Diaz-Ordaz, K.; Keogh, R.H. Increased mortality in community-tested cases of SARS-CoV-2 lineage B. 1.1. 7. *Nature* **2021**, *593*, 270–274. [CrossRef]
14. Oran, D.P.; Topol, E.J. The Proportion of SARS-CoV-2 Infections That Are Asymptomatic: A Systematic Review. *Ann. Intern. Med.* **2021**, *174*, 655–662. [CrossRef]
15. Crits-Christoph, A.; Kantor, R.S.; Olm, M.R.; Whitney, O.N.; Al-Shayeb, B.; Lou, Y.C.; Flamholz, A.; Kennedy, L.C.; Greenwald, H.; Hinkle, A.; et al. Genome sequencing of sewage detects regionally prevalent SARS-CoV-2 variants. *MBio* **2021**, *12*, e02703-20. [CrossRef]
16. Otero, M.C.B.; Murao, L.A.E.; Limen, M.A.G.; Gaite, P.L.A.; Bacus, M.G.; Acaso, J.T.; Corazo, K.; Knot, I.E.; Sajonia, H.; Francis, L.; et al. Wastewater-Based Epidemiology and Whole-Genome Sequencing for Community-Level Surveillance of SARS-CoV-2 in Selected Urban Communities of Davao City, Philippines: A Pilot Study. *medRxiv* **2021**. Available online: https://www.medrxiv.org/content/10.1101/2021.08.27.21262450v1 (accessed on 14 December 2021).
17. Rothman, J.A.; Loveless, T.B.; Kapcia, J., III.; Adams, E.D.; Steele, J.A.; Zimmer-Faust, A.G.; Langlois, K.; Wanless, D.; Griffith, M.; Mao, L.; et al. RNA Viromics of Southern California Wastewater and Detection of SARS-CoV-2 Single-Nucleotide Variants. *Appl. Environ. Microbiol.* **2021**, *87*, e01448-21. [CrossRef]
18. Jahn, K.; Dreifuss, D.; Topolsky, I.; Kull, A.; Ganesanandamoorthy, P.; Fernandez-Cassi, X.; Bänziger, C.; Stachler, E.; Fuhrmann, L.; Jablonski, K.P.; et al. Detection of SARS-CoV-2 variants in Switzerland by genomic analysis of wastewater samples. *medRxiv* **2021**. Available online: https://www.medrxiv.org/content/10.1101/2021.01.08.21249379v1.full.pdf (accessed on 14 December 2021).
19. Wilton, T.; Bujaki, E.; Klapsa, D.; Majumdar, M.; Zambon, M.; Fritzsche, M.; Mate, R.; Martin, J. Rapid increase of SARS-CoV-2 variant B. 1.1. 7 detected in sewage samples from England between October 2020 and January 2021. *Msystems* **2021**, *6*, e00353-21. [CrossRef]
20. Landgraff, C.; Wang, L.Y.; Buchanan, C.; Wells, M.; Schonfeld, J.; Bessonov, K.; Ali, J.; Robert, E.; Nadon, C. Metagenomic sequencing of municipal wastewater provides a near-complete SARS-CoV-2 genome sequence identified as the B. 1.1. 7 variant of concern from a Canadian municipality concurrent with an outbreak. *medRxiv* **2021**. Available online: https://www.medrxiv.org/content/10.1101/2021.03.11.21253409v1.full.pdf (accessed on 14 December 2021).
21. Dharmadhikari, T.; Rajput, V.; Yadav, R.; Boargaonkar, R.; Panse, D.; Kamble, S.; Dastager, S. High throughput sequencing based detection of SARS-CoV-2 prevailing in wastewater of Pune, West India. *medRxiv* **2021**. Available online: https://www.medrxiv.org/content/10.1101/2021.06.08.21258563v1.full.pdf (accessed on 14 December 2021).
22. Medema, G.; Been, F.; Heijnen, L.; Petterson, S. Implementation of environmental surveillance for SARS-CoV-2 virus to support public health decisions: Opportunities and challenges. *Curr. Opin. Environ. Sci. Health* **2020**, *17*, 49–71. [CrossRef]

23. Medema, G.; Heijnen, L.; Elsinga, G.; Italiaander, R.; Brouwer, A. Presence of SARS-Coronavirus-2 RNA in sewage and correlation with reported COVID-19 prevalence in the early stage of the epidemic in the Netherlands. *Environ. Sci. Technol. Lett.* **2020**, *7*, 511–516. [CrossRef]
24. Bivins, A.; North, D.; Ahmad, A.; Ahmed, W.; Alm, E.; Been, F.; Bhattacharya, P.; Bijlsma, L.; Boehm, A.B.; Brown, J.; et al. Wastewater-based epidemiology: Global collaborative to maximize contributions in the fight against COVID-19. *Environ. Sci. Technol.* **2020**, *54*, 7754–7757. [CrossRef]
25. Arora, S.; Nag, A.; Sethi, J.; Rajvanshi, J.; Saxena, S.; Shrivastava, S.K.; Gupta, A.B. Sewage surveillance for the presence of SARS-CoV-2 genome as a useful wastewater based epidemiology (WBE) tracking tool in India. *Water Sci. Technol.* **2020**, *82*, 2823–2836. [CrossRef]
26. Arora, S.; Nag, A.; Rajpal, A.; Tyagi, V.K.; Tiwari, S.B.; Sethi, J.; Sutaria, D.; Rajvanshi, J.; Saxena, S.; Shrivastava, S.K.; et al. Imprints of lockdown and treatment processes on the wastewater surveillance of SARS-CoV-2: A curious case of fourteen plants in northern India. *Water* **2021**, *13*, 2265. [CrossRef]
27. Arora, S.; Nag, A.; Kalra, A.; Sinha, V.; Meena, E.; Saxena, S.; Sutaria, D.; Kaur, M.; Pamnani, T.; Sharma, K.; et al. Successful Application of Wastewater-Based Epidemiology in Prediction and Monitoring of the Second Wave of COVID-19 in India with Fragmented Sewerage Systems-A Case Study of Jaipur (India). *medRxiv* **2021**. Available online: https://www.medrxiv.org/content/10.1101/2021.09.11.21263417v1.full.pdf (accessed on 14 December 2021).
28. Available online: https://www.cdc.gov/ (accessed on 8 January 2022).
29. Available online: www.covid19india.org (accessed on 8 January 2022).
30. Sievers, F.; Higgins, D.G. Clustal Omega, accurate alignment of very large numbers of sequences. In *Multiple Sequence Alignment Methods*; Humana Press: Totowa, NJ, USA, 2014; pp. 105–116.
31. Available online: https://www.ncbi.nlm.nih.gov/nuccore/OK090988 (accessed on 14 December 2021).
32. Available online: https://www.ncbi.nlm.nih.gov/nuccore/OK342100 (accessed on 14 December 2021).
33. Available online: https://www.ncbi.nlm.nih.gov/nuccore/OK342107 (accessed on 14 December 2021).
34. Kannan, S.R.; Spratt, A.N.; Cohen, A.R.; Naqvi, S.H.; Chand, H.S.; Quinn, T.P.; Lorson, C.L.; Byrareddy, S.N.; Singh, K. Evolutionary analysis of the Delta and Delta Plus variants of the SARS-CoV-2 viruses. *J. Autoimmun.* **2021**, *124*, 102715. [CrossRef]
35. Emergence of SARS-CoV-2 B.1.617 Variants in India and Situation in the EU/EEA. Available online: https://www.ecdc.europa.eu/sites/default/files/documents/Emergence-of-SARS-CoV-2-B.1.617-variants-in-India-and-situation-in-the-EUEEA_0 (accessed on 14 December 2021).
36. Assessment, R.R. *Coronavirus Disease 2019 (COVID-19) in the EU/EEA and the UK–Ninth Update*; European Centre for Disease Prevention and Control: Stockholm, Sweden, 2020.
37. Available online: https://indianexpress.com/article/explained/explained-who-classifies-india-variant-as-being-of-global-concern-what-does-it-mean-7310999/ (accessed on 14 December 2021).
38. Volz, E.; Hill, V.; McCrone, J.T.; Price, A.; Jorgensen, D.; O'Toole, Á.; Southgate, J.; Johnson, R.; Jackson, B.; Nascimento, F.F.; et al. Evaluating the effects of SARS-CoV-2 spike mutation D614G on transmissibility and pathogenicity. *Cell* **2021**, *184*, 64–75. [CrossRef]

water

MDPI

Article

Molecular Identification of Human Adenovirus Isolated from Different Wastewater Treatment Plants in Riyadh, Saudi Arabia: Surveillance and Meteorological Impacts

Khalid Maniah [1], Islam Nour [1,*], Atif Hanif [1], Mohamed Taha Yassin [1], Abdulrahman Alkathiri [1], Ibrahim Al-Ashkar [2] and Saleh Eifan [1,*]

1 Botany and Microbiology Department, College of Science, King Saud University, Riyadh 11451, Saudi Arabia; 442106769@student.ksu.edu.sa (K.M.); ahchaudhry@ksu.edu.sa (A.H.); myassin2.c@ksu.edu.sa (M.T.Y.); 441105608@student.ksu.edu.sa (A.A.)

2 Department of Plant Production, College of Food and Agriculture Sciences, King Saud University, Riyadh 11451, Saudi Arabia; ialashkar@ksu.edu.sa

* Correspondence: inour@ksu.edu.sa (I.N.); seifan@ksu.edu.sa (S.E.)

Abstract: Regular water environment monitoring is crucial for minimizing contamination caused by waterborne viruses and reducing health risks. As the human adenovirus (HAdV) is linked to clinical episodes of gastroenteritis in children, the present investigation aimed to detect HAdVs in three wastewater treatment plants in Riyadh, Saudi Arabia (King Saud University (KSU-WWTP), Manfoha (MN-WWTP), and Embassy Quarter (EMB-WWTP)). The impact of seasonal variability and meteorological factors on the prevalence of HAdVs was also investigated. The HAdV hexon sequences of the isolated human adenoviruses were phylogenetically analyzed and revealed that the F species of HAdV, especially serotype 41, dominated. The highest prevalence of HAdV was detected in KSU-WWTP (83.3%), followed by MN-WWTP (75%), and EMB-WWTP (66.6%). Seasonal distribution insignificantly influenced the HAdV prevalence among sampling areas ($p > 0.05$). The highest prevalence of HAdVs (100%) was detected in late Summer and Autumn at temperatures (high: 34–43 °C, low: 18–32 °C) and moderate prevalence of 66.67% in Winter (particularly, in January and February) at lower temperature ranges (high: 26 °C, low: 10 °C–12 °C). The large variation of HAdV prevalence detected at different humidity ranges emphasized the significant impact of relative humidity on HAdV incidence in raw water of WWTPs ($p = 0.009$, $R^2 = 0.419$). In contrast, wind speed was detected to have insignificant influence on HAdV prevalence among different WWTPs ($p > 0.05$, $R^2 = 0.03$). The study provides important data for the incidence of HAdVs in wastewater treatments plants in Riyadh, Saudi Arabia, which enabled the successful management of health hazards of viral diseases transmitted via fecal-oral route. In addition, the non-significant influence of seasonal variability on HAdV prevalence highlights the potentiality of utilizing HAdVs as a potential fecal indicator of wastewater contamination.

Keywords: wastewater treatment, human adenovirus; serotype 41; seasonal variability; temperature

Citation: Maniah, K.; Nour, I.; Hanif, A.; Yassin, M.T.; Alkathiri, A.; Al-Ashkar, I.; Eifan, S. Molecular Identification of Human Adenovirus Isolated from Different Wastewater Treatment Plants in Riyadh, Saudi Arabia: Surveillance and Meteorological Impacts. *Water* **2023**, *15*, 1367. https://doi.org/10.3390/w15071367

Academic Editors: Guangming Jiang, Ryo Honda, Sudipti Arora and Laura Bulgariu

Received: 22 March 2023
Revised: 29 March 2023
Accepted: 31 March 2023
Published: 2 April 2023

1. Introduction

Enteric viruses are significant waterborne pathogens found in wastewater [1–3]. They are often isolated from feces contaminated water and have been connected to a number of waterborne epidemics [4]. Pathogens in this category include adenoviruses, rotaviruses, hepatitis A virus, enteroviruses, and noroviruses [5]. Enteric viruses were detected from recreational beaches and groundwater for public use in the Great Lakes area, suggesting an increased public health risk from drinking or coming into contact with these waters [6,7]. Adenoviruses (HAdVs), which are abundant in wastewater, have been proposed as indicators for viral infections since they meet the majority of the requirements for an ideal indicator [8]. More than 90% of the human population is thought to be positive for one or more adenovirus serotypes [9]. Human adenoviruses are more prevalent in sewage than other enteric viruses and are discharged in high proportions by infected people (up to 10^{11} viral particles/gram of feces) [10]. Transmission mechanisms of adenoviruses include the fecal-oral pathway and inhalation of aerosols [11]. Adenoviruses have been linked with epidemics in numerous contexts, including hospitals [12], day care facilities [13], swimming pools [14], and schools [15]. Furthermore, adenoviruses are one of the main causes of clinical disorders such as gastroenteritis, hemorrhagic cystitis, conjunctivitis, respiratory ailments, and systemic infections [16]. The six genera of the Adenoviridae family that have been classified to date are mastadenovirus, ichtadenovirus, atadenovirus, siadenovirus, aviadenovirus, and testadenovirus [17]. Human adenoviruses are classified into seven species (A–G) and more than 110 types within the Adenoviridae family [18]. Pathogenesis is caused by adenovirus infections in a range of human organs. Adenoviruses B and C are the most prevalent respiratory infections, with Adenoviruses species A affecting the respiratory system in immunodeficient individuals, and Adenoviruses F, including serotypes 40 and 41, recognized as one of the main viruses causing infantile gastroenteritis [19]. Human adenovirus is an enteric, non-enveloped, and icosahedral particle with a double-stranded, linear DNA genome of 34–36 kb [20]. They are now classified into seven HAdV species, A-G, as well as newly found adenovirus types that are constantly appearing [21]. Adenoviruses, like the majority of enteric viruses, are more resilient to environmental conditions and even sewage treatment techniques than the fecal indicator bacteria now in use [20]. Moreover, direct sequencing of PCR products, as well as sequence analysis of cloned PCR products, have been routinely employed to study adenoviruses in aquatic settings [22,23]. The research regarding the molecular detection of HAdVs in Saudi Arabia is scanty; hence, the aim of the current investigation was to detect the prevalence of HAdVs in wastewater samples collected from wastewater treatment plants at three different locations in Riyadh, Saudi Arabia. Furthermore, the impact of environmental conditions on the prevalence of HAdVs was also detected. Finally, the sequences of the detected HAdVs were phylogenetically analyzed to detect their correlation with reference strains in GenBank.

2. Materials and Methods

2.1. Sample Collection

From January 2022 to December 2022, 36 untreated wastewater samples were collected monthly at a rate of three samples per month from the inlets of the three different wastewater treatment plants (WWTPs) in Riyadh, Saudi Arabia. The sampling locations were King Saud University wastewater treatment plants (KSU-WWTP), Manfoha wastewater treatment plants (MN-WWTP), and Embassy Quarter wastewater treatment plants (EMB-WWTP) in Riyadh, Saudi Arabia. The latitude and longitude of the sampling locations are shown in Figure 1. In sterile 200-mL plastic bottles, samples were collected and transferred to the laboratory on dry ice. The data for temperature, relative humidity, and wind speed were collected on each sample day in order to study the influence of the weather on viral persistence.

2.2. Viral Concentration

Adenovirus (Adv) was concentrated using the polyethylene glycol (PEG) precipitation process [24]. Briefly, 200 mL of wastewater was combined with 25 mL of glycine buffer (0.05 M glycine and 0.3 g/L beef extract) to detach virions bound to the organic material, and the pH was adjusted to 9.6 with 1 M NaOH before centrifugation at 8000× g for 30 min. Afterwards, the supernatant was withdrawn using sterile syringe then filtered using Millipore filter (0.22 µm) for the removal of bacterial cell and the other unfavorable debris. The filtrate was then treated with PEG 8000 (80 g/L) and sodium chloride (17.5 g/L) followed by stirring by a magnetic stirrer overnight (100 rpm) at room temperature for viral precipitation. Centrifugation of the filtrate was then performed at 13,000× g for harvesting viral particles then the pellet was eluted in 1 mL phosphate buffer saline (PBS) and finally stored at −80 °C for further experimentations.

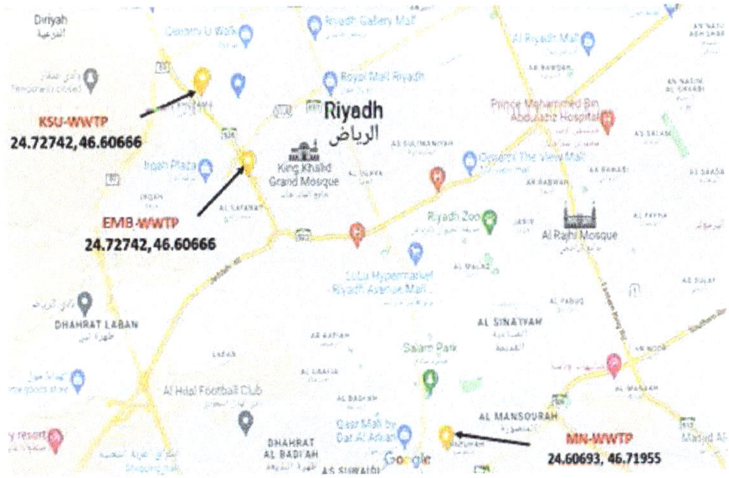

Figure 1. Latitude and longitude of sampling locations in Riyadh, Saudi Arabia.

2.3. DNA Extraction and Specific PCR Detection

Extraction of genomic DNA of HAdVs was conducted for the concentrated 36 wastewater samples using a DNeasy PowerWater Kit (Qiagen GmbH, Hilden, Germany). The viral DNA of Human adenoviruses (HAdV) was directly detected in a 20 µL reaction mixture composed of 2 µL DNA template, 300 nM AdFhex-F: 5′-GCCACCGATACCTACTTCAGCCTG-3′ as forward primer and 300 nM AdFhex-R: 5′-GGCAGTGCCGGAGTAGGGTTTAAA-3′ as a reverse primer targeting hexon gene and 1 X Phusion Master Mix under the following reaction conditions: 98 °C for 30 s, followed by 40 cycles of 98 °C for 10 s and 72 °C for 30 s each, and final extension at 72 °C for 5 min. Positive control was acquired for HAdV virus which was obtained from the Virology Unit, King Khalid University, Hospital, Riyadh, Saudi Arabia.

2.4. Amplicon Purification and Sequencing

The 261-bp amplicons for HAdV were excised and purified using the Wizard SV Gel and PCR Clean-Up System (Promega Co., Madison, WI, USA), in accordance with the manufacturer's instructions. Afterwards, the attained amplicons were then sequenced using the ABI genetic analyzer 3130Xl (Applied Biosystems®, Carlsbad, CA, USA).

2.5. Phylogenetic Analysis

The identified HAdV nucleotide sequences were examined using the MEGA X program and compared to BLAST sequences. ClustalW was used to create the sequence alignments. The phylogenetic tree was built using the minimal Bayesian information criteria and the best-fit model of nucleotide substitution. The bootstrapping of 1000 replicates was used to evaluate the reliability of the phylogenetic tree. The Kimura three-parameter approach was used to calculate genetic distances according to the best fitting substitution model (Table S1).

2.6. Statistical Analysis

Pearson's correlation coefficient matrix was used to investigate possible correlations between the various sample locations during a one-year period. To investigate the relevance of the influence of meteorological parameters (including temperature, wind speed, and relative humidity (RH%)) on HAdV prevalence independent of sample region, a one-way analysis of variance was performed. Linear curve fitting was used to match the connections between distinct sample sites, as dependent variables, and climatic conditions, as independent variables. The XL-STAT statistics package software was used for all statistical studies (Ver. 2019, Excel Add-ins soft SARL, New York, NY, USA).

3. Results

3.1. Prevalence of HAdVs in Sampling Areas

Out of 36 wastewater samples tested for HAdVs, 27 (75%) were found to be positive, as denoted by the detection of the 261-bp amplicon (Figure 2). The highest HAdV prevalence was detected in KSU-WWTP (83.3%), followed by MN-WWTP (75%), and EMB-WWTP (66.6%) (Table 1). The molecular characterization of amplicon sequences by Sanger sequencing showed the presence of 20 sequences; however, seven sequences were not characterized owing to overlapped electropherograms. Moreover, these obtained sequences underwent phylogenetic analysis to define the possible serotype, detect any sequence variation, and to check any possible imported strains.

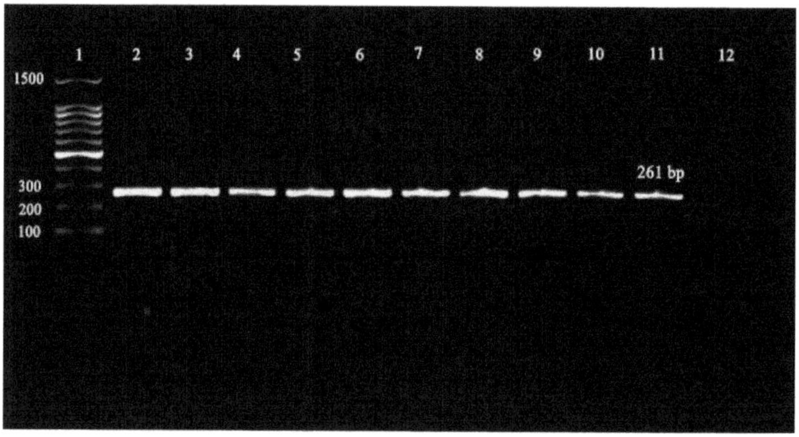

Figure 2. Gel image of the PCR product of HAdV. Lane 1, DNA marker (100–1500 bp); lane 12, negative control; lane 11, positive control; lanes 2 to 10, 261-bp HAdV ampilcons.

Table 1. HAdV prevalence in different water sample locations.

Sampling Area	HAdVs +ve	HAdV Prevalence %
EMB-WWTP	8	66.6%
KSU-WWTP	10	83.3%
MN-WWTP	9	75%

3.2. Predominance of HAdV Serotype 41

The phylogenetic tree illustrated the typical relationship of the HAdV hexon sequences obtained from all sites with serotype 41 followed by serotype 40 of HAdV (type F) (Figure 3). Pairwise distancing analysis uncovered five different HAdV isolates, and most of them (4/5) were collected from KSU-WWTP (Table S2). Brazilian HAdV isolates were the closest to our recovered sequences from the entire locations. However, HAdV recovered sequences from KSU-WTTP, in particular K19-2B64 and K20-2B64, showed the closest relationship with multiple Brazilian isolates as each showed a delicate clustering with specific Brazilian sequences. For instance, K19-2B64 was closely related to five Brazilian HAdV sequences including AD5764, AD6026, AD4578, AD01-71218, and AD102 ($d = 0.0000$, Table S3) than HAdV sequences isolated from irrigation water in Riyadh, Saudi Arabia ($d = 0.0038467$). Interestingly, K20-2B64 shared the same sequence identity as the HAdV sequence (IW isolate) previously recovered from irrigation water in the same area of KSU-WWTP (d = 0.0000, Table S3) rather than that from the landfill wastewater or Wadi Hanifa in Riyadh (d = 0.00384). Since the Brazilian sequences, similar to our sequences, were recovered at different time periods, the time-dependent molecular divergence was taken into account and was displayed in the timetree denoted by Figure S1. The timetree showed that K23-2B64, K26-2B64, and 10UA-2B64 could have earlier molecular divergence than the other sequences that could be the origin of the other sequences and provide interpretation of the distant relationship to sequences previously recovered from the same region or from even the Brazilian sequences. On the contrary, K19-2B64 and K20-2B64 showed later molecular divergence in 2007 and 2003, respectively and consequently a closer relationship to these Brazilian sequences in each cluster. However, the large node height error bar could indicate that the molecular divergence could even have occurred earlier.

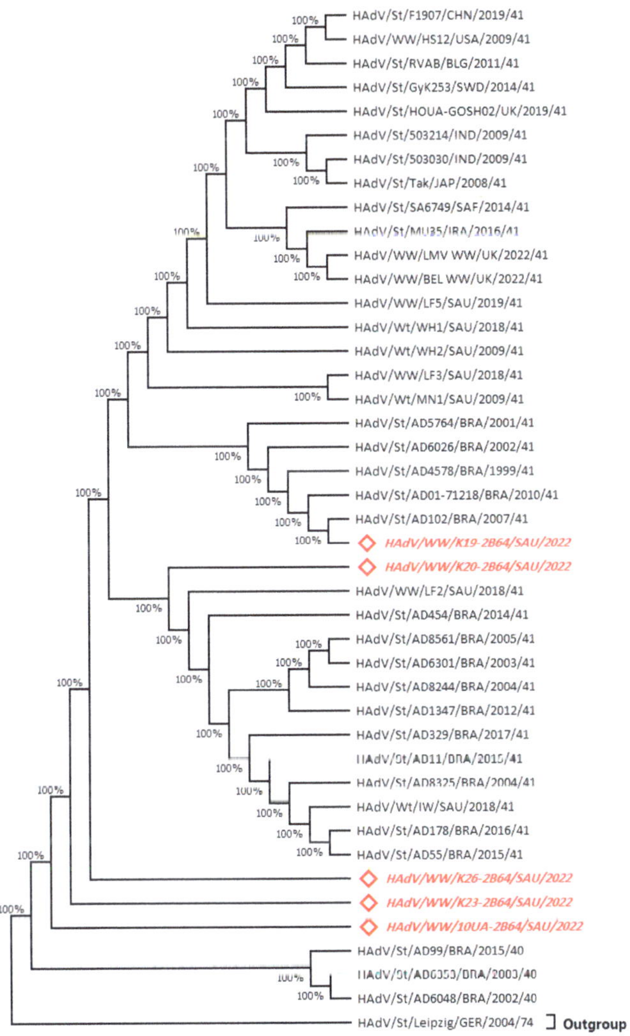

Figure 3. Phylogenetic tree for the HAdV hexon-derived sequences constructed by the maximum likelihood method and Tamura three-parameter model.

The evolutionary history was inferred by using the maximum likelihood method and Tamura 3-parameter model. The tree with the highest log likelihood (−742.94) is shown. Initial tree(s) for the heuristic search were obtained automatically by applying the Neighbor-Join and BioNJ algorithms to a matrix of pairwise distances estimated using the maximum composite likelihood (MCL) approach, and then selecting the topology with a superior log likelihood value. The proportion of sites where at least one unambiguous base is present in at least one sequence for each descendent clade is shown next to each internal node in the tree. This analysis involved 43 nucleotide sequences. The rate variation model was permitted to be evolutionarily uniform for several sites, according to the best fitting substitution model validation (Table S1). The horizontal distance connecting two HAdV sequences is proportional to the genetic distance between these two HAdV sequences. The distance is expressed as the number of nucleotide substitutions per site. HAdV serotype 74 was used as an outgroup. Accession numbers of sequences used for phylogenetic analysis are displayed in Table S4. The red italicized sequences denote the current study sequences.

3.3. Seasonal Distribution of HAdV

HAdV showed various distributions over the different seasons. Overall, the highest HAdV prevalence of 22.22% was observed in Autumn season. However, HAdV recorded the least prevalence (11.11%) in Spring. In the same manner, the highest HAdV prevalence (100%) was recorded in all sampling locations in the Autumn season (Figure 4). Seasonal distribution insignificantly influenced the HAdV prevalence in all sampling areas ($p > 0.05$).

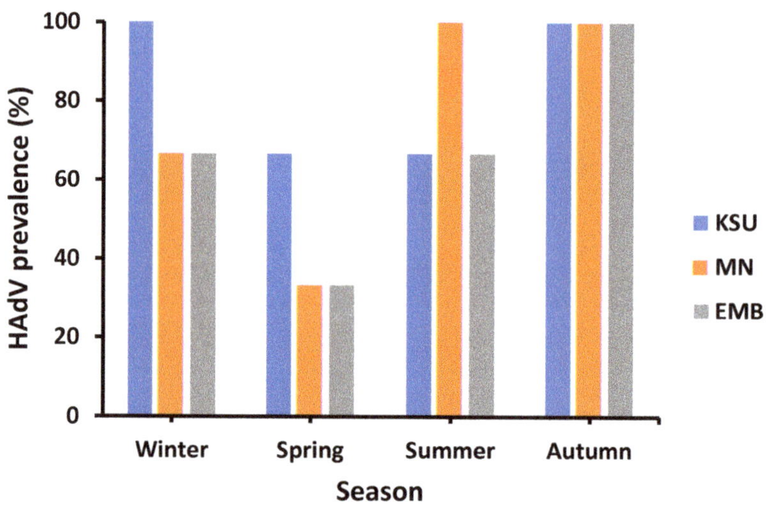

Figure 4. Seasonal prevalence of HAdVs among sampling areas.

3.4. Effect of Environmental Conditions on HAdV Prevalence

3.4.1. Temperature Variation Influence on HAdV Incidence

The lowest prevalence (33.33%) generally occurred in Spring and early Summer at temperatures (high: 30–44 °C, low: 20–30 °C), particularly during March to June (Figure 5). Notably, HAdV prevalence displayed a different pattern with the highest prevalence of 100% in late Summer and Autumn at temperatures (high: 34–43 °C, low: 18–32 °C) and moderate prevalence of 66.67% in Winter (particularly, in January and February) at lower temperature ranges (high: 26 °C, low: 10–12 °C). Consequently, the high temperature ranges were found to have a potentially significant influence on the prevalence of HAdV ($p = 0.001$, $R^2 = 0.689$; Table 2). On the other hand, the HAdV prevalence was insignificantly influenced by low temperature ranges ($p > 0.05$). Likewise, the segregation of sampling areas depicted lack of significant impact of low or high temperature on prevalence of

HAdV ($p > 0.05$). However, the highest HAdV prevalence was mostly favored in at ≥28 °C in MN-WWTP in the low temperature range, which is equivalent to ≥41 °C in the high temperature range (Figure 6). Surprisingly, the HAdV prevalence detected at EMB-WWTP was the highest at lower temperature ranges (low: 16–21 °C, high: 26–30 °C).

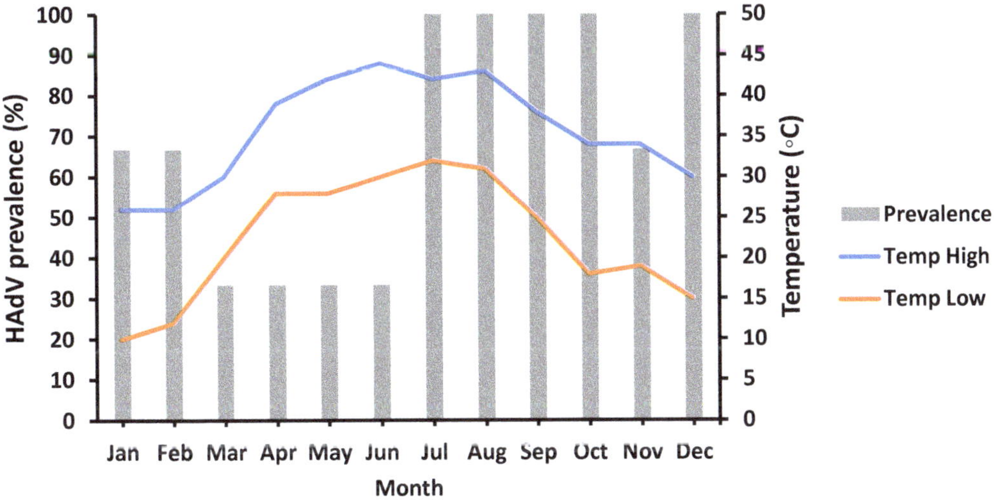

Figure 5. Temperature variation influence on the HAdV prevalence in wastewater samples. "Temp Low" refers to the average low temperature and "Temp High" refers to the average high temperature.

Table 2. Impact significance of environmental factors on the HAdV prevalence.

Environmental Factor	R^2	RMSE	Equation
High temperature (T_H)	**0.689 ***	4.575	$Prev_{HAdV} = -6.15 + 0.79 \times T_H$
Low temperature (T_L)	0	4.571	$Prev_{HAdV} = 18.055556$
Relative humidity (RH%)	**0.419 ****	8.439	$Prev_{HAdV} = 25.926 - 0.525 \times RH\%$
Wind speed (WS)	0.03	17.413	$Prev_{HAdV} = 16.0067 + 0.358 \times WS$

Notes: *Prev* denotes the prevalence of virus. RMSE denotes the root mean squared error. *: significant at $p = 0.001$, **: significant at $p = 0.009$.

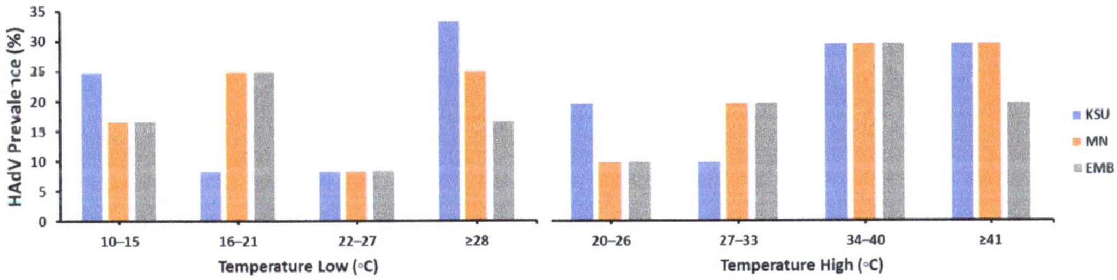

Figure 6. Temperature variation impact on HAdV prevalence in different sampling areas.

3.4.2. Humidity Variation Influenced HAdV Incidence

HAdV showed various distributions among the relative humidity ranges, favoring the lowest humidity range (6–14%) at all sampling locations with a prevalence of 33.3% (Figure 7). However, no HAdV prevalence (0%) was detected at humidity ranges of 24–32% and 33–41%, at KSU-/EMB-WWTP and MN-/EMB-WWTP, respectively. The large variation in HAdV prevalence observed at different humidity ranges highlighted the significant influence of relative humidity on HAdV occurrence in raw water of WWTPs ($p = 0.009$, $R^2 = 0.419$; Table 2).

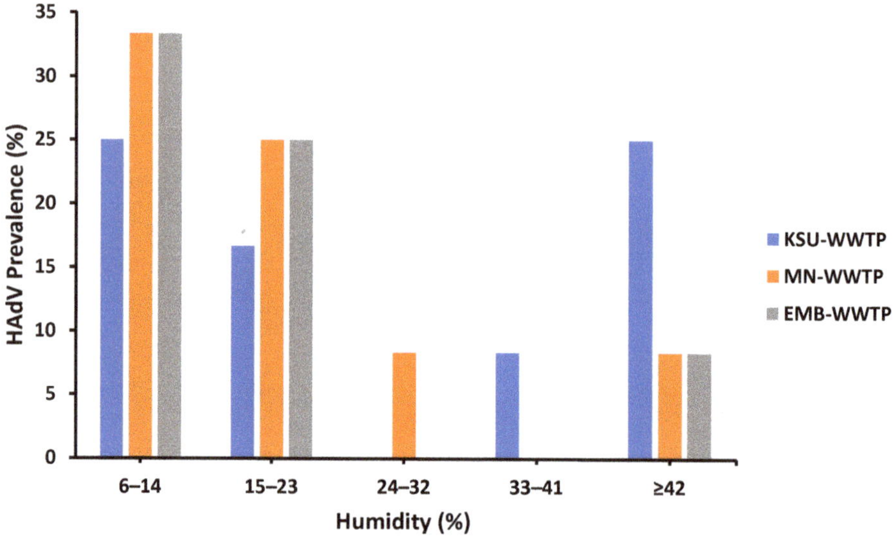

Figure 7. Impact of humidity variation on the HAdV prevalence (%) in different sampling areas.

3.4.3. Wind Speed Influence on HAdV Incidence

HAdV showed a different occurrence frequency at the various ranges of wind speed. The highest HAdV prevalence was detected at a relatively high wind speed range (15–21 km/h). Whereas the wind speed ranges of 8–14 km/h and ≥22 km/h were associated with no HAdV incidence in raw water of KSU-/EMB-WWTP and KSU-WWTP, respectively (Figure 8). Despite the HAdV occurrence pattern, wind speed was found to have insignificant influence on HAdV prevalence in the entire WWTPs ($p > 0.05$, $R^2 = 0.03$, Table 2).

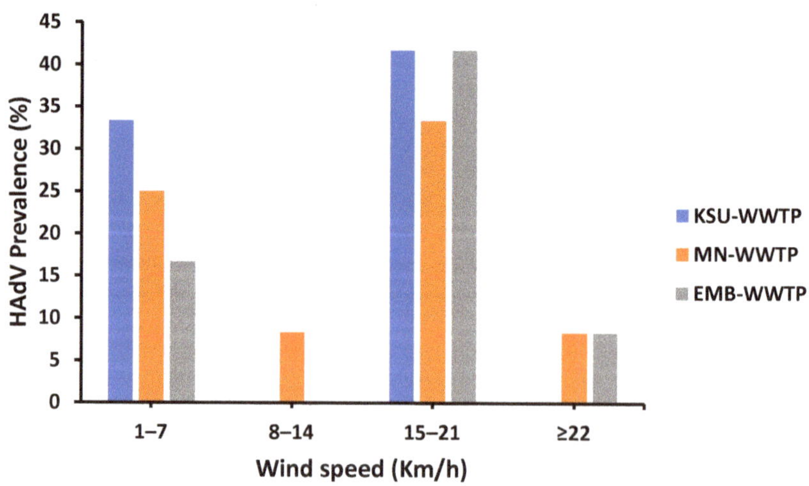

Figure 8. Impact of wind speed ranges (km/h) on the HAdV prevalence (%) in different sampling areas.

4. Discussion

Human adenoviruses are considered important indicators of fecal contamination owing to their wide detection in all types of water throughout the year and also due to their ability to resist the sewage treatment process [17]. Moreover, HAdVs are a major cause of clinical infections, including gastroenteritis, ocular, respiratory diseases, conjunctivitis, hemorrhagic cystitis, and chronic systemic infections in immunocompromised patients [25]. The continuous surveillance of human adenoviruses is of significant importance because these viruses account for 2–10% of diarrheal cases [26]. Hence, the current investigation focused on the detection of HAdVs in wastewater treatment plants in three different locations at Riyadh, Saudi Arabia. The highest prevalence of HAdVs was detected in KSU-WWTP, recording a relative percentage of 83.3%, followed by MN-WWTP (75%), and EMB-WWTP (66.6%). Our results were in accordance with those of a previous study, which reported the detection of HAdVs in 84.4% of raw wastewater samples collected from the Zenin wastewater treatment plant (WWTP), followed by treated sewage (50%) [22]. The HAdV prevalence in Manfoha wastewater treatment was higher than that detected in a prior study at the same location, which emphasizes the importance of continuous surveillance of HAdVs in wastewater samples as these viruses are correlated with gastroenteritis cases among children in hospitals [18].

The various species of adenoviruses are found in distinct tissue tropism, resulting in equally diverse clinical symptoms [27–29]. In this setting, species F and G (serotypes 40 and 41) affect the gastrointestinal tract whereas, species B, C, and E (serotypes 3, 5, and 7) primarily cause respiratory infections [30]. The current study revealed that the HAdVs belong to species F which was previously reported to be associated with gastroenteritis in children [31,32]. Phylogenetic analysis revealed that the isolated HAdV cluster with serotype 41, which had previously been found in children suffering from gastroenteritis in Riyadh, Jeddah, and Mecca in Saudi Arabia [33].

A previous study confirmed that the AdV prevalence was largely stable throughout the year, showing no seasonal effect on the removal of these viruses in wastewater treatment [13]. This finding was consistent with that study's assertion that the seasonal variability insignificantly influenced the HAdV prevalence in the locations under investigation ($p > 0.05$). The occurrence of HAdV in sewage with little or insignificant seasonal variability supported the hypothesis mentioned elsewhere that HAdV occurrence in the aquatic environment is most likely owing to potential contamination with untreated or poorly treated human sewage [34]. Moreover, the lack of seasonal impact on HAdV prevalence implicates the potentiality of HAdV as a quality indicator of water resources [35,36]. However, the highest HAdV prevalence of 22.22% was observed in the Autumn season while the lowest prevalence was detected in Spring (11.11%). This finding was in conflict with that of previous investigations conducted in China, Korea, and Egypt, which found that the highest prevalence of HAdVs was detected in Summer rather than Autumn [22,37,38]. Interestingly, HAdV prevalence was 100% in late Summer and Autumn and of moderate prevalence in Winter, with the lowest prevalence recorded in Spring and early Summer, particularly during March to June. However, a recent research found that HAdVs were more often detected on days with low temperatures (20 °C), mostly during the Winter and Spring seasons [39]. The discrepancy of our results with previous findings could be owing to geographical and temporal considerations. The HAdV prevalence in our study was insignificantly influenced by low temperature ranges, which implicates the stability of HAdV with no favorable distribution pattern even at low temperatures, which agrees a previous study that recorded twice more HAdV prevalence during high temperature seasons [37]; similar to the higher HAdV prevalence recorded in China [38]. On the contrary, a study conducted in Saudi Arabia, reported the highest HAdV prevalence at 5/6 locations favoring lower temperature ranges (22–25 °C) compared to our study [18]. The difference could be owing to the sample source since the later study was concerned with treated water, irrigation water, and surface water (lakes) as well as the temporal variation. Moreover, meteorological factors including temperature, wind speed, and humidity were reported to have no significant impact on HAdV prevalence with few exceptions [37]. However, the HAdV distribution varied across relative humidity ranges, preferring the lowest humidity range (6–14%) at all sample sites, with a prevalence of 33.3%. Nevertheless, no HAdV prevalence (0%) was identified in humidity ranges of 24–32% and 33–41%, at KSU-/EMB-WWTP and MN-/EMB-WWTP, respectively. This finding was contrasted with that of a previous report which demonstrated that humidity had no influence on HAdV prevalence [40]. The highest HAdV prevalence was identified at rather high wind speeds (15–21 km/h); however, wind speed was shown to have an insignificant influence on HAdV prevalence over the investigated wastewater treatment plants, and this coincides with the prior report, which indicated that the virus was detected in low concentrations at extremely low wind speeds [41].

5. Conclusions

The present investigation found that species F, serotype 41 predominated among the identified HAdV strains in various wastewater treatment facilities. Since this serotype has been connected to gastroenteritis in children, ongoing monitoring of HAdVs gives novel data that are critical for fecal contamination management. The research confirmed that HAdVs were unaffected by seasonal variation and were abundant in wastewater samples, suggesting their potential use as a fecal pollution indicator.

Supplementary Materials: The following supporting information can be downloaded at: https://www.mdpi.com/article/10.3390/w15071367/s1, Figure S1. Timetree for HAdV evolutionary analysis by Maximum Likelihood method; Table S1. Maximum Likelihood fits of 24 different nucleotide substitution models; Table S2. Estimates of evolutionary divergence between HAdV sequences recovered from WWTPs raw water in Riyadh, Saudi Arabia; Table S3. Evolutionary divergence estimates between human Adenovirus sequences; Table S4. Full description of sequences used for phylogenetic analysis of human adenovirus isolates.

Author Contributions: Conceptualization, I.N., A.H. and S.E.; methodology, K.M., I.N. and I.A.-A.; validation, K.M., I.N. and A.A.; formal analysis, K.M., I.N., I.A.-A. and M.T.Y.; investigation, K.M. and I.N.; resources, S.E.; data curation, I.N.; writing—original draft preparation, K.M., I.N. and M.T.Y.; writing—review and editing, K.M., I.N., A.H., M.T.Y., A.A., I.A.-A. and S.E.; supervision, I.N., A.H. and S.E.; funding acquisition, S.E. All authors have read and agreed to the published version of the manuscript.

Funding: This research was funded by Researchers Supporting Project, number (RSPD2023R797).

Data Availability Statement: The sequences used in this study for phylogenetic analysis are openly available in NCBI GenBank repository with accession numbers mentioned in Table S4.

Acknowledgments: The authors extend their appreciation to the Researchers Supporting Project, number (RSPD2023R797), King Saud University, Riyadh, Saudi Arabia, for funding this work.

Conflicts of Interest: The authors declare no conflict of interest.

References

1. Ibrahim, Y.; Ouda, M.; Kadadou, D.; Banat, F.; Naddeo, V.; Alsafar, H.; Yousef, A.F.; Barceló, D.; Hasan, S.W. Detection and Removal of Waterborne Enteric Viruses from Wastewater: A Comprehensive Review. *J. Environ. Chem. Eng.* **2021**, *9*, 105613. [CrossRef]
2. Teixeira, P.; Costa, S.; Brown, B.; Silva, S.; Rodrigues, R.; Valerio, E. Quantitative PCR Detection of Enteric Viruses in Wastewater and Environmental Water Sources by the Lisbon Municipality: A Case Study. *Water* **2020**, *12*, 544. [CrossRef]
3. Salvador, D.; Caeiro, M.F.; Serejo, F.; Nogueira, P.; Carneiro, R.N.; Neto, C. Monitoring Waterborne Pathogens in Surface and Drinking Waters. Are Water Treatment Plants (Wtps) Simultaneously Efficient in the Elimination of Enteric Viruses and Fecal Indicator Bacteria (Fib)? *Water* **2020**, *12*, 2824. [CrossRef]
4. Walker, D.I.; Adriaenssens, E.M.; McDonald, J.E.; Hillary, L.S.; Malham, S.K.; Jones, D.L. Viral Indicators for Tracking Domestic Wastewater Contamination in the Aquatic Environment. *Water Res.* **2020**, *181*, 115926.
5. Gholipour, S.; Ghalhari, M.R.; Nikaeen, M.; Rabbani, D.; Pakzad, P.; Miranzadeh, M.B. Occurrence of Viruses in Sewage Sludge: A Systematic Review. *Sci. Total Environ.* **2022**, *824*, 153886. [CrossRef] [PubMed]

6. Fong, T.-T.; Mansfield, L.S.; Wilson, D.L.; Schwab, D.J.; Molloy, S.L.; Rose, J.B. Massive Microbiological Groundwater Contamination Associated with a Waterborne Outbreak in Lake Erie, South Bass Island, Ohio. *Environ. Health Perspect.* **2007**, *115*, 856–864. [CrossRef]
7. Xagoraraki, I.; Kuo, D.H.-W.; Wong, K.; Wong, M.; Rose, J.B. Occurrence of Human Adenoviruses at Two Recreational Beaches of the Great Lakes. *Appl. Environ. Microbiol.* **2007**, *73*, 7874–7881. [CrossRef]
8. Jiang, S.C.; Bischel, H.N.; Goel, R.; Rosso, D.; Sherchan, S.P.; Whiteson, K.L.; Yan, T.; Solo-Gabriele, H.M. Integrating Virus Monitoring Strategies for Safe Non-Potable Water Reuse. *Water* **2022**, *14*, 1187. [CrossRef]
9. Sant'Anna, T.B.; Araujo, N.M. Adeno-Associated Virus Infection and Its Impact in Human Health: An Overview. *Virol. J.* **2022**, *19*, 173. [CrossRef]
10. Lanrewaju, A.A.; Enitan-Folami, A.M.; Sabiu, S.; Edokpayi, J.N.; Swalaha, F.M. Global Public Health Implications of Human Exposure to Viral Contaminated Water. *Front. Microbiol.* **2022**, *13*, 981896. [CrossRef]
11. Fong, T.-T.; Phanikumar, M.S.; Xagoraraki, I.; Rose, J.B. Quantitative Detection of Human Adenoviruses in Wastewater and Combined Sewer Overflows Influencing a Michigan River. *Appl. Environ. Microbiol.* **2010**, *76*, 715–723. [CrossRef] [PubMed]
12. Chmielewicz, B.; Benzler, J.; Pauli, G.; Krause, G.; Bergmann, F.; Schweiger, B. Respiratory Disease Caused by a Species B2 Adenovirus in a Military Camp in Turkey. *J. Med. Virol.* **2005**, *77*, 232–237. [CrossRef] [PubMed]
13. Katayama, H.; Haramoto, E.; Oguma, K.; Yamashita, H.; Tajima, A.; Nakajima, H.; Ohgaki, S. One-Year Monthly Quantitative Survey of Noroviruses, Enteroviruses, and Adenoviruses in Wastewater Collected from Six Plants in Japan. *Water Res.* **2008**, *42*, 1441–1448. [CrossRef]
14. Bonadonna, L.; La Rosa, G. A Review and Update on Waterborne Viral Diseases Associated with Swimming Pools. *Int. J. Environ. Res. Public Health* **2019**, *16*, 166. [CrossRef]
15. Li, D.; Zhou, J.-N.; Li, H.; He, C.-Y.; Dai, Q.-S.; Li, X.-L.; He, J.-F.; He, H.; Li, M.-B.; Jiang, L.I.-L. An Outbreak of Epidemic Keratoconjunctivitis Caused by Human Adenovirus Type 8 in Primary School, Southwest China. *BMC Infect. Dis.* **2019**, *19*, 624. [CrossRef] [PubMed]
16. Mostafa-Hedeab, G.; Allayeh, A.K.; Elhady, H.A.; Eledrdery, A.Y.; Mraheil, M.A.; Mostafa, A. Viral Eco-Genomic Tools: Development and Implementation for Aquatic Biomonitoring. *Int. J. Environ. Res. Public Health* **2022**, *19*, 7707. [CrossRef]
17. Allayeh, A.K.; Al-Daim, S.A.; Ahmed, N.; El-Gayar, M.; Mostafa, A. Isolation and Genotyping of Adenoviruses from Wastewater and Diarrheal Samples in Egypt from 2016 to 2020. *Viruses* **2022**, *14*, 2192. [CrossRef] [PubMed]
18. Nour, I.; Hanif, A.; Zakri, A.M.; Al-Ashkar, I.; Alhetheel, A.; Eifan, S. Human Adenovirus Molecular Characterization in Various Water Environments and Seasonal Impacts in Riyadh, Saudi Arabia. *Int. J. Environ. Res. Public Health* **2021**, *18*, 4773. [CrossRef] [PubMed]
19. Montasser, K.A.; Youssef, M.I.; Ghandour, A.A.; Kamal, M. Infection with Adenovirus, Rotavirus, and Coinfection among Hospitalized Children with Gastroenteritis in an Egyptian University Hospital. *J. Med. Virol.* **2022**, *94*, 4950–4958. [CrossRef]
20. Gallardo, J.; Pérez-Illana, M.; Martín-González, N.; San Martín, C. Adenovirus Structure: What Is New? *Int. J. Mol. Sci.* **2021**, *22*, 5240. [CrossRef]
21. Ferreyra, L.J.; Giordano, M.O.; Martinez, L.C.; Barril, P.A.; Masachessi, G.; Isa, M.B.; Poma, R.; Rajal, V.; Biganzoli, P.; Nates, S.V. Tracking Novel Adenovirus in Environmental and Human Clinical Samples: No Evidence of Endemic Human Adenovirus Type 58 Circulation in Cordoba City, Argentina. *Epidemiol. Infect.* **2015**, *143*, 1427–1431. [CrossRef]
22. Elmahdy, E.M.; Ahmed, N.I.; Shaheen, M.N.F.; Mohamed, E.-C.B.; Loutfy, S.A. Molecular Detection of Human Adenovirus in Urban Wastewater in Egypt and among Children Suffering from Acute Gastroenteritis. *J. Water Health* **2019**, *17*, 287–294. [CrossRef] [PubMed]
23. Iaconelli, M.; Valdazo-González, B.; Equestre, M.; Ciccaglione, A.R.; Marcantonio, C.; Della Libera, S.; La Rosa, G. Molecular Characterization of Human Adenoviruses in Urban Wastewaters Using Next Generation and Sanger Sequencing. *Water Res.* **2017**, *121*, 240–247. [CrossRef] [PubMed]
24. Bibby, K.; Peccia, J. Identification of Viral Pathogen Diversity in Sewage Sludge by Metagenome Analysis. *Environ. Sci. Technol.* **2013**, *47*, 1945–1951. [CrossRef] [PubMed]
25. Zhang, S.-Y.; Luo, Y.-P.; Huang, D.-D.; Fan, H.; Lu, Q.-B.; Wo, Y.; Chen, G.; Zhang, X.-A.; Li, Y.; Tong, Y.-G. Fatal Pneumonia Cases Caused by Human Adenovirus 55 in Immunocompetent Adults. *Infect. Dis.* **2016**, *48*, 40–47. [CrossRef] [PubMed]
26. Kumthip, K.; Khamrin, P.; Ushijima, H.; Maneekarn, N. Enteric and Non-Enteric Adenoviruses Associated with Acute Gastroenteritis in Pediatric Patients in Thailand, 2011 to 2017. *PLoS ONE* **2019**, *14*, e0220263. [CrossRef] [PubMed]
27. Ismail, A.M.; Lee, J.S.; Dyer, D.W.; Seto, D.; Rajaiya, J.; Chodosh, J. Selection Pressure in the Human Adenovirus Fiber Knob Drives Cell Specificity in Epidemic Keratoconjunctivitis. *J. Virol.* **2016**, *90*, 9598–9607. [CrossRef]
28. Yao, L.; Wang, C.; Wei, T.; Wang, H.; Ma, F.; Zheng, L. Human Adenovirus among Hospitalized Children with Respiratory Tract Infections in Beijing, China, 2017–2018. *Virol. J.* **2019**, *16*, 78. [CrossRef]
29. Pscheidt, V.M.; Gregianini, T.S.; Martins, L.G.; da Veiga, A.B.G. Epidemiology of Human Adenovirus Associated with Respiratory Infection in Southern Brazil. *Rev. Med. Virol.* **2021**, *31*, e2189. [CrossRef]
30. Kheder Alatawi, S.; Alyahyawi, H.E.; Akhter, N.; Alharbi, R.A.; AM Ahmed, W.; Saleh, R.; Almalki, S. Prevalence of Adenoviruses as Ocular Disease Causatives in Saudi Arabia. *Saudi J. Biol. Sci.* **2022**, *29*, 2026–2032. [CrossRef]
31. Lee, B.; Damon, C.F.; Platts-Mills, J.A. Pediatric Acute Gastroenteritis Due to Adenovirus 40/41 in Low-and Middle-Income Countries. *Curr. Opin. Infect. Dis.* **2020**, *33*, 398. [CrossRef] [PubMed]

32. Do Nascimento, L.G.; Fialho, A.M.; de Andrade, J.d.S.R.; de Assis, R.M.S.; Fumian, T.M. Human Enteric Adenovirus F40/41 as a Major Cause of Acute Gastroenteritis in Children in Brazil, 2018 to 2020. *Sci. Rep.* **2022**, *12*, 11220. [CrossRef] [PubMed]
33. Tayeb, H.T.; Dela Cruz, D.M.; Al-Qahtani, A.; Al-Ahdal, M.N.; Carter, M.J. Enteric Viruses in Pediatric Diarrhea in Saudi Arabia. *J. Med. Virol.* **2008**, *80*, 1919–1929. [CrossRef] [PubMed]
34. Jiang, S.C. Human adenoviruses in water: Occurrence and health implications: A critical review. *Environ. Sci. Technol.* **2006**, *40*, 7132–7140. [CrossRef]
35. Nagarajan, V.; Chen, J.S.; Hsu, B.M.; Hsu, G.J.; Wang, J.L.; Hussain, B. Prevalence, distribution, and genotypes of adenovirus and norovirus in the Puzi River and its tributaries and the surrounding areas in Taiwan. *GeoHealth.* **2021**, *5*, e2021GH000465. [CrossRef]
36. Pina, S.; Puig, M.; Lucena, F.; Jofre, J.; Girones, R. Viral pollution in the environment and in shellfish: Human adenovirus detection by PCR as an index of human viruses. *Appl. Environ. Microbiol.* **1998**, *64*, 3376–3382. [CrossRef] [PubMed]
37. Wang, Z.; Shin, H.; Jung, S.; Yeo, D.; Park, H.; Shin, S.; Seo, D.J.; Park, K.H.; Choi, C. Effects of Weather and Environmental Factors on the Seasonal Prevalence of Foodborne Viruses in Irrigation Waters in Gyeonggi Province, Korea. *Microorganisms* **2020**, *8*, 1224. [CrossRef]
38. Xie, L.; Zhang, B.; Xiao, N.; Zhang, F.; Zhao, X.; Liu, Q.; Xie, Z.; Gao, H.; Duan, Z.; Zhong, L. Epidemiology of Human Adenovirus Infection in Children Hospitalized with Lower Respiratory Tract Infections in Hunan, China. *J. Med. Virol.* **2019**, *91*, 392–400. [CrossRef]
39. Bortagaray, V.; Lizasoain, A.; Piccini, C.; Gillman, L.; Berois, M.; Pou, S.; Díaz, M.d.P.; Tort, F.L.; Colina, R.; Victoria, M. Microbial Source Tracking Analysis Using Viral Indicators in Santa Lucía and Uruguay Rivers, Uruguay. *Food Environ. Virol.* **2019**, *11*, 259–267. [CrossRef]
40. Sundell, N.; Andersson, L.-M.; Brittain-Long, R.; Lindh, M.; Westin, J. A Four Year Seasonal Survey of the Relationship between Outdoor Climate and Epidemiology of Viral Respiratory Tract Infections in a Temperate Climate. *J. Clin. Virol.* **2016**, *84*, 59–63. [CrossRef]
41. Colston, J.M.; Zaitchik, B.F.; Badr, H.S.; Burnett, E.; Ali, S.A.; Rayamajhi, A.; Satter, S.M.; Eibach, D.; Krumkamp, R.; May, J.; et al. Associations Between Eight Earth Observation-Derived Climate Variables and Enteropathogen Infection: An Independent Participant Data Meta-Analysis of Surveillance Studies with Broad Spectrum Nucleic Acid Diagnostics. *GeoHealth* **2022**, *6*, e2021GH000452. [CrossRef] [PubMed]

MDPI
St. Alban-Anlage 66
4052 Basel
Switzerland
www.mdpi.com

Water Editorial Office
E-mail: water@mdpi.com
www.mdpi.com/journal/water

www.ingramcontent.com/pod-product-compliance
Lightning Source LLC
Chambersburg PA
CBHW061555210926
51850CB00009B/2108
* 9 7 8 3 7 2 5 8 0 6 3 8 6 *